Statistical Inference

Paul H. Garthwaite
University of Aberdeen

Ian T. Jolliffe
University of Aberdeen

Byron Jones
De Montfort University

PRENTICE HALL

London New York Toronto Sydney Tokyo Singapore
Madrid Mexico City Munich

First published 1995 by
Prentice Hall Europe
Campus 400, Maylands Avenue
Hemel Hempstead
Hertfordshire, HP2 7EZ
A division of
Simon & Schuster International Group

Typeset in 10/12pt Times
by PPS Limited, London Road, Amesbury, Wilts.

Printed and bound in Great Britain by
Redwood Books, Trowbridge, Wiltshire

Library of Congress Cataloging-in-Publication Data

Garthwaite, Paul H.
Statistical inference / Paul H. Garthwaite, Ian T. Jolliffe, Byron Jones.
p. cm.
Includes bibliographical references (p. —) and index.
ISBN 0–13–847260–2 (acid-free)
1. Mathematical statistics. I. Jolliffe, I. T. II. Jones, Byron, 1951– . III. Title.
QA276.G36 1995
519.5′4–dc20 95–7038
CIP

British Library Cataloguing in Publication Data

A catalogue record for this book is available from the British Library

ISBN 0-13-847260-2

2 3 4 5 99 98 97

To Pamela, Jean and Hilary

Contents

Preface

This book is designed for courses in statistical inference for final year undergraduate or postgraduate students in statistics. It should also be useful to the working statistician as a reference for the basic ideas in statistical inference, as well as providing information on newer developments. There are a number of excellent, comprehensive texts on the theory of inference, including Bickel and Doksum (1977), Cox and Hinkley (1974), Stuart and Ord (1991), and Zacks (1971). The present text differs from these books in two ways. Firstly, it is less detailed than most of them. The emphasis is on explaining what various techniques do, with mathematical proofs kept to a minimum. Secondly, it includes some recent advances in inference. For example, there are chapters on computationally intensive methods and on inference associated with generalized linear models. Recent work on topics such as conditional inference is also included.

The book gives most prominence to the frequentist approach to statistics, but the Bayesian approach and other competing theories also receive substantial space. We take a neutral stance in the controversies between the different approaches, simply describing how the various techniques are implemented and reporting the advantages and drawbacks of each. For a good discussion of the philosophical arguments underlying various schools of inference the reader may consult Barnett (1982).

Some prior knowledge of probability and statistics is required. In probability, knowledge is assumed of marginal and conditional distributions and associated expectations and variances, while in statistics one or two courses on statistical methods, including an introduction to linear models, should be adequate. The mathematical background required is standard calculus and a passing acquaintance with matrix algebra, but little more.

Most of the material in the book can be covered in a course of about 48 one-hour lectures. Shorter courses could be based on Chapters 2–6, which contain what we regard as core material, although Sections 3.3, 4.7 and 6.5 might be omitted. The instructor could then select topics from subsequent chapters. These chapters can largely be taught independently of each other, although Section 9.3 assumes Section 8.2.1 has previously been covered and Secton 9.6 requires familiarity with much of the material in Chapters 6 and 7. Each chapter, except the first, includes a selection of exercises of varying difficulty. Some of these augment material in the text, but the purpose of most is to help the student consolidate ideas and results.

Acknowledgements

We are grateful to a number of friends and colleagues who have commented on earlier drafts of parts of the book, and whose comments have served to improve the final product. The analyses in Chapter 10 were done using GLIM 4 (Francis et al., 1993) and Genstat (Payne et al., 1993). Dave Collett, John Hinde, Jim Lindsey and Martin Ridout provided us with useful advice, and/or pieces of Genstat or GLIM code, and Chris Gilligan supplied one of the data sets. Jim Dickey made many useful comments on Chapters 6 and 7. An anonymous reviewer made many helpful points throughout the book. Kay Penny was responsible for the excellent figures in the first nine chapters.

Most of the text, in various drafts, was expertly typed by Mavis Swain and Louise Thomson. We are extremely grateful to them both for their efficiency and patience.

Large parts of the book have evolved from courses that we have taught at the Universities of Kent and Aberdeen over a number of years. Some of the Exercises and Examples have been in our notes for so long that we have lost track of their original sources. We therefore acknowledge the debt we owe to colleagues and ex-colleagues and to the other anonymous originators of these problems.

Finally, we are each grateful to our families for their patience and understanding throughout the gestation period of the book.

Chapter One

Introduction

1.1 Background

Statistics is a very broad subject, with applications in a vast number of different fields. At heart it is an applied subject, but it would be unsatisfactory if each application had to be tackled separately, devising fresh techniques for each new problem. For this reason, a body of common statistical theory has been developed so that most problems can be tackled within the same framework.

There are three main subdivisions within statistics: efficient summarization, tabulation and graphical display of data; design of experiments; and statistical inference. Data summarization was historically the first major statistical activity. Experimental design is of crucial importance before data are collected. However, it is statistical inference which has seen most research and practical application in recent years, and it is inference which forms the subject of this book. As already noted, the range of applications of statistical inference is vast, as a glance through issues of journals such as *Applied Statistics*, *Biometrics*, the *Journal of the American Statistical Association*, and the *Journal of Applied Statistics* will reveal. Agricultural research provided an impetus for many statistical techniques which have long been 'standard'. Medical statistics, or more broadly biostatistics, is currently an area where much new statistical methodology is being developed. But advances in inferential techniques have come from many other disciplines, ranging from chemistry to climatology and from image processing to quality control, to name but a few.

This introductory chapter starts with a brief discussion of the various approaches to inference. We then give an overview of the contents of the book and continue by establishing some basic ideas and notation which will be used throughout the following chapters.

In statistical inference we use a sample of data to draw inferences about some aspect of the population (real or hypothetical) from which the data were taken. Often the inference concerns the value of one or more unknown parameters, which describe some attribute of the population such as its location or spread.

There are three main types of inference, namely point estimation, interval estimation and hypothesis testing. In point estimation, for each unknown parameter of interest a single value is computed from the data, and used as an estimate of that parameter.

For example, suppose that in a study of food-storing behaviour in birds, 1 out of 4 feeders contains food, and 14 out of 20 birds visit the correct feeder. We could use the proportion of birds making the correct choice, 0.7, as an estimate of the probability, p, that birds of that species choose the correct feeder in similar circumstances. Such an estimate on its own gives no information about the precision of the estimate, although estimates of precision can be computed as well.

Instead of producing a single estimate of a parameter, interval estimation provides a range of values which have a predetermined high probability of including the true, but unknown, value of the parameter. In our food-storing example we might, for example, state with a high degree of confidence that p lies between 0.5 and 0.9.

Hypothesis testing sets up specific hypotheses regarding the parameters of interest and assesses the plausibility of any specified hypothesis by seeing whether the observed data support or refute that hypothesis. Returning to our food-storing birds, if birds were choosing among the 4 feeders at random, then p should be $\frac{1}{4}$. We could test whether this value of p is plausible, given our data. Also, we may collect data for birds of a different species and wish to test whether p takes the same value for the two species. Although hypothesis testing can often be artificial in the sense that none of the proposed hypotheses will be exactly correct (for example, exact equality of p for two species of birds is unlikely), it is often a convenient way to proceed and underlies a substantial part of scientific research.

As well as the different types of inference there are also different underlying philosophies. The most widely used approach is the frequentist, classical or sampling theory approach. The theory makes the assumption that we can randomly take repeated samples of data, under the same conditions as hold for our single observed sample, from the same population. Properties of point estimators, interval estimators, and tests of hypothesis are all derived under this repeated sampling assumption. Frequentist inference works reasonably well in many circumstances, but in complicated (and sometimes very simple) situations it can break down and produce silly results. We shall see a number of examples of this behaviour throughout the book.

The main rival to frequentist inference is the Bayesian approach. Here there is no need to invoke the possibility of repeated sampling, but we must specify a probability distribution for the parameters of interest, rather than treating them as fixed but unknown. Bayesian inference removes many of the mathematical and logical problems suffered by the frequentist approach. However, it suffers from practical difficulties, in particular how to choose the probability distribution for the parameters.

A third approach to inference is based on decision theory, in which an inference is viewed as making a decision (between competing estimators or hypotheses). As in Bayesian inference, a probability distribution is assigned to the parameters of interest, but in addition a loss function or utility function is needed. Such a function measures the desirability or otherwise of each possible decision (estimator, hypothesis) for each set of possible values of the parameters.

In this book we do not discuss in detail the philosophical differences between these various approaches to inferences. The reader may refer to Barnett (1982) for such comparisons, and also for some other philosophies within statistical inference. Instead we concentrate on how the approaches are implemented, together with some of the advantages and disadvantages of each.

1.2 Plan of the book

Chapters 2–5 are concerned with the frequentist approach to inference, with Chapters 2, 3 covering point estimation, Chapter 4 on hypothesis testing and Chapter 5 dealing with interval estimation.

In Chapter 2 properties of estimators are described. Unbiasedness, consistency, and especially efficiency and sufficiency are emphasized. The concept of minimum variance unbiased estimators is discussed and exponential families of distributions are introduced. Chapter 3 looks at methods of finding point estimators, and discusses such estimators with respect to the properties of Chapter 2. The emphasis is on maximum likelihood estimators which play a central role in inference. Various modifications and extensions of maximum likelihood estimators are included, as well as a number of alternatives to the maximum likelihood approach.

For hypothesis testing, both properties of tests and methods of constructing tests are dealt with in Chapter 4. Starting with the relatively straightforward case of two simple hypotheses, the chapter moves on to the composite case and describes some general methods of test construction, and the properties which they possess.

Chapter 5, covering interval estimation, is shorter than the three previous chapters. This is because much of the theory follows on from that of point estimation and hypothesis testing, and does not reflect any lack of importance attached to interval estimation. As with point estimation and hypothesis testing, properties of interval estimates and methods for their construction are both described. At various points in Chapters 2–5 problems which can arise with the various techniques are noted, but their positive qualities are also discussed.

Chapters 6 and 7 cover, respectively, the decision theory and Bayesian approaches to inference. All three types of inference, point estimation, interval estimation and hypothesis testing, are discussed in each chapter. The problems associated with choosing a (prior) distribution for a parameter, and a loss function or utility function, are described. Some discussion of sequential testing procedures is included in Chapter 6, while Chapter 7 describes empirical Bayes as well as 'pure' Bayes procedures.

Most inferential techniques assume that the probability distribution from which the data are drawn is known, apart from the values of a small number of parameters. Chapter 8 deals mainly with two approaches to the case where the assumption of a known distribution cannot safely be made. In nonparametric, or more accurately distribution-free, inference, techniques are devised which make minimal assumptions about the underlying distribution and so are valid for a wide range of distributions. Robust inference, on the other hand, has a working assumption about the underlying probability distribution. Procedures are therefore sought which have good properties under this assumption, but which are not much inferior for a range of alternatives to the assumption. Semiparametric methods are also outlined in Chapter 8.

Modern computer power has obviously had an impact on statistical inference and in Chapter 9 we discuss some important methods that require extensive calculations. For some problems the methods provide the only viable means of solution and, for some, they make fewer distributional assumptions than alternative methods. For example, permutation and randomization tests, introduced in Chapter 8, are discussed further. These make comparatively few assumptions and there are occasions when they seem the only suitable tests. We also consider simulation methods and give examples which show their versatility and usefulness for examining complex problems. Methods that involve repeated subsampling or resampling are also given, such as cross-validation and the bootstrap. In addition we describe the Gibbs sampler, which is important in Bayesian statistics for estimating parameters when integrals, which would otherwise have to be evaluated, are intractable.

Finally, in Chapter 10 we consider the analysis of data using generalized linear models. This chapter differs from the others in describing a specific class of models, rather than discussing general approaches to inference. However, such models have played an important role in recent developments in inference, and have widespread applicability. A separate chapter devoted to them therefore seems appropriate. Generalized linear models can be applied to data whose distribution comes from the exponential family. Therefore, the methods described can be applied not only to data sampled from a normal distribution, but also to data from other distributions such as the binomial and Poisson. Extensions beyond the exponential family also exist. We describe how maximum likelihood estimates of the model parameters can be obtained using an iterative algorithm, as well as explaining how the goodness of fit of a model can be evaluated. Hypothesis tests for model parameters and ways of checking the model assumptions are described. The method of quasi-likelihood estimation is introduced as a way of fitting a model to data whose distribution is incompletely specified. A number of data sets are used to illustrate the fitting and testing.

1.3 Notation and terminology

A large proportion of our notation and terminology will be introduced as needed. However, a few concepts which run right through inference will be established now. Our objective in choosing notation is to be consistent with general usage, whilst keeping things as simple as possible. Any departures from these general principles will be explained.

In much of inference we have available to us n observations $x_1, x_2, \ldots, x_n$, on the basis of which we draw conclusions about the value of one or more parameters θ. The quantities $x_1, x_2, \ldots, x_n$ are considered to be particular values of the random variables $X_1, X_2, \ldots, X_n$ respectively. Often, though not always, $X_1, X_2, \ldots, X_n$ are assumed to be independent, identically distributed (i.i.d.) with a distribution whose probability function (discrete random variables) or probability density function (continuous random variables) is $f_X(x; \theta)$, or more briefly $f(x; \theta)$. In this case we say that we have a random sample of n observations from the distribution with probability density function $f(x;\,\theta)$. We will also abuse terminology slightly by simply saying that the observations have distribution $f(x;\,\theta)$.

In what follows we generally treat $X_1, X_2, \ldots, X_n$ as continuous random variables, although much of what we say holds equally for discrete random variables. In this chapter, we write θ as a scalar, even though it may represent a vector of several parameters. Later, we sometimes use $\boldsymbol{\theta}$ to distinguish a vector of parameters $\boldsymbol{\theta} = (\theta_1, \theta_2, \ldots, \theta_k)^{\mathrm{T}}$. Here the superscript T denotes transpose.

Four useful abbreviations which will be adopted are i.i.d. (as above), p.d.f. (probability density function), c.d.f. (cumulative distribution function) and random variable (r.v.). The notation $\mathrm{E}[\cdot]$, $\mathrm{Var}[\cdot]$ will denote the expected value and variance respectively of the expression in brackets; $\Pr[\cdot]$ will be the probability of the bracketed event.

A quantity which appears throughout inference is the likelihood function. Suppose that we have our usual random sample of n observations, with p.d.f. $f(x;\,\theta)$. Then the joint p.d.f. of $X_1, X_2, \ldots, X_n$ is

$$f(\mathbf{x};\,\theta) = \prod_{i=1}^{n} f(x_i;\,\theta).$$

Here $\mathbf{x} = (x_1, x_2, \ldots, x_n)^{\mathrm{T}}$, the vector of observations. In its most usual, and simplest, form the likelihood function $L(\theta;\,\mathbf{x})$ is identical to $f(\mathbf{x};\,\theta)$, i.e. it is the joint p.d.f. of $X_1, X_2, \ldots, X_n$, but viewed as a function of θ. In some circumstances this definition of L needs to be extended (see Example 3.2 and Cheng and Iles, 1987), but for most purposes this definition is adequate.

In many circumstances it is more convenient to work with the natural logarithm of $L(\theta;\mathbf{x})$ rather than the likelihood itself. We denote this log-likelihood by $l(\theta;\,\mathbf{x}) = \ln[L(\theta;\,\mathbf{x})]$. Note that this L, l notation is probably the most commonly used, but that the reverse notation, l for likelihood and L for its logarithm, is also in frequent use.

For some purposes it is useful to look at the order statistics, denoted $X_{(1)}, X_{(2)}, \ldots, X_{(n)}$, for an i.i.d. set of r.v.s. These are the same as the original r.v.s., except that they have been rearranged in ascending order, so that $X_{(1)} \leqslant X_{(2)} \leqslant \cdots \leqslant X_{(n)}$.

1.3.1 *Some standard distributions*

We shall use the following abbreviations for standard distributions:

- $N(\mu, \sigma^2)$ normal distribution with mean μ, variance σ^2.
- $B(n, p)$ binomial distribution with n trials, probability of success p.
- $P(\theta)$ Poisson distribution with mean θ.
- $U[0, \theta]$ uniform distribution on the interval $[0, \theta]$.
- $\Gamma(\alpha, \beta)$ the gamma distribution with parameters α, β, where the parameterization is that given in Table 2.2.

t_ν	the t-distribution with ν degrees of freedom.
χ^2_ν	the χ^2-distribution with ν degrees of freedom.
F_{ν_1,ν_2}	the F-distribution with ν_1, ν_2 degrees of freedom.

The symbol $\sim$ will be used to denote 'distributed as'; for example $X \sim N(\mu, \sigma^2)$ means that X has a normal distribution with mean μ and σ^2. Similarly, $\dot{\sim}$ means 'approximately distributed as'.

We shall use Z to denote an r.v. with the standard normal distribution, $N(0, 1)$, and z_α will denote the critical value of Z, defined so that $\Pr[Z > z_\alpha] = \alpha$. Similarly, $t_{\nu;\alpha}$, $\chi^2_{\nu;\alpha}$ will denote critical values cutting off probability α in the upper tails of the t and χ^2 distributions respectively, each with ν degrees of freedom.

Chapter Two

Properties of estimators

2.1 Introduction

Suppose that we have a random sample $(x_1, x_2, \ldots, x_n)$ from a probability distribution with p.d.f. $f(x; \theta)$, and that we wish to use the values $x_1, x_2, \ldots, x_n$ to estimate θ, which is unknown. In particular, let $\hat{\theta}(x_1, x_2, \ldots, x_n)$ be a function of $x_1, x_2, \ldots, x_n$ which we use as a (point) estimate of θ; the corresponding function $\hat{\theta}(X_1, X_2, \ldots, X_n)$ of the r.v.s $X_1, X_2, \ldots, X_n$, which is itself an r.v., is an *estimator* for θ.

In any situation there will be a variety of possible estimators, though some may be more obvious than others, and we need some way of choosing between them. In this chapter we shall look at a number of desirable properties which we might like estimators to possess – unbiasedness, consistency, efficiency and sufficiency. These might be termed 'classical' properties of estimators. Further properties, such as admissibility and minimum expected risk or robustness, will be examined in subsequent chapters.

2.2 Unbiasedness

Definition 2.1

$\hat{\theta}$ is an *unbiased estimator* for θ if $\mathrm{E}[\hat{\theta}] = \theta$; otherwise it is *biased*. The *bias* of $\hat{\theta}$ is defined to be $\mathrm{bias}(\hat{\theta}) = \mathrm{E}[\hat{\theta}] - \theta$.

Intuitively, this means that the distribution of $\hat{\theta}$ is centred at θ, and there is no persistent tendency to under- or overestimate θ.

Example 2.1

Estimating the mean of the normal distribution $N(\mu, \sigma^2)$. The obvious estimator is $\bar{X} = (1/n)\Sigma_{i=1}^{n} X_i$, which is unbiased (as it is for distributions other than the normal). However, there are many other unbiased estimators, e.g. median, mid-range, X_1, so unbiasedness is not necessarily conclusive in choosing an estimator.

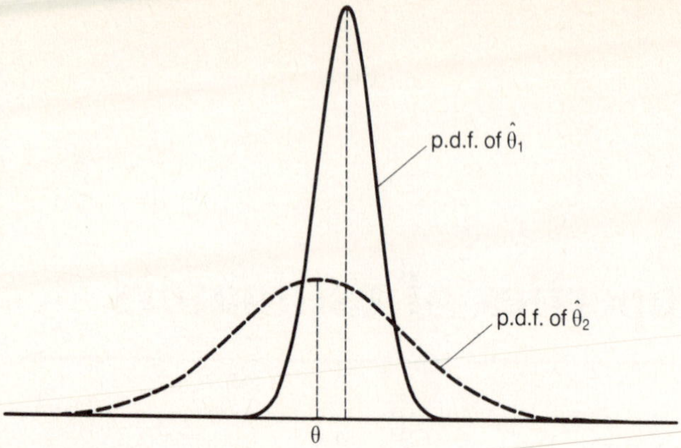

Fig. 2.1 Probability density functions for two estimators $\hat{\theta}_1$, $\hat{\theta}_2$.

Example 2.2

Consider a binomial experiment with n trials, and probability of success p. X is the number of successes. An obvious estimator for p is $\hat{p} = X/n$, the proportion of successes. Now $\mathrm{E}[X] = np$, so $\mathrm{E}[\hat{p}] = p$, and $\hat{p}$ is an unbiased estimator for p.

As well as not necessarily distinguishing very well between estimators (see Example 2.1 above) unbiasedness is by no means an essential property, for consider Figure 2.1, which shows the p.d.f.s for two estimators $\hat{\theta}_1$, $\hat{\theta}_2$ of a parameter θ. $\hat{\theta}_2$ is unbiased, $\hat{\theta}_1$ is biased, but $\hat{\theta}_1$ may well be preferred to $\hat{\theta}_2$ because of its much smaller variance. Although $\hat{\theta}_1$ is biased, it is never far from θ; $\hat{\theta}_2$ is unbiased but may be a long way from θ. A situation where biased estimators may be preferred to unbiased estimators is in multiple regression when there are large correlations between the regressor variables – see, for example, Vinod and Ullah (1981). There can also be theoretical objections to any insistence on unbiasedness – see, for example, De Groot (1986, Section 7.7). In much of this chapter we leave aside these objections, but we return to them in Section 2.8.

2.3 Consistency

Although some bias may be acceptable in an estimator, we would like the bias to tend to 0 as the sample size, n, tends to ∞. In addition we would like variance to tend to 0 as n tends to ∞. These requirements are related to the idea of consistency.

Definition 2.2

An estimator $\hat{\theta}$ for θ is (*weakly*) *consistent* if $\Pr[|\hat{\theta} - \theta| > \varepsilon] \to 0$ as $n \to \infty$ for any $\varepsilon > 0$ i.e. the p.d.f. of $\hat{\theta}$ becomes increasingly concentrated around θ for large n.

Alternative terminology for this form of consistency is that $\hat{\theta}$ converges to θ in probability. Strong consistency corresponds to convergence with probability 1 – see Cox and Hinkley (1974, pp. 287–288), where a third type of consistency, called Fisher consistency is also noted.

An estimator which is not consistent will rarely be acceptable, except occasionally on the grounds of ease of computation, robustness, etc.; fortunately most 'sensible-looking' estimators are consistent.

It may be difficult to prove consistency using the definition above, but it turns out that a sufficient (though not necessary – see Exercise 2.12) condition for consistency is that bias$(\hat{\theta}) \to 0$ and var$[\hat{\theta}] \to 0$ as $n \to \infty$.

To prove this we first define the mean square error (MSE), of $\hat{\theta}$. As we shall see, MSE$(\hat{\theta})$ is a property of an estimator $\hat{\theta}$ which takes account of both its bias and variance.

Definition 2.3 $\mathrm{MSE}(\hat{\theta}) = \mathrm{E}[(\hat{\theta} - \theta)^2]$.

From this definition it follows immediately that

$$\begin{aligned}\mathrm{MSE}(\hat{\theta}) &= \mathrm{E}[\{(\hat{\theta} - \bar{\theta}) + (\bar{\theta} - \theta)\}^2] \text{ where } \bar{\theta} = \mathrm{E}[\hat{\theta}] \\ &= \mathrm{E}[(\hat{\theta} - \bar{\theta})^2] + \mathrm{E}[(\bar{\theta} - \theta)^2] + 2(\bar{\theta} - \theta)\mathrm{E}[(\hat{\theta} - \bar{\theta})] \\ &= \mathrm{var}(\hat{\theta}) + [\mathrm{bias}(\hat{\theta})]^2 (+0).\end{aligned}$$

If bias $\to 0$ and variance $\to 0$ as $n \to \infty$ then MSE $\to 0$, as $n \to \infty$.

Now consider $\Pr[|\hat{\theta} - \theta| > \varepsilon]$ and assume that $\hat{\theta}$ is continuous, with a well-behaved p.d.f. $f(\hat{\theta})$ (a similar argument, replacing integrals by summations can easily be produced for discrete $\hat{\theta}$).

$$\begin{aligned}\Pr[|\hat{\theta} - \theta| > \varepsilon] &= \int\limits_{|\hat{\theta} - \theta| > \varepsilon} f(\hat{\theta})\mathrm{d}\hat{\theta} \\ &\leqslant \int\limits_{|\hat{\theta} - \theta| > \varepsilon} \frac{(\hat{\theta} - \theta)^2}{\varepsilon^2} f(\hat{\theta})\mathrm{d}\hat{\theta} \\ &\leqslant \int_{-\infty}^{\infty} \frac{(\hat{\theta} - \theta)^2}{\varepsilon^2} f(\hat{\theta})\mathrm{d}\hat{\theta}\end{aligned}$$

since the integrand $\geqslant 0$ everywhere

$$= \frac{1}{\varepsilon^2} \mathrm{MSE}(\hat{\theta}).$$

So $\Pr[|\hat{\theta} - \theta| > \varepsilon] \to 0$ as $n \to \infty$ if MSE$(\hat{\theta}) \to 0$ as $n \to \infty$. Thus bias$(\hat{\theta}) \to 0$ and var$(\hat{\theta}) \to 0$ as $n \to \infty$ implies that $\hat{\theta}$ is weakly consistent.

Although the converse is not true, it holds in all but somewhat pathological situations.

Example 2.1 (*continued*)

Of the four estimators mentioned above for μ in $N(\mu, \sigma^2)$, three are consistent. Variances are proportional to $1/n$ for $\bar{X}$ and for the median, and proportional to $1/[\log n]$ for the mid-range – see Kendall *et al.* (1987, pp. 331, 332, 467). Since each is unbiased and has a variance which $\to 0$ as $n \to \infty$ these three estimators are consistent. For X_1, $\Pr[|X_1 - \mu| > \varepsilon]$ is the same regardless of n, and certainly does not tend to 0 as $n \to \infty$ for any $\varepsilon > 0$. Hence X_1 is not a consistent estimator for μ.

Example 2.2 (*continued*)

From standard results for the binomial distribution, we know that $\text{var}(X) = np(1-p)$, so that $\text{var}(\hat{p}) = p(1-p)/n$. $\hat{p}$ has a zero bias and a variance which tends to 0 as $n \to \infty$, and is therefore a consistent estimator for p.

Example 2.3

Consider the (one-parameter) Cauchy distribution with p.d.f.

$$f(x; \theta) = [\pi(1 + (x - \theta)^2)]^{-1} \qquad -\infty < x < \infty.$$

The distribution is symmetric about θ, so $\bar{X}$ seems a sensible estimator for θ. However, the distribution of $\bar{X}$ is exactly the same as that of a single X (see Kendall *et al.* (1987) pp. 347–9, for a proof). Hence $\Pr[|\bar{X} - \theta| > \varepsilon]$ is exactly the same, however large the sample size, and $\Pr[|\bar{X} - \theta| > \varepsilon] \not\to 0$ for any $\varepsilon > 0$ as $n \to \infty$, so $\bar{X}$ is not a consistent estimator for θ.

Here the median is an unbiased estimator for θ with variance $\pi^2/4n$ which tends to 0 as $n \to \infty$. (See, for example, Gibbons, 1985, Section 2.6, for general results regarding order statistics and in particular the median. The Cauchy result may be obtained as a special case.) The median is therefore a consistent estimator for θ.

2.4 Efficiency

Using unbiasedness and consistency may still leave a (possibly infinite) number of candidate estimators. How can we choose between them (if we feel that it is necessary to make a choice)? One fairly obvious strategy is to try to find estimators which minimize MSE; however, this is often difficult to do and it is not true in general that estimators with uniformly minimum MSE actually exist. This is illustrated by the trivial, and usually silly, estimator $\hat{\theta} = \theta_0$, where θ_0 is a constant which ignores the data. Since this estimator has zero MSE when $\theta = \theta_0$, no other estimator can be uniformly best unless it has zero MSE everywhere. This is clearly an impossibility in general. We could therefore just try to minimize variance, and since it does not make sense to compare estimators with different biases with respect to variance alone ($\hat{\theta} = \theta_0$ is always best in this case), we compare only estimators with the same bias.

Further, to keep things simple, in much of what follows we restrict ourselves to unbiased estimators and look for minimum variance unbiased estimators (MVUEs).

A lot of work has been done on the theory surrounding MVUEs, and we shall cover the major aspects of this theory in the remainder of the present chapter. Towards the end of the chapter, however, we shall devote some space to criticisms of the whole approach to estimation based on MVUEs.

The words 'efficient' and 'efficiency', when applied to estimators, refer to the variances of the estimators. The lower the variance of an unbiased estimator, the more efficient it is.

Definitions 2.4

An unbiased estimator is said to be *efficient* if it has the minimum possible variance; the *efficiency* of an unbiased estimator is the ratio of the minimum possible variance to the variance of the estimator.

The *relative efficiency* of two (unbiased) estimators is the reciprocal of the ratio of their variances.

Since efficiencies may vary with sample sizes, the *asymptotic efficiencies* and *asymptotic relative efficiencies* (as $n \to \infty$) are often used as once-and-for-all measures.

We have talked in this group of definitions about a 'minimum possible variance' with which we compare variances of estimators. The definitions are incomplete until this 'minimum possible variance' is itself defined. It is usual to take as this 'minimum variance' benchmark a well-known lower bound to the variance of unbiased estimators.

2.4.1 *The Cramér–Rao inequality (and lower bound)*

Suppose that $X_1, X_2, \ldots, X_n$ form a random sample from the distribution with p.d.f. $f(x; \theta)$. Subject to certain regularity conditions on $f(x; \theta)$ we have that for any unbiased estimator $\hat{\theta}$ for θ,

$$\text{var}[\hat{\theta}] \geqslant I_\theta^{-1}$$

where

$$I_\theta = \mathrm{E}\left[\left(\frac{\partial \ln[L(\theta;\, \mathbf{x})]}{\partial \theta}\right)^2\right] = \mathrm{E}\left[\left(\frac{\partial l}{\partial \theta}\right)^2\right]$$

$L(\theta; \mathbf{x})$ is the likelihood function defined in Section 1.3, and $l = \ln(L)$. I_θ^{-1} is known as the Cramér–Rao lower bound, and the corresponding inequality is the Cramér–Rao inequality.

Comments

(1) I_θ is sometimes known as the (Fisher) information about θ in the observations. Clearly the lower is the attainable variance of $\hat{\theta}$, the more 'information', in an

intuitive sense, we have about θ, and the larger is I_θ. The inequality itself is sometimes known as the 'information inequality'.

(2) We shall not give technical details of the required regularity conditions. The main reason for their presence is to be able to interchange integration and differentiation (see Zacks, 1971, p. 182, for details) at various points in the proof – see below. One particular condition is that the range of values of X must not depend on θ so, for example, the result cannot be applied when θ is the parameter in the uniform distribution $U[0, \theta]$.

(3) The proof which follows goes through for discrete r.v.s if we replace integrals by summations, and the regularity conditions will be somewhat less restrictive in this case (it is easier to interchange summation and differentiation).

Proof

$\mathrm{E}[\hat{\theta}] = \int \hat{\theta}(x_1, x_2, \ldots, x_n)L(\theta; \mathbf{x})d\mathbf{x}$ (a multiple integral with respect to $x_1, x_2, \ldots, x_n$; L is the joint p.d.f. of the X_i as well as the likelihood function). $\hat{\theta}$ is an unbiased estimator, so $\mathrm{E}[\hat{\theta}] = \int \hat{\theta}(\mathbf{x})L(\theta; \mathbf{x})d\mathbf{x} = \theta$.

Differentiating both sides of this equation with respect to θ, and interchanging the order of integration and differentiation, gives

$$\int \hat{\theta}\, \frac{\partial L}{\partial \theta}\, d\mathbf{x} = 1 \qquad (\hat{\theta} \text{ does not depend on } \theta).$$

Now

$$\frac{\partial l}{\partial \theta} = \frac{\partial \ln (L)}{\partial \theta} = \frac{1}{L}\frac{\partial L}{\partial \theta}, \qquad \text{so} \qquad \frac{\partial L}{\partial \theta} = L\,\frac{\partial (\ln L)}{\partial \theta} = L\,\frac{\partial l}{\partial \theta}$$

Thus

$$1 = \int \hat{\theta}\, \frac{\partial l}{\partial \theta}\, L\, d\mathbf{x} = \mathrm{E}\left[\hat{\theta}\, \frac{\partial l}{\partial \theta}\right].$$

We next use the well-known result that for any two r.v.s U and V, $(\mathrm{Cov}[U, V])^2 \leqslant \mathrm{Var}[U]\,\mathrm{Var}[V]$, and let $U = \hat{\theta}$, and $V = \partial l/\partial \theta$.

Now

$$\begin{aligned}\mathrm{E}[V] &= \int \frac{\partial l}{\partial \theta}\, L\, d\mathbf{x} = \int \frac{\partial L}{\partial \theta}\, d\mathbf{x} \\ &= \frac{\partial}{\partial \theta}\left[\int L\, d\mathbf{x}\right] \qquad \text{(using regularity conditions)} \\ &= \frac{\partial}{\partial \theta}[1] = 0.\end{aligned}$$

Hence

$$\mathrm{Cov}[U, V] = \mathrm{E}[UV] = 1.$$

Also

$$\mathrm{Var}[V] = \mathrm{E}[V^2] = E\left[\left(\frac{\partial l}{\partial \theta}\right)^2\right] = I_\theta$$

so

$$\mathrm{Var}[U] = \mathrm{Var}[\hat{\theta}] \geqslant \frac{\{\mathrm{Cov}[U, V]\}^2}{\mathrm{Var}[V]} = 1^2/I_\theta = I_\theta^{-1}, \text{ as required.} \qquad \square$$

Definition 2.4 (*continued*)

As pointed out above, our definitions of efficiency and efficient were incomplete, since we did not give a definition of the 'minimum variance' with respect to which we were making comparisons. It is conventional to take the Cramér–Rao lower bound as this minimum variance, and define the efficiency of an estimator $\hat{\theta}$ as the ratio $I_\theta^{-1}/\mathrm{Var}[\hat{\theta}] = [I_\theta \mathrm{Var}(\hat{\theta})]^{-1}$. This definition is sometimes known as Bahadur efficiency (Zacks, 1971, p. 14), and has some drawbacks. The Cramér–Rao lower bound is not always applicable – we need regularity conditions on $f(x; \theta)$. Also, it is only a lower bound and may not be attainable. It is, however, frequently attainable asymptotically, and because asymptotic efficiencies are often of greatest interest, the definition is relevant in many cases. Note, though, that Mikulski (1982) defines efficiency with respect to the maximum likelihood estimator (see Section 3.2).

Extensions to the Cramér–Rao inequality

(1) If $\hat{\theta}$ is an estimator for θ with bias($\hat{\theta}$) now denoted by $b(\theta)$, then

$$\mathrm{Var}[\hat{\theta}] \geqslant \left(1 + \frac{\partial b}{\partial \theta}\right)^2 I_\theta^{-1}.$$

The proof of this is very similar to that given above for unbiased estimators. The left-hand side of the first equation in that proof is $\theta + b(\theta)$, and this has derivative $(1 + \partial b/\partial \theta)$.

(2) A lower bound for unbiased estimators of a function

$$g(\theta) \text{ of } \theta \text{ is } \left(\frac{\partial g}{\partial \theta}\right)^2 I_\theta^{-1}.$$

(3) We leave the extension to vector $\boldsymbol{\theta}$ until slightly later.

In the statement of the Cramér–Rao inequality, we defined Fisher information as $I_\theta = \mathrm{E}[(\partial l/\partial \theta)^2]$. An alternative form, which is often more convenient for computation is given by the following lemma.

Lemma 2.1

Under the same regularity conditions as for the Cramér–Rao inequality,

$$I_\theta = -\mathrm{E}\left[\frac{\partial^2 l}{\partial \theta^2}\right].$$

Proof

$$\frac{\partial^2 l}{\partial \theta^2} = \frac{\partial}{\partial \theta}\left\{\frac{1}{L}\frac{\partial L}{\partial \theta}\right\} = -\frac{1}{L^2}\left(\frac{\partial L}{\partial \theta}\right)^2 + \frac{1}{L}\frac{\partial^2 L}{\partial \theta^2}$$

$$= -\left(\frac{\partial l}{\partial \theta}\right)^2 + \frac{1}{L}\left(\frac{\partial^2 L}{\partial \theta^2}\right).$$

If we can show that the second term has zero expectation, then we have the required result. But

$$\mathrm{E}\left[\frac{1}{L}\left(\frac{\partial^2 L}{\partial \theta^2}\right)\right] = \int \frac{1}{L}\frac{\partial^2 L}{\partial \theta^2} L \,\mathrm{d}\mathbf{x} = \int \frac{\partial^2 L}{\partial \theta^2}\,\mathrm{d}\mathbf{x}$$

$$= \frac{\partial^2}{\partial \theta^2}\left[\int L \,\mathrm{d}\mathbf{x}\right]$$

using regularity conditions to interchange the order of the differentiation and integration,

$$= \frac{\partial^2}{\partial \theta^2}[1] = 0.$$ □

We mentioned above, the fact that the Cramér–Rao lower bound is often attainable asymptotically. We now give a lemma which specifically allows us to investigate when it is attainable.

Lemma 2.2

Under the same regularity conditions as before, there exists an unbiased estimator $\hat{\theta}$ which attains the Cramér–Rao lower bound if and only if

$$\frac{\partial l}{\partial \theta} = I_\theta(\hat{\theta} - \theta).$$

Proof

In deriving the bound we used the inequality $(\mathrm{Cov}[U, V])^2 \leqslant \mathrm{Var}[U]\,\mathrm{Var}[V]$, and the bound will be attained if and only if equality is achieved here. However, equality occurs if and only if there is an exact linear relationship between U and V (corresponding to a correlation of ± 1 between U and V).

Remember that $U = \hat{\theta}$, $V = \partial l/\partial \theta$, so we have $\partial l/\partial \theta = c + d\hat{\theta}$, where c, d are constants.

Taking expectations in this equation gives $0 = c + \mathrm{d}\theta$ i.e. $c = -\mathrm{d}\theta$, so

$$\frac{\partial l}{\partial \theta} = \mathrm{d}(\hat{\theta} - \theta).$$

Now multiply the last equation by $(\partial l/\partial\theta)$ and take expectations.

$$\text{LHS} = \text{E}\left[\left(\frac{\partial l}{\partial\theta}\right)^2\right] = I_\theta$$

$$\text{RHS} = \text{dE}\left[\frac{\partial l}{\partial\theta}\,\hat{\theta}\right] - \text{d}\theta\ \text{E}\left[\frac{\partial l}{\partial\theta}\right]$$

$$= \text{d},$$

using

$$\text{E}\left[\frac{\partial l}{\partial\theta}\right] = 0 \quad \text{and} \quad \text{E}\left[\frac{\partial l}{\partial\theta}\,\hat{\theta}\right] = 1$$

from the proof of the inequality. □

Next we give two examples which illustrate Lemmas 2.1 and 2.2, as well as the Cramér–Rao inequality itself.

Example 2.1 (*continued*)

As before, suppose that $X_1, X_2, \ldots, X_n$ form a random sample from $N(\mu, \sigma^2)$ with σ^2 known.

$$L = \prod_{i=1}^{n} f(x_i; \mu) = \prod_{i=1}^{n} (2\pi\sigma^2)^{-1/2} \exp\left\{-\frac{1}{2\sigma^2}(x_i - \mu)^2\right\}$$

$$l = -\frac{n}{2}\ln(2\pi\sigma^2) - \frac{1}{2\sigma^2}\sum_{i=1}^{n}(x_i - \mu)^2$$

$$\frac{\partial l}{\partial\mu} = \frac{1}{\sigma^2}\sum_{i=1}^{n}(x_i - \mu) = \frac{n}{\sigma^2}(\bar{x} - \mu)$$

$$I_\theta = \text{E}\left[\left(\frac{\partial l}{\partial\mu}\right)^2\right] = \text{E}\left[\frac{n^2}{\sigma^4}(\bar{x} - \mu)^2\right].$$

But using the well-known result, $\bar{X} \sim N(\mu, \sigma^2/n)$, we have

$$\text{E}[(\bar{X} - \mu)^2] = \text{Var}[\bar{X}] = \sigma^2/n,$$

and

$$I_\theta = n/\sigma^2.$$

The lower bound is thus $I_\theta^{-1} = \sigma^2/n$, which is attained by $\bar{X}$, i.e. $\bar{X}$ is an MVUE for μ.

Confirmation that the bound is attained by $\bar{X}$ could alternatively have come from Lemma 2.2. With $I_\theta = n/\sigma^2$, $\hat{\theta} = \bar{X}$, and $\theta = \mu$, $\partial l/\partial\mu$ can be written as $I_\theta(\hat{\theta} - \theta)$, which is the form required by Lemma 2.2.

We could also have calculated I_θ using the alternative form given by Lemma 2.1.

When we looked at this example earlier with respect to unbiasedness and consistency there was nothing to choose between $\bar{X}$, the median, and the mid-range. However, if we are looking for MVUEs we now know that $\bar{X}$ cannot be improved upon by either of these estimators or by any other unbiased estimator. The asymptotic efficiencies of the other two estimators are $2/\pi$ and 0 respectively, and although their efficiencies improve for small sample sizes they are less than 1 except at $n = 2$, when all three estimators are identical. These results are, of course, for the normal distribution; the mean may not be the best estimator of the 'centre' of other symmetric distributions – see, for example, De Groot (1986, pp. 567–569) for an illustration of this point.

Example 2.4

Suppose that $X_1, X_2, \ldots, X_n$ form a random sample from the Poisson distribution $P(\theta)$.

$$L = \prod_{i=1}^{n} f(x_i; \theta) = e^{-n\theta}\theta^{\sum_{i=1}^{n} x_i} \Big/ \prod_{i=1}^{n} x_i!$$

$$l = -n\theta + \sum_{i=1}^{n} x_i \ln \theta - \ln\left(\prod_{i=1}^{n} x_i!\right)$$

$$\frac{\partial l}{\partial \theta} = -n + \sum_{i=1}^{n} x_i/\theta = \frac{n}{\theta}(\bar{x} - \theta)$$

$$\frac{\partial^2 l}{\partial \theta^2} = -\sum_{i=1}^{n} x_i/\theta^2$$

$$I_\theta = -\mathrm{E}\left[\frac{\partial^2 l}{\partial \theta^2}\right] = \frac{n\theta}{\theta^2} = \frac{n}{\theta}.$$

As with the previous example, it is not difficult to find I_θ directly from the definition $I_\theta = \mathrm{E}[(\partial l/\partial \theta)^2]$ using the fact that for the Poisson distribution, $\bar{X}$ has mean θ and variance θ/n. In other examples, the alternative formula for I_θ may be much easier to compute.

Again we can see that $\bar{X}$ attains the lower bound, either from Lemma 2.1 or directly from knowledge of its variance.

Example 2.5

We have skimmed over the regularity conditions which are required for the Cramér–Rao inequality. However, we should not lose sight of the fact that some restrictions exist on the use of the inequality. To illustrate this, we compute $I_\theta = \mathrm{E}[(\partial l/\partial \theta)^2]$ in a case where the regularity conditions do not hold. We see I_θ does not then provide a lower bound to the variance of unbiased estimators.

One of the regularity conditions is that the range of values of X must not depend on θ. Consider a random sample $X_1, X_2, \ldots, X_n$ from $U[0, \theta]$,

$$L = \begin{cases} \theta^{-n}, & 0 \leqslant x_{(1)} \leqslant x_{(2)} \leqslant \cdots \leqslant x_{(n)} \leqslant \theta, \\ 0, & \text{elsewhere} \end{cases}$$

where $x_{(i)}$ is the ith-order statistic. If we differentiate l, in the region where L is non-zero, we find

$$I_\theta = \mathrm{E}\left[\left(\frac{\partial l}{\partial \theta}\right)^2\right] = \frac{n^2}{\theta^2}.$$

However, it can be shown (see Exercise 2.3) that if $\hat{\theta} = [(n+1)/n]X_{(n)}$, then $\hat{\theta}$ is an unbiased estimator for θ, and

$$\mathrm{var}(\hat{\theta}) = \frac{\theta^2}{n(n+2)} < I_\theta^{-1}.$$

Furthermore, $-\mathrm{E}[\partial^2 l/\partial \theta^2]$ is different from $\mathrm{E}[(\partial l/\partial \theta)^2]$, and is actually negative. Therefore we should not try to use the inequality where it is inappropriate.

Extensions to the Cramér–Rao inequality (continued)

(3) Suppose now that we have a vector of k parameters $\boldsymbol{\theta}$. Then we define the (Fisher) information matrix as the matrix whose (i, j)th element is

$$I_{ij} = \mathrm{E}\left[\frac{\partial l}{\partial \theta_i}\frac{\partial l}{\partial \theta_j}\right]$$

where θ_i, θ_j are the ith and jth elements of $\boldsymbol{\theta}$.

Let I^{ij} be the (i, j)th element of the inverse of the information matrix. Then, under appropriate regularity conditions, we have for any unbiased estimator $\hat{\theta}_i$ of θ_i, $\mathrm{var}(\hat{\theta}_i) \geqslant I^{ii}$. Of course, we already know from the single-parameter Cramér–Rao inequality that $\mathrm{var}(\hat{\theta}_i) \geqslant I_{ii}^{-1}$, but $I^{ii} \geqslant I_{ii}^{-1}$ with strict inequality (and hence an improved lower bound) unless I_θ is diagonal (Rao, 1973, p. 74). To aid in calculation, we have an analogue to Lemma 2.1 which tells us that

$$I_{ij} = -\mathrm{E}\left[\frac{\partial^2 l}{\partial \theta_i \partial \theta_j}\right].$$

In fact, the inequality $\mathrm{var}(\hat{\theta}_i) \geqslant I^{ii}$ follows from a more general result, namely that $\mathrm{var}(\hat{\boldsymbol{\theta}}) - I_\theta^{-1}$ is positive semi-definite, where $\mathrm{var}(\hat{\boldsymbol{\theta}})$ is the covariance matrix of a vector of estimators $\hat{\boldsymbol{\theta}}$ for $\boldsymbol{\theta}$ – see Silvey (1970, pp. 41–42), who also introduces the following example.

Example 2.6

Consider the trinomial distribution, where X_1, X_2, X_3 has probability function

$$f(x_1, x_2, x_3; \theta_1, \theta_2, \theta_3) = \frac{n!}{x_1!x_2!x_3!}\theta_1^{x_1}\theta_2^{x_2}\theta_3^{x_3}.$$

We have the constraints $x_1 + x_2 + x_3 = n$, $\theta_1 + \theta_2 + \theta_3 = 1$, so the probability function can be written

$$f(\mathbf{x};\boldsymbol{\theta}) = \frac{n!}{x_1!x_2!(n-x_1-x_2)!}\theta_1^{x_1}\theta_2^{x_2}(1-\theta_1-\theta_2)^{n-x_1-x_2}$$

and we have two unknown parameters θ_1, θ_2.

$$L(\boldsymbol{\theta};\mathbf{x}) = f(\mathbf{x};\boldsymbol{\theta})$$

and

$$l = \ln\left[\frac{n!}{x_1!x_2!(n-x_1-x_2)!}\right] + x_1\ln\theta_1 + x_2\ln\theta_2 + (n-x_1-x_2)\ln(1-\theta_1-\theta_2)$$

$$\frac{\partial^2 l}{\partial\theta_1\partial\theta_2} = -\frac{(n-x_1-x_2)}{(1-\theta_1-\theta_2)^2}$$

$$\mathrm{E}[X_i] = n\theta_i \qquad i = 1, 2$$

so

$$-\mathrm{E}\left[\frac{\partial^2 l}{\partial\theta_1\partial\theta_2}\right] = \frac{n}{(1-\theta_1-\theta_2)} = I_{12}.$$

Similarly

$$I_{11} = n\left[\frac{1}{\theta_1} + \frac{1}{(1-\theta_1-\theta_2)}\right]$$

and

$$I_{22} = n\left[\frac{1}{\theta_2} + \frac{1}{(1-\theta_1-\theta_2)}\right].$$

Inverting the information matrix, we find, for example, that $I^{11} = \theta_1(1-\theta_1)/n$, and this gives a lower bound to the variance of any unbiased estimator for θ_1. But we know from properties of the binomial distribution that if $\hat{\theta}_1 = X_1/n$, then $\mathrm{var}(\hat{\theta}_1) = \theta_1(1-\theta_1)/n$, so $\hat{\theta}_1$ attains the lower bound.

Finally in this example, let us compare I^{11} with I_{11}^{-1}, the bound derived from the single-parameter version of the inequality

$$I_{11}^{-1} = \frac{1}{n}\left[\frac{1}{\theta_1} + \frac{1}{(1-\theta_1-\theta_2)}\right]^{-1} = \frac{\theta_1(1-\theta_1-\theta_2)}{n(1-\theta_2)}.$$

But $(1-\theta_1-\theta_2) < (1-\theta_1)(1-\theta_2)$ whenever $0 < \theta_1, \theta_2 < 1$, so

$$I_{11}^{-1} < \frac{\theta_1(1-\theta_1)}{n} = I^{11}.$$

2.5 Sufficiency

Although we introduce sufficiency as a property of point estimators, the idea is important throughout all branches of inference. We will give a Bayesian definition of sufficiency in Chapter 7, but here we give the more usual definition. In fact the two definitions turn out to be equivalent – see Lemma 7.1.

Definition 2.5

As usual, suppose that $X_1, X_2, \ldots, X_n$ form a random sample from $f(x; \theta)$. Suppose further that $t(x_1, x_2, \ldots, x_n)$ is a function of the observations $x_1, x_2, \ldots, x_n$, and not of θ, and that $T(X_1, X_2, \ldots, X_n)$ is the corresponding r.v.

T is then a *statistic*, and T is *sufficient* for θ – a *sufficient statistic* for θ – if the conditional distribution of the X_i given the value of T, does not depend on θ.

Sometimes we talk of the set of distributions with p.d.f.s $f(x; \theta)$ for some value of θ as a family of distributions, indexed by θ. If T satisfies the above property, then T is said to be *sufficient for the family of distributions* $\{f(x; \theta): \theta \in \Theta\}$.

What this definition means is that if T is sufficient for θ, then it contains all the information about θ which is contained in the X_i; once the value of T is known we can squeeze no more information out of the X_i regarding θ.

Example 2.7

Consider a binomial experiment with n trials and probability of success p, and define T to be the number of successes. We can write

$$T = \sum_{i=1}^{n} X_i, \text{ where } X_i = \begin{cases} 1, & \text{if the } i\text{th trial is a success} \\ 0, & \text{if the } i\text{th trial is a failure.} \end{cases}$$

The complete description of the outcome of the experiment is given by the sequence $X_1, X_2, \ldots, X_n$.

Once the value of T is known there is no way in which we can add to our knowledge about p – the only other information available is the order of successes and failures and this cannot help us. To prove the sufficiency of T formally, we proceed as follows.

Let $f(x_1, x_2, \ldots, x_n)$, $h(t)$ denote marginal probability functions for $(X_1, X_2, \ldots, X_n)$ and T respectively, with the conditional probability function for $(X_1, X_2, \ldots, X_n)$, given $T = t$, denoted by $g(x_1, x_2, \ldots, x_n \mid t)$. Then

$$\begin{aligned} g(x_1, x_2, \ldots, x_n \mid t) &= \frac{f(x_1, x_2, \ldots, x_n)}{h(t)} \\ &= \prod_{i=1}^{n} p^{x_i}(1-p)^{1-x_i} \bigg/ \binom{n}{t} p^t(1-p)^{n-t} \\ &= p^t(1-p)^{n-t} \bigg/ \binom{n}{t} p^t(1-p)^{n-t} \end{aligned}$$

$$= \left[\binom{n}{t}\right]^{-1}$$

which is independent of p, so T is sufficient for p or for the family of binomial distributions indexed by p.

In the above verification of sufficiency, note that the joint distribution $f(x_1, x_2, \ldots, x_n)$ has no term $\binom{n}{t}$ since all the different orderings of the x_i leading to the same value t are treated distinctly.

Despite our notation above, a sufficient statistic need not be a scalar; it could be a vector and this leads to a problem of definition, because if we add another element to our (vector) sufficient statistic, then it remains sufficient. To overcome this we introduce the idea of minimal sufficiency.

Definition 2.6

A statistic is *minimal sufficient* if it can be expressed as a function of every other sufficient statistic. Shortly we show that T in the example above is minimal sufficient, but first we look at an alternative interpretation of sufficiency and minimal sufficiency in general.

Consider the sample space defined by all possible values of $X_1, X_2, \ldots, X_n$; any statistic corresponds to a partitioning of this space, according to the different values of the statistic. A statistic which is not sufficient gives a partitioning which is too coarse to give all possible information about θ; a statistic which is sufficient, but not minimal sufficient, corresponds to a partition which is finer than necessary. Furthermore, taking a (1–1) function of a statistic does not change the partition, so a (1–1) function of a (minimal) sufficient statistic is also (minimal) sufficient. Thus even the requirement of minimal sufficiency does not lead to a unique statistic.

Example 2.8

Consider a binomial experiment with four trials and probability of success p. Let S denote success, and F failure. Then the sample space, and three possible partitions of this sample space are shown in Figure 2.2.

The three partitions correspond to three different statistics as follows:

(1) the solid lines correspond to $Y_1 = \{\text{outcome of first trial}\}$ – this is not sufficient and the partition is too coarse;
(2) the large dashed lines correspond to the statistic $Y_2 = \{\text{Number of successes}\}$ – this is minimal sufficient;
(3) the small dashed lines correspond to $Y_3 = (Y_1, Y_2)$ – this is sufficient, but not minimal sufficient, and the partition is finer than it need be.

Note that in (2) the same partition would be given by any statistic which is a (1–1) function of Y_2; such statistics would therefore also be minimal sufficient. For example, $\frac{1}{4}Y_2$, the proportion of successes, is minimal sufficient.

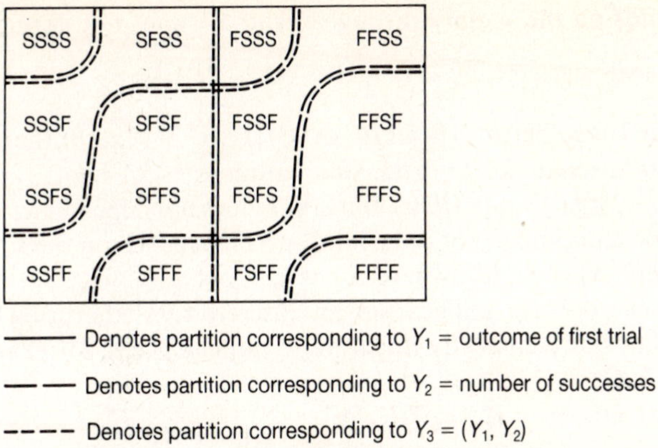

Fig. 2.2 Sample space and three partitions for a binomial experiment with four trials.

Example 2.7 (*continued*)

Returning to our general binomial example, suppose that T, defined earlier, is not minimal sufficient, but that some statistic U is. Then U is a function of T, such that $U(t_1) = U(t_2) = u$ for some $t_1 \neq t_2$. Consider the conditional distribution of the Xs given $U = u$ (with $T = t_i$) – this is

$$g(x_1, x_2, \ldots, x_n \mid u) = \frac{p^{t_i}(1-p)^{n-t_i}}{\binom{n}{t_1}p^{t_1}(1-p)^{n-t_1} + \binom{n}{t_2}p^{t_2}(1-p)^{n-t_2}}.$$

Clearly $g(\mathbf{x} \mid u)$ depends on p, so U is not sufficient. Hence T, already shown to be sufficient, is minimal sufficient.

Here we have assumed equality of U at only two values of T; similar reasoning will clearly hold if there are further equalities between the $U(t_i)$.

2.5.1 *Finding sufficient and minimal sufficient statistics*

The basic definitions of sufficiency and minimal sufficiency are often tedious to verify, so we introduce two results, Theorem 2.1 and Lemma 2.3, which provide more convenient ways of finding sufficient or minimal sufficient statistics. The first is:

Theorem 2.1 (The factorization theorem/criterion)

Suppose, as usual, that $X_1, X_2, \ldots, X_n$ form a random sample from $f(x; \theta)$. Then $T(X_1, X_2, \ldots, X_n)$ is a sufficient statistic for θ if and only if there exist two non-negative functions K_1, K_2 such that the likelihood function $L(\theta; \mathbf{x})$ can be written

$$L(\theta; \mathbf{x}) = K_1[t(x_1, x_2, \ldots, x_n); \theta]\ K_2[x_1, x_2, \ldots, x_n]$$

where K_1 depends on the x_i only through t, and K_2 does not depend on θ.

Proof

(a) Suppose that T is sufficient. Then $L(\theta; \mathbf{x}) = f(x_1, x_2, \ldots, x_n; \theta)$, the joint probability function (for discrete X) or p.d.f. (for continuous X) for $X_1, X_2, \ldots, X_n$, is $g(x_1, x_2, \ldots, x_n \mid t)h(t) = g(\mathbf{x} \mid t)h(t)$. But $g(\mathbf{x} \mid t)$ does not depend on θ if T is sufficient, and $h(t)$ is the marginal p.f. or p.d.f. for T, which depends on the x_i only through t.
   Hence $L(\theta; \mathbf{x}) = K_1 K_2$, where $K_1 \equiv h$, $K_2 \equiv g$.

(b) We prove the converse implication for discrete r.v.s; the proof for continuous r.v.s is similar, but technically much more difficult – see, for example, Lehmann (1986, pp. 54–6).
Suppose that $L(\theta; \mathbf{x}) = K_1[t; \theta]K_2[\mathbf{x}]$.
Now

$$h(t) = \sum_{\mathbf{x} \in \tau} L(\theta; \mathbf{x}), \text{ where } \tau = \{\mathbf{x}\colon T(\mathbf{x}) = t\}$$

$$= \sum_{\mathbf{x} \in \tau} K_1[t; \theta]K_2[\mathbf{x}] = K_1[t; \theta] \sum_{\mathbf{x} \in \tau} K_2[\mathbf{x}]$$

since K_1 takes the same value for all $\mathbf{x} \in \tau$. Thus

$$g(x_1, x_2, \ldots, x_n \mid t) = \frac{L(\theta; \mathbf{x})}{h(t)} = \frac{K_2[\mathbf{x}]}{\Sigma_{\mathbf{x} \in \tau} K_2[\mathbf{x}]}$$

where K_1 has cancelled out from the numerator and denominator. This last expression does not depend on θ, so T is sufficient. □

Before looking at examples of the factorization theorem note that it, and the concept of sufficiency in general, can be easily extended to the case of more than one parameter, i.e. if $\boldsymbol{\theta}$ is a vector of k parameters, and the statistic $\mathbf{T}$ is a vector of m scalar statistics, then $\mathbf{T}$ is sufficient for $\boldsymbol{\theta}$ if and only if $g(X_1, X_2, \ldots, X_n \mid \mathbf{T})$ does not depend on $\boldsymbol{\theta}$. An equivalent condition is that $L(\boldsymbol{\theta}; \mathbf{x})$ may be written

$$L(\boldsymbol{\theta}; \mathbf{x}) = K_1[\mathbf{t}; \boldsymbol{\theta}]K_2[\mathbf{x}], \qquad \text{with obvious notation.}$$

Note that m, the number of elements in $\mathbf{T}$, need not be the same as k, the number of parameters. However, if we are interested in *minimal* sufficiency, we will usually have $m = k$ although $m < k$, $m > k$ can each occasionally occur – see Cox and Hinkley (1974), p. 29.

Example 2.9

Consider a random sample of size n from $N(\mu, \sigma^2)$, so the likelihood function is

$$L = (2\pi\sigma^2)^{-n/2} \exp\left\{-\frac{1}{2\sigma^2} \sum_{i=1}^{n} (x_i - \mu)^2\right\}$$

$$= (2\pi\sigma^2)^{-n/2} \exp\left\{-\frac{1}{2\sigma^2}\sum_{i=1}^{n} x_i^2 + \frac{\mu}{\sigma^2}\sum_{i=1}^{n} x_i - n\mu^2/2\sigma^2\right\}.$$

Consider three possibilities:

(a) μ is unknown, σ^2 is known. If

$$t = \sum_{i=1}^{n} x_i, \qquad K_1[t;\mu] = \exp\{\mu t/\sigma^2 - n\mu^2/2\sigma^2\}$$

$$K_2[\mathbf{x}] = (2\pi\sigma^2)^{-n/2} \exp\left\{-\frac{1}{2\sigma^2}\sum_{i=1}^{n} x_i^2\right\}$$

then the factorization theorem shows that T or, equivalently $\bar{X}$, is sufficient for μ.

(b) μ, σ^2 both unknown. Let

$$t_1 = \sum_{i=1}^{n} x_i, \qquad t_2 = \sum_{i=1}^{n} x_i^2, \qquad \mathbf{t} = \begin{pmatrix} t_1 \\ t_2 \end{pmatrix}, \qquad K_1[\mathbf{t};\boldsymbol{\theta}] = L,\ K_2 \equiv 1.$$

Then the factorization theorem shows that $\mathbf{T}$ is sufficient for $\boldsymbol{\theta} = \begin{pmatrix} \mu \\ \sigma^2 \end{pmatrix}$. In fact any (1–1) function of $\mathbf{T}$ is also sufficient, such as $\begin{pmatrix} \bar{X} \\ S^2 \end{pmatrix}$, where

$$S^2 = \frac{1}{(n-1)}\sum_{i=1}^{n} (X_i - \bar{X})^2.$$

Note also that a factorization need not be unique. For example, here we could use $K_2 = (2\pi)^{-n/2}$ rather than $K_2 \equiv 1$.

(c) σ^2 unknown, μ known.

Let $t = \sum_{i=1}^{n} (x_i - \mu)^2$, $K_1 = L$, $K_2 \equiv 1$; then the factorization theorem shows that T is sufficient for σ^2.

Notice that the single sufficient statistic for σ^2 is not one of the two statistics which are jointly sufficient for μ and σ^2. Neither of these two statistics is, on its own, sufficient for σ^2, though clearly together they both are. However, the single sufficient statistic for μ, in (a), can be one of the pair found in (b).

Example 2.10

Consider a random sample of size n from $U[0, \theta]$. Now

$$L(\theta;\mathbf{x}) = \begin{cases} \theta^{-n}, & 0 \leqslant x_{(1)} \leqslant x_{(2)} \leqslant \cdots \leqslant x_{(n)} \leqslant \theta \\ 0, & \text{elsewhere} \end{cases}$$

At first sight this appears not to depend on any function of the data, but there *is* dependency since for some values of $\mathbf{x}$, $L = 0$. We can rewrite L, throughout the range of values of $\mathbf{x}$, as

$$L(\theta; \mathbf{x}) = \theta^{-n} H[\theta - x_{(n)}] H[x_{(1)}]$$

where

$$H(z) = \begin{cases} 1, & z \geqslant 0 \\ 0, & z < 0 \end{cases}.$$

If we put $K_1 = \theta^{-n} H[\theta - x_{(n)}]$, $K_2 = H[x_{(1)}]$, we see from the factorization theorem that $X_{(n)}$ is sufficient for θ.

Example 2.11

Let $X_1, X_2, \ldots, X_n$ be a random sample from the one-parameter Cauchy distribution, so

$$L(\theta; \mathbf{x}) = \prod_{i=1}^{n} \{1 + (x_i - \theta)^2\}^{-1} \times \pi^{-n}.$$

However hard you try, it is not possible to get the terms involving θ into a form which involves fewer than n functions of the x_i alone. The minimal sufficient statistic here consists of the vector $(X_{(1)}, X_{(2)}, \ldots, X_{(n)})$ of order statistics, as we see below.

The factorization theorem will provide us with a sufficient statistic, T, but we cannot be certain that T is minimal sufficient. The following result enables minimal sufficient statistics to be found.

Lemma 2.3

Consider the partition of the sample space of $X_1, X_2, \ldots, X_n$ which is defined by putting $\mathbf{x}$ and $\mathbf{y}$ into the same class of the partition if and only if $L(\boldsymbol{\theta}; \mathbf{x})/L(\boldsymbol{\theta}; \mathbf{y})$ does not depend on $\boldsymbol{\theta}$. Then any statistic corresponding to this partition is minimal sufficient.

Proof

We first prove sufficiency of $\mathbf{T}$ corresponding to the suggested partition, and then minimal sufficiency. As with the factorization theorem, the proof is done for discrete X; the proof for continuous X is identical in spirit, but technically more difficult – see Lehmann and Scheffé (1950).

$$g(\mathbf{x} \mid \mathbf{t}) = \frac{f(x_1, x_2, \ldots, x_n)}{h(\mathbf{t})} = \frac{L(\boldsymbol{\theta}; \mathbf{x})}{\Sigma_{\mathbf{y} \in \tau} L(\boldsymbol{\theta}; \mathbf{y})},$$

where τ is the set of $\mathbf{y}$s such that $\mathbf{T} = \mathbf{t}$, i.e. such that $L(\boldsymbol{\theta}; \mathbf{y})/L(\boldsymbol{\theta}; \mathbf{x}) = m(\mathbf{x}, \mathbf{y})$, a function which does not depend on $\boldsymbol{\theta}$.

Thus

$$g(\mathbf{x} \mid \mathbf{t}) = \frac{L(\boldsymbol{\theta}; \mathbf{x})}{\Sigma_{\mathbf{y} \in \tau} L(\boldsymbol{\theta}; \mathbf{x}) m(\mathbf{x}, \mathbf{y})} = \frac{L(\boldsymbol{\theta}; \mathbf{x})}{L(\boldsymbol{\theta}; \mathbf{x}) \, \Sigma_{\mathbf{y} \in \tau} m(\mathbf{x}, \mathbf{y})}$$

$$= [\Sigma_{\mathbf{y} \in \tau} m(\mathbf{x}, \mathbf{y})]^{-1}$$

which does not depend on $\boldsymbol{\theta}$. Hence **T** is sufficient.

Now suppose that **U** is any (other) sufficient statistic and that $\mathbf{U}(\mathbf{x}) = \mathbf{U}(\mathbf{y})$ for some pair of values of **X**, $(\mathbf{x}, \mathbf{y})$. If we can show that $\mathbf{U}(\mathbf{x}) = \mathbf{U}(\mathbf{y})$ implies $\mathbf{T}(\mathbf{x}) = \mathbf{T}(\mathbf{y})$, then **T** defines a partition at least as coarse as any other sufficient statistic, and so must be minimal sufficient.

Since **U** is sufficient, we have $L(\boldsymbol{\theta}; \mathbf{x}) = K_1[\mathbf{u}(\mathbf{x}); \boldsymbol{\theta}]K_2[\mathbf{x}]$, $L(\boldsymbol{\theta}; \mathbf{y}) = K_1[\mathbf{u}(\mathbf{y}); \boldsymbol{\theta}]\, K_2[\mathbf{y}]$, and since $\mathbf{u}(\mathbf{x}) = \mathbf{u}(\mathbf{y})$,

$$\frac{L(\boldsymbol{\theta};\ \mathbf{x})}{L(\boldsymbol{\theta};\ \mathbf{y})} = \frac{K_2[\mathbf{x}]}{K_2[\mathbf{y}]}$$

which does not depend on $\boldsymbol{\theta}$.

Hence, by our definition of **T**, $\mathbf{T}(\mathbf{x}) = \mathbf{T}(\mathbf{y})$ so the result is proved. □

Example 2.9 (*continued*)

Consider $X_1, X_2, \ldots, X_n$ from $N(\mu, \sigma^2)$ with μ, σ^2 both unknown. We have seen that we can get a pair of statistics which are jointly sufficient for μ, σ^2, but are they minimal sufficient?

$$\frac{L(\boldsymbol{\theta};\ \mathbf{x})}{L(\boldsymbol{\theta};\ \mathbf{y})} = \exp\left\{-\frac{1}{2\sigma^2}\sum_{i=1}^{n}[(x_i - \mu)^2 - (y_i - \mu)^2]\right\}$$

$$= \exp\left\{-\frac{1}{2\sigma^2}\left[\sum_{i=1}^{n}(x_i - \bar{x})^2 - \sum_{i=1}^{n}(y_i - \bar{y})^2 + n(\bar{x} - \mu)^2 - n(\bar{y} - \mu)^2\right]\right\},$$

using $\Sigma_{i=1}^n (x_i - \mu)^2 = \Sigma_{i=1}^n (x_i - \bar{x})^2 + n(\bar{x} - \mu)^2$ with a similar expression for the y_i.

Now in order for this ratio of likelihoods to be independent of μ, we must have $\bar{x} = \bar{y}$ – the only term depending on μ is $(n/\sigma^2)(\bar{x} - \bar{y})\mu$ in the exponent, and the coefficient of μ must be zero. For the ratio also to be independent of σ^2 we must have, in addition,

$$\sum_{i=1}^{n}(x_i - \bar{x})^2 = \sum_{i=1}^{n}(y_i - \bar{y})^2.$$

Thus, the partition we require for minimal sufficiency has $\bar{X}$ and $\Sigma_{i=1}^n (X_i - \bar{X})^2$ constant throughout any class of the partition, so $[\bar{X}, \Sigma_{i=1}^n (X_i - \bar{X})^2]$, or any (1–1) function of this pair, forms a minimal sufficient statistic for (μ, σ^2).

Example 2.11 (*continued*)

$X_1, X_2, \ldots, X_n$ are a random sample from the one-parameter Cauchy distribution; we said earlier that it was impossible to get fewer than n functions of $X_1, X_2, \ldots, X_n$, not depending on θ, in our likelihood. Let us verify this.

$$\frac{L(\theta;\ \mathbf{x})}{L(\theta;\ \mathbf{y})} = \frac{\prod_{i=1}^{n}[1 + (y_i - \theta)^2]}{\prod_{i=1}^{n}[1 + (x_i - \theta)^2]}.$$

The ratio can only have no dependence on θ if coefficients of all powers of θ, $(\theta, \theta^2, \ldots, \theta^{2n})$ are the same in the numerator and denominator. This will occur only if the *x*s and *y*s are the same (apart from a possible permutation of their order.) Thus the order statistics $(X_{(1)}, X_{(2)}, \ldots, X_{(n)})$ form a minimal sufficient statistic for θ. This assumes that the $X_{(i)}$ are all distinct, as will hold almost surely.

Example 2.12

Consider the general linear model

$$E[\mathbf{Y}] = X\boldsymbol{\beta},$$

where $\mathbf{Y} = (Y_1, Y_2, \ldots, Y_n)^T$ is a vector of n random variables, with $\text{var}(Y_i) = \sigma^2, i = 1, 2, \ldots, n$, $\text{Cov}[Y_i, Y_j] = 0, i \neq j$, X is a $(n \times p)$ matrix of fixed constants and $\boldsymbol{\beta}$ is a vector of unknown parameters. If the elements of $\mathbf{Y}$ are assumed to be normally distributed, then the likelihood function is

$$L(\boldsymbol{\beta}, \sigma^2; \mathbf{y}) = (2\pi\sigma^2)^{-n/2} \exp\left\{-\frac{1}{2\sigma^2}(\mathbf{y} - X\boldsymbol{\beta})^T(\mathbf{y} - X\boldsymbol{\beta})\right\}.$$

Now

$$(\mathbf{y} - X\boldsymbol{\beta})^T(\mathbf{y} - X\boldsymbol{\beta}) = [(\mathbf{y} - X\hat{\boldsymbol{\beta}}) + X(\hat{\boldsymbol{\beta}} - \boldsymbol{\beta})]^T[(\mathbf{y} - X\hat{\boldsymbol{\beta}}) + X(\hat{\boldsymbol{\beta}} - \boldsymbol{\beta})], \quad (2.1)$$

where $\hat{\boldsymbol{\beta}} = \hat{\boldsymbol{\beta}}(\mathbf{y}) = (X^TX)^{-1}X^T\mathbf{y}$. The estimate $\hat{\boldsymbol{\beta}}$ is the standard one for $\boldsymbol{\beta}$, and is derived by the method of least squares – see Section 3.4.2.

Expanding equation (2.1), the cross-product disappears because $X^TX\hat{\boldsymbol{\beta}} = X^T\mathbf{y}$, and we have

$$L(\boldsymbol{\beta}, \sigma^2; \mathbf{y}) = (2\pi\sigma^2)^{-n/2} \exp\left\{-\frac{1}{2\sigma^2}[(\mathbf{y} - X\hat{\boldsymbol{\beta}})^T(\mathbf{y} - X\hat{\boldsymbol{\beta}}) + (\hat{\boldsymbol{\beta}} - \boldsymbol{\beta})^T X^T X(\hat{\boldsymbol{\beta}} - \boldsymbol{\beta})]\right\}.$$

If $\mathbf{y}$, $\mathbf{z}$ are two realizations of the vector of random variables $\mathbf{Y}$, then

$$\frac{l(\boldsymbol{\beta}, \sigma^2; \mathbf{y})}{l(\boldsymbol{\beta}, \sigma^2; \mathbf{z})} = \exp\left\{-\frac{1}{2\sigma^2}[(\mathbf{y} - X\hat{\boldsymbol{\beta}}(\mathbf{y}))^T(\mathbf{y} - X\hat{\boldsymbol{\beta}}(\mathbf{y})) - (\mathbf{z} - X\hat{\boldsymbol{\beta}}(\mathbf{z}))^T(\mathbf{z} - X\hat{\boldsymbol{\beta}}(\mathbf{z})) + (\hat{\boldsymbol{\beta}}(\mathbf{y}) - \boldsymbol{\beta})^T X^T X(\hat{\boldsymbol{\beta}}(\mathbf{y}) - \boldsymbol{\beta}) - (\hat{\boldsymbol{\beta}}(\mathbf{z}) - \boldsymbol{\beta})^T X^T X(\hat{\boldsymbol{\beta}}(z) - \boldsymbol{\beta})]\right\}.$$

In order for this ratio not to depend on $\boldsymbol{\beta}$, the last two terms in the exponent must sum to zero, which only occurs if $\hat{\boldsymbol{\beta}}(\mathbf{y}) = \hat{\boldsymbol{\beta}}(\mathbf{z})$, so that $\hat{\boldsymbol{\beta}}$ is minimal sufficient for $\boldsymbol{\beta}$. If, in addition, σ^2 is unknown, then we must also have

$$(\mathbf{y} - X\hat{\boldsymbol{\beta}}(\mathbf{y}))^T(\mathbf{y} - X\hat{\boldsymbol{\beta}}(\mathbf{y})) = (\mathbf{y} - X\hat{\boldsymbol{\beta}}(\mathbf{z}))^T(\mathbf{y} - X\hat{\boldsymbol{\beta}}(\mathbf{z}))$$

in order that the likelihood ratio does not depend on σ^2. Hence, the pair of statistics $[\hat{\boldsymbol{\beta}}, (\mathbf{y} - X\hat{\boldsymbol{\beta}})^T(\mathbf{y} - X\hat{\boldsymbol{\beta}})]$ is minimal sufficient for $(\boldsymbol{\beta}, \sigma^2)$. But these are precisely the statistics which are generally used to make inferences about $\boldsymbol{\beta}$ and σ^2 in the standard approach to the general linear model.

2.5.2 *The role of sufficiency in finding MVUEs*

We noted at the beginning of Section 2.4 that much of the present chapter would be concerned with finding MVUEs. The introduction of sufficiency may have seemed like something of a digression, but we now show that sufficiency has an important role to play in the search for MVUEs. This follows from the corollary to a theorem which is now discussed.

Theorem 2.2 (The Rao–Blackwell Theorem)

Let $X_1, X_2, \ldots, X_n$ be a random sample of observations from a distribution with p.d.f. $f(x; \theta)$. Suppose that T is a sufficient statistic for θ and that $\hat{\theta}$ is any unbiased estimator for θ. Define $\hat{\theta}_T = \mathrm{E}[\hat{\theta} \mid T]$. Then

(a) $\hat{\theta}_T$ is a function of T alone;
(b) $\mathrm{E}[\hat{\theta}_T] = \theta$;
(c) $\operatorname{var} \hat{\theta}_T \leqslant \operatorname{var} \hat{\theta}$.

Proof

We shall not give the proof, although it is not too difficult – (a) is fairly obvious and (b), (c) follow from standard equalities and inequalities concerning expectations (see, for example, Cox and Hinkley, 1974, p. 258, Silvey, 1970 p. 28). □

Comments

(1) The theorem appears in several slightly different forms in different texts. For example, De Groot (1986, pp. 373, 374) looks at versions in which the MSE and mean absolute error are shown to be reduced for $\hat{\theta}_T$ compared to $\hat{\theta}$, which is no longer restricted to be unbiased.
(2) Examples are given in some texts (see, for example, Hogg and Craig (1970, p. 224) of starting with T and $\hat{\theta}$, and then deriving $\hat{\theta}_T$, but this procedure is rarely used in practice. Rather it is a corollary of the theorem that is more important than the theorem itself.

Corollary to Theorem 2.2

If an MVUE for θ exists, then there must be a function of the minimal sufficient statistic for θ which is an MVUE.

Proof

Suppose $\hat{\theta}$ is an MVUE and T is minimal sufficient, so T is certainly sufficient. Then by the Rao–Blackwell theorem, $\hat{\theta}_T$ (which might be identical to $\hat{\theta}$, though we have not ruled out the possibility of more than one MVUE), which is a function only of T, is unbiased and has a variance no larger than $\hat{\theta}$. Hence it is also an MVUE. We shall return to MVUEs later to show that often there is a unique function of T which is an MVUE, but we leave the subject again for the moment. □

2.6 Exponential families of distributions

At this point we introduce a family, or class of families, of probability distributions, which has found widespread application. The immediate purpose is to use the distributions in examining when a sufficient statistic of a specified dimension exists. Later in the chapter we link the families to the existence of MVUEs. More generally, the families are widely used in practice, for instance in the large area of generalized linear models – see Chapter 10 and McCullagh and Nelder (1989).

Definition 2.7

We say that a random variable X belongs to the *k-parameter exponential family of distributions* if its p.d.f., or probability function, can be written in the form

$$f(x; \boldsymbol{\theta}) = \exp\left\{\sum_{j=1}^{k} A_j(\boldsymbol{\theta})B_j(x) + C(x) + D(\boldsymbol{\theta})\right\}$$

where $A_1(\boldsymbol{\theta}), A_2(\boldsymbol{\theta}), \ldots, A_k(\boldsymbol{\theta}), D(\boldsymbol{\theta})$ are functions of $\boldsymbol{\theta}$ alone, and $B_1(x), B_2(x), \ldots, B_k(x), C(x)$ are well-behaved functions of x alone. Note that, as with many other definitions, there are some alternatives. A common one is to take $C(x)$ and $D(\boldsymbol{\theta})$ outside the exponential so that we have

$$f(x; \boldsymbol{\theta}) = C^*(x)D^*(\boldsymbol{\theta}) \exp\left\{\sum_{j=1}^{k} A_j(\boldsymbol{\theta})B_j(x)\right\}$$

with $C^*(x) = \exp\{C(x)\}$, $D^*(\boldsymbol{\theta}) = \exp\{D(\boldsymbol{\theta})\}$ in the above notation.

Table 2.1 Some well-known members of the one-parameter exponential family of distributions

Distribution	$f(x; \theta)$	$A(\theta)$	$B(x)$	$C(x)$	$D(\theta)$
Binomial	$\binom{n}{x}p^x(1-p)^{n-x}$	$\ln\{(p/(1-p))\}$	x	$\ln\left[\binom{n}{x}\right]$	$n\ln(1-p)$
Poisson	$e^{-\theta}\theta^x/x!$	$\ln\theta$	x	$-\ln(x!)$	$-\theta$
Exponential	$\theta e^{-\theta x}$	$-\theta$	x	0	$\ln\theta$
$N(0, \sigma^2)$	$(2\pi\sigma^2)^{-1/2}\exp\{-x^2/2\sigma^2\}$	$-\dfrac{1}{2\sigma^2}$	x^2	0	$-\frac{1}{2}\ln(2\pi\sigma^2)$
$N(\mu, 1)$	$(2\pi)^{-1/2}\exp\{-\frac{1}{2}(x-\mu)^2\}$	μ	x	$-x^2/2$	$-\frac{1}{2}\ln(2\pi)$ $-\mu^2/2$
Gamma (one parameter)	$\theta^n x^{n-1} e^{-\theta x}/(n-1)!$	$-\theta$	x	$(n-1)\ln(x)$	$n\ln(\theta)$ $-\ln[(n-1)!]$

Tables 2.1 and 2.2 show that many well-known distributions belong to the exponential family. Although we shall see that membership of an exponential family implies a number of desirable properties, there are nevertheless useful probability distributions which lie outside such families. For example, the Cauchy distribution has one parameter but does not belong to the one-parameter exponential family.

Table 2.2 Some members of the two-parameter exponential family of distributions

Distribution	$f(x; \boldsymbol{\theta})$	$A(\boldsymbol{\theta})$	$B(x)$	$C(x)$	$D(\boldsymbol{\theta})$
$N(\mu, \sigma^2)$	$(2\pi\sigma^2)^{-1/2} \exp\left\{-\dfrac{1}{2\sigma^2}(x-\mu)^2\right\}$	$A_1(\boldsymbol{\theta}) = -\dfrac{1}{2\sigma^2}$	$B_1(x) = x^2$	0	$-\frac{1}{2}\ln(2\pi\sigma^2)$
		$A_2(\boldsymbol{\theta}) = \dfrac{\mu}{\sigma^2}$	$B_2(x) = x$		$-\frac{1}{2}\mu^2/\sigma^2$
Gamma (two parameter)	$\dfrac{\beta^\alpha x^{\alpha-1} e^{-\beta x}}{\Gamma(\alpha)}$	$A_1(\boldsymbol{\theta}) = (\alpha - 1)$ $A_2(\boldsymbol{\theta}) = -\beta$	$B_1(x) = \ln(x)$ $B_2(x) = x$	0	$-\ln[\Gamma(\alpha)\beta^{-\alpha}]$

We have seen that sufficiency is a useful property and, in a trivial sense, sufficient statistics always exist, i.e. $(X_1, X_2, \ldots, X_n)$ is always trivially sufficient. However, we may be interested in whether a sufficient statistic $\mathbf{T} = (T_1, T_2, \ldots, T_m)$ exists for $\boldsymbol{\theta} = (\theta_1, \theta_2, \ldots, \theta_k)$ for a specified value of $m < n$ (usually we are interested in $m = k$). Consider the case where the regularity conditions hold which are relevant for the Cramér–Rao lower bound (we shall refer to these as the usual regularity conditions).

Lemma 2.4

If the usual conditions hold, then a vector of k sufficient statistics $\mathbf{T}$ exists for a vector of parameters $\boldsymbol{\theta}$ if and only if the distribution of X belongs to the k-parameter exponential family.

Proof

Given our usual random sample $X_1, X_2, \ldots, X_n$, the likelihood function when our distribution is a member of the k-parameter exponential family is

$$L(\boldsymbol{\theta}; \mathbf{x}) = \prod_{i=1}^{n} f(x_i; \boldsymbol{\theta}) = \prod_{i=1}^{n} \exp\left\{\sum_{j=1}^{k} A_j(\boldsymbol{\theta})B_j(x_i) + C(x_i) + D(\boldsymbol{\theta})\right\}$$

$$= \exp\left\{\sum_{j=1}^{k} A_j(\boldsymbol{\theta})\left(\sum_{i=1}^{n} B_j(x_i)\right) + nD(\boldsymbol{\theta}) + \sum_{i=1}^{n} C(x_i)\right\}.$$

If we set, say,

$$\mathbf{t} = \left(\sum_{i=1}^{n} B_1(x_i), \sum_{i=1}^{n} B_2(x_i), \ldots, \sum_{i=1}^{n} B_k(x_i)\right) = (t_1, t_2, \ldots, t_k)$$

with $\mathbf{T}$ the corresponding random vector, and define

$$K_1[\mathbf{t};\,\boldsymbol{\theta}] = \exp\left\{\sum_{j=1}^{k} A_j(\boldsymbol{\theta})t_j + nD(\boldsymbol{\theta})\right\}$$

and

$$K_2[\mathbf{x}] = \exp\left\{\sum_{i=1}^{n} C(x_i)\right\}$$

then it follows immediately from the factorization theorem that $\mathbf{T}$ is sufficient for $\boldsymbol{\theta}$. □

This proof is, of course, only in one direction. The proof that membership of the k-parameter exponential family is a necessary, as well as sufficient, condition for the existence of k sufficient statistics is much trickier – see, for example, Zacks (1971, Section 2.5).

Lemma 2.4 assumes that the usual regularity conditions hold. In cases where they do not, it may be possible to find alternative results which ensure the existence of a sufficient statistic of specified dimension – see, for example, Exercise 2.13. Returning to the case where the usual regularity conditions hold, we can go further and demonstrate minimal sufficiency for $\mathbf{T}$.

Lemma 2.5

Under the same conditions as Lemma 2.4, $\mathbf{T}$ defined in that lemma is minimal sufficient for $\boldsymbol{\theta}$.

Proof

Using Lemma 2.3, we must investigate the ratio

$$\frac{L(\boldsymbol{\theta};\,\mathbf{x})}{L(\boldsymbol{\theta};\,\mathbf{y})} = \exp\left\{\sum_{j=1}^{k} A_j(\boldsymbol{\theta})\left[\sum_{i=1}^{n} B_j(x_i) - \sum_{i=1}^{n} B_j(y_i)\right] + \sum_{i=1}^{n} [C(x_i) - C(y_i)]\right\}.$$

In order for the ratio not to depend on $\boldsymbol{\theta}$, we must have

$$\sum_{i=1}^{n} B_j(x_i) = \sum_{i=1}^{n} B_j(y_i), \qquad j = 1, 2, \ldots, k$$

which implies minimal sufficiency of $\mathbf{T}$ defined above. □

2.6.1 *Alternative parameterizations in the exponential family*

For many distributions, there is a conventional way in which to express the parameters defining the distributions, but for other distributions more than one parameterization may be in common use. For example, in Table 2.1 we have a fairly natural parameterization for the exponential distribution, but an alternative is to use the mean of the distribution as its single parameter. Then the p.d.f becomes $1/\theta\ e^{-x/\theta}$, $x > 0$, $A(\theta) = -(1/\theta)$ and $D(\theta) = -\ln(\theta)$, with $B(x)$, $C(x)$ remaining the same as in Table 2.1.

The factorization of the likelihood function in Lemma 2.4 suggests that a particular form of parameterization may be convenient.

Definition 2.8

Consider a random variable, whose distribution is a member of the exponential family. Let $\phi_j = A_j(\boldsymbol{\theta})$, $j = 1, 2, \ldots, k$. We can then write the p.d.f of the distribution as

$$f(x; \boldsymbol{\phi}) = \exp\left\{\sum_{j=1}^{k} \phi_j B_j(x) + C(x) + D(\boldsymbol{\phi})\right\}.$$

The parameters $\phi_1, \phi_2, \ldots, \phi_k$ are known as *natural parameters* or *canonical parameters* for this distribution. Such canonical parameterizations play an important role in the analysis of generalized linear models – see Chapter 10 and McCullagh and Nelder (1989).

We can, of course, also re-express the minimal sufficient statistic **T** by taking any (1–1) transformation. In the form given above, we have *natural sufficient statistics.* This natural sufficient statistic has a distribution which is also in a k-parameter exponential family. Its p.d.f. is

$$f(\mathbf{t}; \boldsymbol{\phi}) = \exp\left\{\sum_{j=1}^{k} \phi_j t_j + C(\mathbf{t}) + D(\boldsymbol{\phi})\right\}.$$

Example 2.13

If we look again at Table 2.1 we see that the natural parameter ϕ does not always coincide with the simplest, or most usual, parameterization of the distribution. For example, in the Poisson distribution, $\phi = \ln(\theta)$ is the natural parameter, and in the Binomial distribution the log odds, $\phi = \ln[p/(1-p)]$ is the natural parameter.

2.6.2 *Connections between exponential families and MVUEs*

As we saw earlier with sufficiency, our examination of exponential families of distributions is not entirely a digression from the search for MVUEs. There are various connections between MVUEs and membership of the exponential family. For example, it can be shown (Bickel and Doksum, 1977, p. 130) that if $k = 1$, and $A(\theta)$ is a strictly monotone function of θ, then T is an MVUE under the usual regularity conditions. Conversely, under the same conditions, for the existence of an MVUE for a function of θ we need membership of the exponential family.

2.7 Complete sufficient statistics

We next define a property called completeness, which also helps in our search for MVUEs.

Definition 2.9

Suppose that T is a sufficient statistic for θ, where, as usual, T is a function of X_1, $X_2, \ldots, X_n$, which are independent r.v.s, each with p.d.f. $f(x; \theta)$. Then T is *complete* if, whenever $h(T)$ is a function of T for which $\mathrm{E}[h(T)] = 0$ for all θ, $h(T) \equiv 0$ almost everywhere.

A weaker version of this definition states that T is *boundedly complete* if the property holds for all bounded functions $h(T)$. The reason for including boundedness in the definition is that it may be easier to verify bounded completeness, rather than completeness, and it is often all that is necessary for various results to hold.

The definition of completeness may also be given in terms of the family of distributions for T, indexed by θ, i.e. the family is (boundedly) complete if and only if T is (boundedly) complete.

The main implication of this definition is given by the following lemma.

Lemma 2.6

If T is a complete sufficient statistic for θ, then there is at most one function of T which is an unbiased estimator for θ, i.e. if we find a function $h(T)$ for which $\mathrm{E}[h(T)] = \theta$, then $h(T)$ is unique.

Proof

Suppose that $h_1(T)$, $h_2(T)$ are two such functions. Then

$$\mathrm{E}[h_1(T) - h_2(T)] = \theta - \theta = 0, \qquad \text{for all } \theta.$$

Hence $h_1(T) - h_2(T) \equiv 0$ almost everywhere, i.e. $h_1(T) \equiv h_2(T)$ almost everywhere. □

Lemma 2.7

If there exists an MVUE for θ, and $h(T)$ is an unbiased estimator for θ, where T is a complete minimal sufficient statistic for θ, then $h(T)$ is an MVUE.

Proof

The corollary to the Rao–Blackwell theorem says that there is at least one function of the minimal sufficient statistic T which is an MVUE. If T is complete then there is only one function of T which is unbiased. Thus if we can find a function of T which is unbiased it must be an MVUE.

Note, however, that we have not ruled out the possibility that another MVUE exists which is not a function of T.

Because of Lemma 2.7, completeness is a useful property, but does it often hold? The answer is 'yes', and one result is of particular importance.

Lemma 2.8

If the random variable X has a distribution belonging to the k-parameter exponential family, then, under the usual regularity conditions, the statistic

$$\left(\sum_{i=1}^{n} B_1(X_i), \sum_{i=1}^{n} B_2(X_i), \ldots, \sum_{i=1}^{n} B_k(X_i)\right)$$

which we already know is minimal sufficient, is complete.

Proof

The proof that membership of the exponential family implies completeness is quite difficult and we shall not include it – see, for example, Zacks (1971, Theorem 2.6.1).

Lemma 2.8 tells us that whenever we deal with a member of the exponential family we have complete (minimal) sufficient statistics – minimal is bracketed because sufficiency + completeness implies minimal sufficiency automatically, although the converse is not true – Zacks (1971, p. 73).

Membership of the exponential family is not a necessary condition for completeness, as the following example shows.

Example 2.14

Suppose that $X_1, X_2, \ldots, X_n$ form a random sample from the uniform distribution $U[0, \theta]$. In Example 2.10 we saw that $X_{(n)}$ is sufficient for θ, and it is not difficult to show, using Lemma 2.3, that $X_{(n)}$ is minimal sufficient.

The p.d.f. of $X_{(n)}$ is $f(x; \theta) = \begin{cases} \dfrac{nx^{n-1}}{\theta^n}, & 0 \leqslant x \leqslant \theta \\ 0, & \text{elsewhere} \end{cases}$

and $(n+1)X_{(n)}/n$ is an unbiased estimator for θ (Exercise 2.3).

Consider a function $h[X_{(n)}]$ such that

$$\mathrm{E}[h(X_{(n)})] = 0 = \int_0^\theta h(x) \frac{nx^{n-1}}{\theta^n} \, \mathrm{d}x.$$

Differentiating this equation with respect to θ gives

$$h(\theta) \frac{n\theta^{n-1}}{\theta^n} = 0 = \frac{n}{\theta} h(\theta).$$

Since $n \neq 0$, $\theta < \infty$, we must have $h(\theta) = 0$ for all θ, so $X_{(n)}$ is complete. We know that $(n+1)X_{(n)}/n$ is unbiased for θ, so it follows, from Lemma 2.7, that it must be an MVUE. Although completeness holds in a wide range of situations, it is not universal – see, for example, Exercise 2.15.

2.8 Problems with MVUEs

Much of this chapter has been concerned, directly or indirectly, with the search for MVUEs, and we now have a strategy based on complete sufficient statistics which will often find them. We illustrate this strategy with one further example, but we then discuss some problems associated with MVUEs, and in particular with the concept of unbiasedness.

Example 2.15

Consider a binomial experiment with n trials and probability of success p, and let X be the number of successes observed. Suppose that we wish to find an MVUE for $\theta = p(1-p)/n$, the variance of the usual estimator, $\hat{p} = X/n$, for p. Let us first examine $\hat{p}(1-\hat{p})/n$, as a possible estimator for θ.

Now

$$\mathrm{E}[\hat{p}] = p \qquad \text{and} \qquad \mathrm{E}[\hat{p}^2] = \frac{1}{n}[p + (n-1)p^2]$$

so

$$\mathrm{E}[\hat{p}(1-\hat{p})/n] = \frac{1}{n}[\mathrm{E}(\hat{p}) - \mathrm{E}(\hat{p}^2)] = \frac{1}{n^2}[np - p - np^2 + p^2]$$

$$= (n-1)p(1-p)/n^2.$$

An unbiased estimator of θ is therefore $\hat{\theta} = \hat{p}(1-\hat{p})/(n-1)$. As $p(1-p)/n$ is a variance, the divisor $(n-1)$ is not unexpected. From Example 2.7, $\hat{p}$ is minimal sufficient for p, and hence is sufficient for any function of p. Thus $\hat{p}$ is sufficient for θ and $\hat{p}$ is also complete because of membership of the one-parameter exponential family. Hence $\hat{p}(1-\hat{p})/(n-1)$ must be an MVUE for $p(1-p)/n$.

One problem with insisting on unbiasedness is that unbiased estimators need not necessarily exist, as illustrated in the next example.

Example 2.16

Consider a binomial experiment, as in Example 2.15, but now suppose that we are interested in estimating the odds $\psi = p/(1-p)$, when $0 < p < 1$. Suppose that $\hat{\psi}$ is an unbiased estimator for ψ taking the value $\hat{\psi}_x$ when $X = x$, $x = 0, 1, \ldots, n$. Then $p/(1-p) = \Sigma_{x=0}^{n}\hat{\psi}_x f(x)$ holds for all $0 < p < 1$, where $f(x)$ denotes the probability function for X.

Suppose that $M = \mathrm{Max}_{x=0,1,2,\ldots,n}(\hat{\psi}_x)$, so that $\Sigma_{x=0}^{n} \hat{\psi}_x f(x) \leqslant M$. By taking p sufficiently close to 1, we can obtain $p/(1-p) > M$ for any finite M, so M must be infinite. But if M is infinite, then so is $\Sigma_{x=0}^{n} \hat{\psi}_x f(x)$, because $f(x) > 0$, $x = 0, 1, \ldots, n$ when $0 < p < 1$.

Hence $p/(1-p) \neq \Sigma_{x=0}^{n} \hat{\psi}_x f(x)$ because $p/(1-p)$ is finite, so $\hat{\psi}_x$ is biased. It is therefore apparent that no unbiased estimator can exist in this example.

Even in examples where unbiased estimators, or indeed MVUEs, can be found, they are not always sensible. Consider the following example.

Example 2.17

Suppose that X is a Poisson random variable, with mean θ, and we are interested in estimating $\psi = \mathrm{e}^{-2\theta}$. The quantity to be estimated is the probability of observing zero on two successive occasions, which might be relevant if X were the number of accidents, failures, etc.

X, our single observation, is certainly complete since the Poisson distribution is a member of the one-parameter exponential family, and sufficient, so if we can find a function of X which is unbiased, then it must be an MVUE.

Consider

$$Y = (-1)^X, \qquad \mathrm{E}[Y] = \sum_{x=0}^{\infty} (-1)^x \frac{\theta^x \, \mathrm{e}^{-\theta}}{x!} = \mathrm{e}^{-\theta} \sum_{x=0}^{\infty} \frac{(-\theta)^x}{x!}$$
$$= \mathrm{e}^{-\theta} \, \mathrm{e}^{-\theta} = \mathrm{e}^{-2\theta}.$$

Hence $(-1)^X$ is unbiased, and is, in fact, an MVUE. However, it does not seem sensible since $(-1)^X = \pm 1$ and $0 < \mathrm{e}^{-2\theta} < 1$ for all $0 < \theta < \infty$.

This example has been discussed at length in the literature. Meeden (1987), for instance, argues that it is the concept of unbiasedness itself which causes the problem, although Lehmann (1983) suggests that it is 'inadequacy of information' in a certain sense which is to blame.

Although the example, and other aspects of unbiasedness, gives cause for concern, it does not detract from the general usefulness of MVUEs. It does, however, show that some caution may be necessary in their use.

2.9 Summary

In this chapter we have looked at a number of desirable properties of estimators. A dominant theme is the search for MVUEs. We have defined unbiasedness, and have seen how efficiency can be defined in terms of a lower bound to the variance of unbiased estimators. A condition for attainability of the lower bound was given.

The idea of sufficiency was then introduced. This concept is important throughout statistics and will appear prominently in several of the subsequent chapters. We gave results which enable us to find sufficient, and more particularly minimal sufficient, statistics. The role of sufficient statistics in finding MVUEs was then discussed.

Another important idea, which will reappear frequently in later chapters, is that of exponential families of distributions. These distributions were defined, and their relationship to the existence of sufficient statistics was discussed. Completeness was then defined, leading us finally to a strategy for finding MVUEs based on functions of complete sufficient statistics.

In the last section, two examples illustrated the fact that searching for MVUEs, although often useful, can be problematical.

Exercises

2.1 Obtain the Cramér–Rao lower bound when

$$f(x;\theta) = \begin{cases} \theta \, \mathrm{e}^{-\theta x}, & x > 0, \ \theta > 0 \\ 0, & \text{elsewhere.} \end{cases}$$

Can the bound be attained by an unbiased estimator in this example? Give reasons.

2.2 A random sample, $X_1, X_2, \ldots, X_n$, of n observations is taken from the two-parameter Weibull distribution with probability density function

$$f(x; \alpha, \beta) = \frac{\beta}{\alpha^\beta} x^{\beta-1} \exp\left\{-\left(\frac{x}{\alpha}\right)^\beta\right\}, \qquad x > 0, \quad \alpha > 0, \quad \beta > 0.$$

Assuming β is known, find a single function of $X_1, X_2, \ldots, X_n$ which is sufficient for α.

Show, however, that if α is known there is no single function of $X_1, X_2, \ldots, X_n$ which is sufficient for β.

2.3 Given that $X_1, X_2, \ldots, X_n$ is a random sample from $U[0, \theta]$, find the p.d.f. of $X_{(n)}$, the largest of the X_i.

Show that $2\bar{X}$ and $(n + 1)X_{(n)}/n$ are both consistent estimators of θ and compare their variances.

2.4 $\hat{\theta}_1, \hat{\theta}_2$ are independent unbiased estimators of a parameter θ, with variances σ_1^2, σ_2^2 respectively, and $\tilde{\theta} = k_1\hat{\theta}_1 + k_2\hat{\theta}_2$, where k_1, k_2 are constants. Find the values of k_1, k_2 for which $\tilde{\theta}$ is unbiased and has the smallest possible variance.

2.5 A random sample $X_1, X_2, \ldots, X_n$ is obtained from the distribution with p.d.f.

$$f(x; \lambda) = \left(\frac{\lambda}{2\pi x^3}\right)^{1/2} \exp\left[-\frac{\lambda(x-\mu)^2}{2\mu^2 x}\right], \qquad x > 0, \quad \lambda > 0, \quad \mu > 0.$$

Write down the likelihood function and find a pair of jointly sufficient statistics for the parameters λ and μ.

2.6 (a) Find the Cramér–Rao lower bound to the variance of unbiased estimators of θ, given a random sample $X_1, X_2, \ldots, X_n$ from the distribution with density

$$f(x; \theta) = e^{-(x-\theta)} \exp(-e^{-(x-\theta)}), \qquad -\infty < x < \infty.$$

(b) In the case of a random sample $X_1, X_2, \ldots, X_n$ from the Bernouilli distribution with probability function

$$f(x; \theta) = \theta^x(1-\theta)^{1-x}, \qquad x = 0, 1, \quad 0 \leqslant \theta \leqslant 1,$$

find, by differentiating the log of the likelihood or otherwise, whether the Cramér–Rao bound is attained for

(a) estimators of θ;

(b) estimators of θ^2.

2.7 Find minimal sufficient statistics for samples of size n from

(a) the uniform distribution on $[\theta - \frac{1}{2}, \theta + \frac{1}{2}]$

(b) the uniform distribution on $[-\theta, \theta]$.

2.8 The random variables $X_1, X_2, \ldots, X_n$ are independent with common probability density $\theta x^{\theta-1}(0 < x < 1)$, where the parameter $\theta > 0$ is unknown. Find a sufficient statistic for θ. Given that $-\log X_1$ is an unbiased estimator of θ^{-1}, find another unbiased estimator with smaller variance.

2.9 Find the MVUE, based on a random sample of size n, of the mean of an exponential distribution. Does this attain the Cramér–Rao lower bound?

2.10 A random sample $X_1, X_2, \ldots, X_n$ is obtained from $N(\mu, 1)$, and it is desired to estimate $\Phi(-\mu)$, the probability that an observation is less than zero. Write down a simple statistic depending upon X_1 above which is an unbiased estimator of $\Phi(-\mu)$, and hence or otherwise obtain the MVUE of this parameter.

2.11 Random variables X_1 and X_2 are independently and identically distributed with p.d.f. $\lambda\,e^{-\lambda x}$, $x \geqslant 0$. Given that the p.d.f. of $Z = X_1 + X_2$ is $\lambda^2 z\,e^{-\lambda z}$, show that the distribution of X_1 conditional upon $Z = z$ is the uniform distribution on $(0, z)$. Prove that the minimum variance unbiased estimator of

$$\Pr[X_i > 1] = e^{-\lambda}$$

based on observations X_1 and X_2 is given by

$$T = 0 \qquad \text{if } z \leqslant 1,$$

$$= \frac{z-1}{z} \qquad \text{if } z > 1.$$

2.12 Suppose $\hat{\theta}$ is an estimator for θ with probability function $\Pr[\hat{\theta} = \theta] = (n-1)/n$ and $\Pr[\hat{\theta} = \theta + n] = 1/n$ (and no other values of $\hat{\theta}$ are possible); show that $\hat{\theta}$ is consistent but that bias $(\hat{\theta}) \not\to 0$ as $n \to \infty$.

2.13 Let $X_1, X_2, \ldots, X_n$ denote a random sample from a distribution whose p.d.f. is

$$\begin{cases} f(x;\,\theta), & a \leqslant x \leqslant b(\theta) \\ 0 & x < a \text{ or } x > b(\theta) \end{cases}$$

where a is a constant, $b(\theta)$ is a fixed function of θ, and θ is a parameter to be estimated. Show that if a single sufficient statistic exists then it must be $X_{(n)}$, and that a necessary condition for $X_{(n)}$ to be a sufficient statistic is that $f(x;\,\theta) = g(x)h(\theta)$, where $g(x)$ does not depend on θ, and $h(\theta)$ does not depend on x.

2.14 In a binomial experiment with n trials, p is the probability of success, and X is the number of observed successes. Let $0 < p < 1$. Show that

(a) $X/(n - X)$ is not an unbiased estimator of $p/(1-p)$

(b) there is no unbiased estimator of $p/(1-p)$.

2.15 Consider a binomial experiment with probability of success p in which m (fixed) trials are conducted, resulting in R successes; a further set of trials is then conducted until s (fixed) further successes have occurred. The number of trials necessary in the second set is a random variable, N. By considering the function $U(R, N) = R/m - (s-1)/(N-1)$ show that (R, N) are jointly sufficient for p, but not complete.

2.16 Let $X_1, X_2, \ldots, X_n$ be n independent observations from an exponential distribution, with density

$$f(x;\,\theta) = \theta^{-1}\,e^{-x/\theta}, \qquad x \geqslant 0, \qquad \theta > 0.$$

Consider the following estimators of θ; $T_1 = \Sigma X_i/n$, $T_2 = \Sigma X_i/(n+1)$ and $T_3 = nY$, where $Y = \text{minimum } (X_1, X_2, \ldots, X_n)$. Which of these estimators are unbiased, which are functions of sufficient statistics and which are consistent?

Discuss the relative merits of T_1, T_2 and T_3 with regard to the above and any other relevant criteria.

2.17 A random sample $X_1, X_2, \ldots, X_n$ is selected from $N(\mu, 1)$. Write down the joint distribution of X_1 and $Y = X_2 + \cdots + X_n$, and hence or otherwise obtain the distribution of X_1 conditional upon the observed value $\bar{x}$ of the sample mean $\bar{X}$.

Let a random variable P take the value 1 if X_1 is less than 0, and the value 0 if X_1 exceeds 0. If the parameter μ is unknown, show, by using P or otherwise, that $\Phi(-\bar{X}\sqrt{n}/\sqrt{n-1})$ is an efficient estimator of $\Phi(-\mu)$, where the function Φ is the c.d.f. of a random variable distributed as $N(0, 1)$.

2.18 A random sample $X_1, X_2, \ldots, X_n$ is obtained from a truncated Poisson distribution with probability function

$$p(x;\lambda)=\frac{e^{-\lambda}}{1-e^{-\lambda}}\cdot\frac{\lambda^x}{x!},\qquad x=1,2,\ldots\ \lambda>0.$$

For $i=1,2,\ldots,n$, a random variable Z_i is defined by

$$Z_i=\begin{cases}X_i, & X_i\geqslant 2,\\ 0, & X_i=1.\end{cases}$$

Show that $\Sigma_{i=1}^n Z_i/n$ is an unbiased estimator of λ with efficiency

$$(1-e^{-\lambda})\bigg/\left[1-\left\{\frac{\lambda\, e^{-\lambda}}{1-e^{-\lambda}}\right\}\right]^2.$$

Explain briefly how you could construct a more efficient unbiased estimator. Does there exist an unbiased estimator with efficiency 1?

2.19 A random sample of size n is available from the distribution with density

$$f(x;\lambda)=\begin{cases}\dfrac{\lambda^3x(x+1)}{\lambda+2}\,e^{-\lambda x}, & x\geqslant 0\\ 0, & x<0\end{cases}$$

For what functions of λ do there exist unbiased estimators which attain the Cramér–Rao lower bound? Show that the information function is given by

$$I(\lambda)=\frac{2n(\lambda^2+6\lambda+6)}{\lambda^2(\lambda+2)^2},$$

and verify, for one of the above functions, the fact that it possesses an unbiased estimator attaining the Cramér–Rao lower bound.

2.20 Observations are made of the value of a random variable Y under two experimental conditions and in association with various values of a variable x. For each condition the model proposed for the dependence of Y on x is one of linear regression; observations (y_{ij}, x_{ij}), $j=1,2,\ldots,n_i$ are made under condition i ($i=1,2$), and the model may be written as

$$Y_{ij}=\alpha_i+\beta_i(x_{ij}-\bar{x}_i)+e_{ij}$$

($i=1,2$; $j=1,2,\ldots,n_i$) where $\bar{x}_i=(1/n_i)\,\Sigma_{j=1}^{n_i}x_{ij}$, and the e_{ij} are independent $N(0,\sigma^2)$ random variables. Find expressions for the minimum-variance unbiased estimators of $\alpha_1, \alpha_2, \beta_1, \beta_2$, and σ^2, and identify their joint distribution.

2.21 A random sample of size $n (n\geqslant 3)$ is available from the exponential distribution with density

$$f(x;\lambda)=\begin{cases}\lambda e^{-\lambda x} & x>0\\ 0 & x\leqslant 0\end{cases}.$$

Find the minimum-variance unbiased estimator of λ, and show that it has efficiency $(n-2)/n$. Why would you not expect this estimator to have efficiency 1?

2.22 X_1 and X_2 are independent binary random variables with

$$\left.\begin{aligned}\Pr[X_i = 1] &= p \\ \Pr[X_i = 0] &= 1 - p = q\end{aligned}\right\}, \qquad i = 1, 2.$$

Find a function of X_1 and X_2 which is an unbiased estimator of pq. Hence, or otherwise, find the minimum-variance unbiased estimator of the variance of a binomial distribution with index n ($\geqslant 2$) and parameter p, given a single observation from that distribution.

2.23 Suppose that $X_1, X_2, \ldots, X_n$ form a random sample from the distribution with probability density function

$$f(x; \sigma^2) = \frac{x}{\sigma^2}\,\mathrm{e}^{-x^2/2\sigma^2}, \qquad x \geqslant 0.$$

Determine the Cramér–Rao lower bound for θ when θ is (a) σ^2, (b) σ, and demonstrate whether or not the lower bound is attainable in the two cases.

2.24 The random variable X has a discrete distribution such that $\Pr[X = r] = \theta^{-1}$ for $r = 1, 2, \ldots, \theta$, where θ is an unknown positive integer. Show that Y, the maximum of a sample of n independent observations of X, is a complete sufficient statistic for θ and hence verify that $[Y^{n+1} - (Y-1)^{n+1}]/[Y^n - (Y-1)^n]$ is a minimum-variance unbiased estimator for θ.

2.25 Let $X_1, X_2, \ldots, X_n$ be a random sample from a Poisson distribution with mean $\mu > 0$. Find the minimum-variance unbiased estimator for

$$\Pr[X \leqslant 1] = (1 + \mu)\mathrm{e}^{-\mu}.$$

Does this estimator attain the Cramér–Rao lower bound?

2.26 Suppose that $X_1, X_2, \ldots, X_n$ are independent random variables, each with the inverse Gaussian distribution whose probability density function is

$$f(x; \theta_1, \theta_2) = \sqrt{\frac{\theta_1}{2\pi x^3}}\exp\left\{\frac{-\theta_1(x - \theta_2)^2}{2\theta_2^2 x}\right\}.$$

Show that this distribution is a member of the two-parameter exponential family of distributions. Hence, or otherwise, find a minimal sufficient statistic for (θ_1, θ_2). Find also a minimal sufficient statistic

(a) for θ_2 when θ_1 is known;

(b) for θ_1 when θ_2 is known.

2.27 Suppose $X_1, X_2, \ldots, X_n$ form a random sample from the normal distribution with unknown variance σ^2. Show that the sample variance

$$S^2 = \sum_{i=1}^{n} (X_i - \bar{X})^2/(n-1)$$

does not attain the Cramér–Rao lower bound for finite n, but does so as n tends to infinity. For what value of c does the estimator

$$c\sum_{i=1}^{n} (X_i - \bar{X})^2$$

of σ^2 have the lowest mean square error?

2.28 Consider a random sample of observations $X_1, X_2, \ldots, X_n$ from the Rayleigh distribution with probability density function

$$f(x;\theta) = \frac{x}{\theta}\exp(-x^2/2\theta), \qquad x > 0, \quad \theta > 0.$$

Show that this distribution belongs to the one-parameter exponential family of distributions, and deduce a minimal sufficient statistic, T, for θ, based on $X_1, X_2, \ldots, X_n$.

2.29 Observations made on random variables $X_1, X_2, \ldots, X_n$ are independent and identically distributed, each with a beta distribution whose p.d.f. is

$$f(x;\alpha,\beta) = \frac{1}{B(\alpha,\beta)}x^{\alpha-1}(1-x)^{\beta-1}, \qquad 0 < x < 1,$$

where α, β are positive parameters, and $B(\alpha, \beta)$ is a beta function. Write down minimal sufficient statistics for (α, β).

Chapter Three

Maximum likelihood and other methods of estimation

3.1 Introduction

The previous chapter was concerned with properties of estimators. It also included a strategy for constructing MVUEs, which constitutes one possible method for finding estimators. The present chapter examines several other, more general, techniques for finding estimators.

A great deal of the chapter is taken up with the method of maximum likelihood (ML) and various extensions or modifications of ML. The dominant role played by ML in this chapter is a fair reflection of its importance in recent work in statistical inference. However, like all other statistical inference procedures that we know of, ML has its faults. Other methods of estimation are therefore also discussed in this chapter, although we leave some approaches, such as those based on decision theory and the Bayesian approach until later chapters.

3.2 Maximum likelihood estimation

This is the best known, most widely used, and most important of the methods of estimation. All that is done is to write down the likelihood function $L(\boldsymbol{\theta}; \mathbf{x})$, and then find the value $\hat{\boldsymbol{\theta}}$ of $\boldsymbol{\theta}$ which maximizes $L(\boldsymbol{\theta}; \mathbf{x})$.

Definition 3.1

The *maximum likelihood estimator* (*MLE*) *is the value* $\hat{\boldsymbol{\theta}}$ *which* maximizes $L(\boldsymbol{\theta}; \mathbf{x})$.

This seems a sensible procedure – we are selecting the value of $\boldsymbol{\theta}$ for which our given set of observations has maximum probability. It turns out that MLEs are often those that might be suggested by 'common sense', as we shall see in the examples below. In addition, MLEs usually have 'fairly good' properties, as we shall discuss later.

To find the maximum of $L(\boldsymbol{\theta}; \mathbf{x})$ is an optimization problem, and we often need iterative procedures to solve the problem – see Section 3.2.2 below. For simple cases we can find closed-form expressions for $\hat{\boldsymbol{\theta}}$, but before illustrating this by examples, two general comments need to be made.

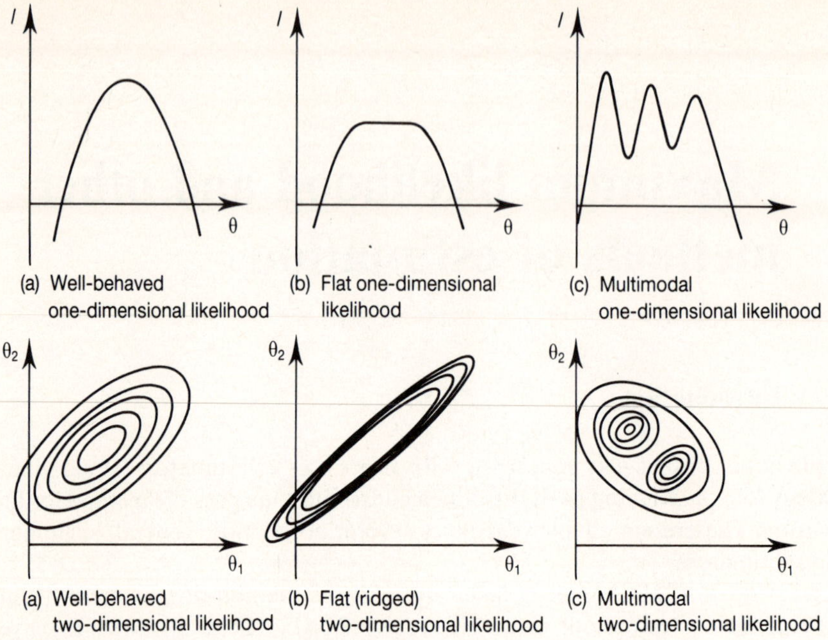

Fig. 3.1 Some illustrative plots of log-likelihood functions.

First, it is frequently the case that maximization of $l(\boldsymbol{\theta}; \mathbf{x}) = \ln[L(\boldsymbol{\theta}; \mathbf{x})]$ is easier than direct maximization of $L(\boldsymbol{\theta}; \mathbf{x})$. Because ln(.) is a monotone function, the same value $\hat{\boldsymbol{\theta}}$ maximizes both $l(\boldsymbol{\theta}; \mathbf{x})$ and $L(\boldsymbol{\theta}; \mathbf{x})$, and we usually choose to work with $l(\boldsymbol{\theta}; \mathbf{x})$.

Second, whenever it is feasible (i.e. if $\boldsymbol{\theta}$ consists of only one or two parameters) it is very often a good idea to plot $L(\boldsymbol{\theta}; \mathbf{x})$ or $l(\boldsymbol{\theta}; \mathbf{x})$. This will give information regarding possible problems in finding MLEs, caused by such phenomena as multiple maxima, or a 'flat' likelihood in the neighbourhood of $\hat{\boldsymbol{\theta}}$. Figure 3.1 gives illustrative plots of log likelihoods which are 'well behaved', 'flat', and 'multimodal', for both the one-parameter and two-parameter cases.

Example 3.1

Suppose that $X_1, X_2, \ldots, X_n$ is a random sample from the exponential distribution with p.d.f.

$$f(x; \theta) = \begin{cases} \theta \, e^{-\theta x} & x > 0 \\ 0 & \text{elsewhere} \end{cases}.$$

$$L(\theta; \mathbf{x}) = \theta^n \, e^{-\theta \sum_{i=1}^n x_i}, \qquad \text{so} \qquad l(\theta; \mathbf{x}) = n \ln(\theta) - \theta \sum_{i=1}^{n} x_i$$

$$\frac{\partial l}{\partial \theta} = \frac{n}{\theta} - \sum_{i=1}^{n} x_i \quad .$$

Equating $\partial l/\partial \theta$ to zero gives

$$\theta = \frac{1}{\bar{x}}.$$

$$\frac{\partial^2 l}{\partial \theta^2} = -\frac{n}{\theta^2} < 0,$$

so $\hat{\theta}$ does indeed correspond to a maximum, i.e. $1/\bar{X}$ is the MLE.

Example 3.2.

Consider an extension to the previous example, in which $X_1, X_2, \ldots, X_n$ are again exponentially distributed random variables. However, the values of $X_1, X_2, \ldots, X_n$ are censored at time T, so that values $X_1, X_2, \ldots, X_m$ are observed, but all we know about $X_{m+1}, X_{m+2}, \ldots, X_n$ is that they exceed T. Data of this sort arise when industrial components are tested, and time to failure is observed, but no component is observed beyond time T. The likelihood function is now

$$L(\theta; \mathbf{x}) = \prod_{i=1}^{m} f(x_i; \theta) \prod_{i=m+1}^{n} (1 - F(T; \theta))$$

where $F(T; \theta) = \Pr[X_i \leqslant T]$ is the (cumulative) distribution function for the X_i.
Now

$$1 - F(T; \theta) = \int_T^\infty \theta \, \mathrm{e}^{-\theta x} \, \mathrm{d}x$$

$$= \mathrm{e}^{-\theta T}.$$

Thus

$$L(\theta; \mathbf{x}) = \prod_{i=1}^{m} \theta \, \mathrm{e}^{-\theta x_i} \prod_{i=m+1}^{n} \mathrm{e}^{-\theta T}$$

and

$$l = \ln(L) = m \ln(\theta) - \theta \sum_{i=1}^{m} x_i - (n-m)\theta T$$

$$\frac{\partial l}{\partial \theta} = \frac{m}{\theta} - \left[(n-m)T + \sum_{i=1}^{m} x_i \right].$$

Equating this derivative to zero gives

$$\hat{\theta} = \frac{m}{(n-m)T + \Sigma_{i=1}^{m} x_i}.$$

Taking the second derivative of l shows that $\hat{\theta}$ corresponds to a maximum, so the MLE for θ is

$$\hat{\theta} = \frac{m}{(n-m)T + \sum_{i=1}^{m} X_i}.$$

3.2.1 Maximum likelihood estimation and exponential families of distributions

When a distribution in the k-parameter exponential family is written in terms of its natural parameterization (see Section 2.6), equations for the MLEs of the natural parameters are readily found, as seen in the following lemma.

Lemma 3.1

Suppose that $X_1, X_2, \ldots, X_n$ form a random sample from a distribution which is a member of the k-parameter exponential family, and that the p.d.f. of the distribution is written in terms of its natural parameterization:

$$f(x; \boldsymbol{\phi}) = \left\{ \sum_{j=1}^{k} \phi_j B_j(x) + C(x) + D(\boldsymbol{\phi}) \right\}.$$

If $T_j = \Sigma_{i=1}^{n} B_j(X_i)$, $j = 1, 2, \ldots, k$, then the MLEs of $\phi_1, \phi_2, \ldots, \phi_k$ are found by solving the equations

$$t_j = \mathrm{E}[T_j], \quad j = 1, 2, \ldots, k.$$

Proof

The likelihood function is

$$L(\boldsymbol{\phi}; \mathbf{x}) = \exp\left\{ \sum_{j=1}^{k} \phi_j t_j + \sum_{i=1}^{n} C(x_i) + nD(\boldsymbol{\phi}) \right\}$$

$$l = \ln(L) = \text{constant} + \sum_{j=1}^{k} \phi_j t_j + nD(\boldsymbol{\phi})$$

$$\frac{\partial l}{\partial \phi_j} = t_j + n \frac{\partial D(\boldsymbol{\phi})}{\partial \phi_j}.$$

Now recall the proof of the Cramér–Rao inequality (Section 2.4.1) where we saw that

$$\mathrm{E}\left[\frac{\partial l}{\partial \phi_j}\right] = 0, \quad \text{so } \mathrm{E}[T_j] = -n \frac{\partial D(\boldsymbol{\phi})}{\partial \phi_j}.$$

Hence

$$\frac{\partial l}{\partial \phi_j} = t_j - \mathrm{E}[T_j],$$

and so solving $\partial l/\partial\phi_j = 0$ is equivalent to $t_j = \mathrm{E}[T_j]$. □

In fact it can be shown (Bickel and Doksum, 1977, p. 106), that if these equations have a solution, then it is the *unique* MLE. Thus, we need not check second derivatives to verify that any solution corresponds to a maximum.

This is an important simplification in many cases where we find MLEs in exponential families. Of course, we may want MLEs for parameterizations other than the natural parameterization. This poses no additional problems because of the invariance property of MLEs which is discussed in Section 3.2.3 below.

Example 3.1 (*continued*)

For the exponential distribution, $T = \Sigma_{i=1}^{n} X_i$, $\mathrm{E}[T] = n\,\mathrm{E}[X_i] = n/\theta$.

Setting $t = \mathrm{E}[T]$ gives

$$\sum_{i=1}^{n} x_i = n/\theta$$

and, solving for θ, confirms $\hat{\theta} = 1/\bar{X}$ as the MLE.

3.2.2 *Computation of MLEs*

We have seen that for members of the exponential family it is easy to find equations whose solutions give MLEs. However, such equations may not have explicit solutions, and iterative methods are then needed to solve the equations. Indeed, once we get beyond the simple illustrative examples of text books such as this, and into 'real-world' problems involving ML estimation, it is common to need iterative optimization methods.

This section discusses those general optimization methods which have been found most useful in computing MLEs. The Newton–Raphson technique, the method of scoring, the simplex method and the EM algorithm are each described. Further information on these techniques, and others, can be found in Everitt (1987). A more recent, alternative, method of solving the equations needed to obtain the MLEs is described by Mak (1993).

The Newton–Raphson method

Suppose, as usual, that $l(\boldsymbol{\theta};\mathbf{x})$ is the log-likelihood function, which we wish to maximize. Let $\mathbf{g}(\boldsymbol{\theta})$ denote the vector of first derivatives of $l(\boldsymbol{\theta};\mathbf{x})$, and let $\mathbf{H}(\boldsymbol{\theta})$ denote the matrix of second derivatives, so that the ith element of $\mathbf{g}(\boldsymbol{\theta})$ is $\partial l(\boldsymbol{\theta};\mathbf{x})/\partial\theta_i$, and the (i, j)th element of $\mathbf{H}(\boldsymbol{\theta})$ is $\partial^2 l(\boldsymbol{\theta};\mathbf{x})/\partial\theta_i\partial\theta_j$, $i, j = 1, 2, \ldots, k$. The matrix $\mathbf{H}(\boldsymbol{\theta})$ is known as the *Hessian matrix.*

Suppose that $\boldsymbol{\theta}_0$ is an initial estimate of $\boldsymbol{\theta}$, and that $\hat{\boldsymbol{\theta}}$ is the MLE. We can expand $\mathbf{g}(\boldsymbol{\theta})$ about $\boldsymbol{\theta}_0$ using a Taylor series expansion to give

$$\mathbf{g}(\boldsymbol{\theta}) = \mathbf{g}(\boldsymbol{\theta}_0) + (\boldsymbol{\theta} - \boldsymbol{\theta}_0)^{\mathrm{T}}\mathbf{H}(\boldsymbol{\theta}_0) + \cdots \tag{3.1}$$

At $\boldsymbol{\theta} = \hat{\boldsymbol{\theta}}$ we know that $\mathbf{g}(\hat{\boldsymbol{\theta}}) = \mathbf{0}$, so substituting in (3.1) gives

$$\mathbf{0} = \mathbf{g}(\boldsymbol{\theta}_0) + (\hat{\boldsymbol{\theta}} - \boldsymbol{\theta}_0)^{\mathrm{T}}\mathbf{H}(\boldsymbol{\theta}_0) + \cdots$$

and $\hat{\boldsymbol{\theta}}$ is approximated by

$$\boldsymbol{\theta}_1 = \boldsymbol{\theta}_0 - \mathbf{g}(\boldsymbol{\theta}_0)\mathbf{H}^{-1}(\boldsymbol{\theta}_0). \tag{3.2}$$

If we now substitute $\boldsymbol{\theta}_1$ for $\boldsymbol{\theta}_0$ in (3.1), we get an improved estimate

$$\boldsymbol{\theta}_2 = \boldsymbol{\theta}_1 - \mathbf{g}(\boldsymbol{\theta}_1)\mathbf{H}^{-1}(\boldsymbol{\theta}_1). \tag{3.3}$$

This procedure is continued to find $\boldsymbol{\theta}_3, \boldsymbol{\theta}_4, \ldots$ until convergence is achieved. Provided that $\boldsymbol{\theta}_0$ is not too far from $\hat{\boldsymbol{\theta}}$, the procedure will converge to $\hat{\boldsymbol{\theta}}$ and will do so rapidly. If, however, $\boldsymbol{\theta}_0$ is not close to $\hat{\boldsymbol{\theta}}$ the method may fail to converge, for example because $\mathbf{H}(\boldsymbol{\theta}_0)$ may not be positive definite – for further details see Everitt (1987, Section 3.3).

Fisher's method of scoring

The method of scoring is a simple modification of the Newton–Raphson method. In (3.2), (3.3) and subsequent steps the Hessian matrices $\mathbf{H}(\boldsymbol{\theta}_0)$, $\mathbf{H}(\boldsymbol{\theta}_1)$, ... are replaced by their expectations. Note that $\mathrm{E}[\mathbf{H}(\boldsymbol{\theta})] = -I_{\theta}$, where I_{θ} is Fisher's information matrix – see Section 2.4. This use of $\mathrm{E}[\mathbf{H}(\boldsymbol{\theta})]$ rather than $\mathbf{H}(\boldsymbol{\theta})$ has two advantages over the Newton–Raphson method. First, because

$$\mathrm{E}\left[\frac{\partial^2 l(\boldsymbol{\theta};\, \mathbf{x})}{\partial\theta_i\partial\theta_j}\right] = -\mathrm{E}\left[\left(\frac{\partial l}{\partial\phi_i}\right)\left(\frac{\partial l}{\partial\theta_j}\right)\right]$$

under the usual regularity conditions (Lemma 2.1) we need only calculate first, rather than second, derivatives of l. The vector of first derivatives is known as the *score vector* or *score function*, hence the terminology 'method of scoring'. A second advantage is that $\mathrm{E}[\mathbf{H}(\boldsymbol{\theta})]$ is guaranteed to be positive-definite, thus eliminating the possible non-convergence problems of the Newton–Raphson method. Convergence may, however, be slower than for the Newton–Raphson method, and it can be advisable to switch to the latter procedure once a few iterations of the method of scoring have produced a value of $\boldsymbol{\theta}$ reasonably close to $\hat{\boldsymbol{\theta}}$. A thorough discussion of the properties of the method of scoring is given by Osborne (1992).

Note that for a distribution from the k-parameter exponential family with the natural parameterization, the method of scoring and the Newton–Raphson method are equivalent. This follows because

$$l(\boldsymbol{\theta};\, \mathbf{x}) = \text{constant} + \sum_{j=1}^{k} \phi_j t_j + nD(\boldsymbol{\phi})$$

$$\frac{\partial l(\boldsymbol{\phi};\, \mathbf{x})}{\partial\phi_j} = t_j + n\frac{\partial D(\boldsymbol{\phi})}{\partial\phi_j}$$

and

$$\frac{\partial^2 l(\boldsymbol{\phi}; \mathbf{x})}{\partial \phi_i \partial \phi_j} = n \frac{\partial^2 D(\boldsymbol{\phi})}{\partial \phi_i \partial \phi_j}.$$

As $D(\boldsymbol{\phi})$ does not depend on **x**, **H** and E[**H**] are identical.

Example 3.3

Suppose that $X_1, X_2, \ldots, X_n$ form a random sample from the two-parameter Cauchy distribution with p.d.f.

$$f(x; \boldsymbol{\theta}) = \frac{\theta_2}{\pi[\theta_2^2 + (x - \theta_1)^2]}, \qquad -\infty < \theta_1 < \infty, \quad \theta_2 > 0.$$

The log-likelihood function is

$$l(\boldsymbol{\theta}; \mathbf{x}) = n \ln(\theta_2) - \sum_{i=1}^{n} \ln[\theta_2^2 + (x_i - \theta_1)^2] - n \ln(\pi)$$

$$\mathbf{g}'(\boldsymbol{\theta}) = \left(2 \sum_{i=1}^{n} \frac{(x_i - \theta_1)}{[\theta_2^2 + (x_i - \theta_1)^2]}, \frac{n}{\theta_2} - 2 \sum_{i=1}^{n} \frac{\theta_2}{[\theta_2^2 + (x_i - \theta_1)^2]} \right)$$

and

$$\mathbf{H}(\boldsymbol{\theta}) = \begin{bmatrix} 2 \displaystyle\sum_{i=1}^{n} \frac{(x_i - \theta_1)^2 - \theta_2^2}{[\theta_2^2 + (x_i - \theta_1)^2]^2} & -4\theta_2 \displaystyle\sum_{i=1}^{n} \frac{(x_i - \theta_1)}{[\theta_2^2 + (x_i - \theta_1)^2]^2} \\ -4\theta_2 \displaystyle\sum_{i=1}^{n} \frac{(x_i - \theta_1)}{[\theta_2^2 + (x_i - \theta_1)^2]^2} & -\dfrac{n}{\theta_2^2} + 2 \displaystyle\sum_{i=1}^{n} \frac{\theta_2^2 - (x_i - \theta_1)^2}{[\theta_2^2 + (x_i - \theta_1)^2]^2} \end{bmatrix}.$$

Taking expectations of **H** gives

$$\mathrm{E}[\mathbf{H}] = \begin{bmatrix} -\dfrac{n}{2\theta_2^2} & 0 \\ 0 & -\dfrac{n}{2\theta_2^2} \end{bmatrix}$$

so the method of scoring is much easier to implement than the Newton–Raphson method in this case. Further discussion of this example, including a numerical illustration, is given by Morgan (1978).

The simplex method

The method of scoring and the Newton–Raphson method are examples of *gradient methods*, so-called because they use the derivatives of $l(\boldsymbol{\theta}; \mathbf{x})$ to decide the direction of change between successive estimates of $\boldsymbol{\theta}$. The simplex method is an example of a direct-search method which does not require the computation of derivatives. At each stage of the method, the value of $l(\boldsymbol{\theta}; \mathbf{x})$ is calculated at the vertices of a simplex (a triangle when $k = 2$, a tetrahedron when $k = 3$). These values then determine the position and size of the next simplex. The position will be adjacent to that of the previous simplex, in a direction which appears likely to increase the value of $l(\boldsymbol{\theta}; \mathbf{x})$. The size of the simplex will shrink as a maximum is approached, thus ensuring convergence. Further details, together with an example, are given by Everitt (1987, Chapter 4).

All iterative maximization methods can encounter problems when there are multiple maxima, because of possible convergence to a local, rather than global, maximum. Gates (1993) suggests that, in an example from spatial statistics, the simplex method is less likely to become trapped at a local maximum than are gradient methods.

The EM algorithm

This algorithm, which was popularized by Dempster *et al.* (1977), is particularly useful when some of the data are missing but, as noted in Dempster *et al.*'s (1977) paper, other types of problem can also be solved by the same general method. For example, Dempster *et al.* (1977) applied the algorithm in several contexts, including grouped data, mixtures of distributions, variance components and factor analysis. More recently, Green (1990) used a modification of the EM algorithm to maximize penalized likelihoods (see Section 3.3.2), and demonstrated the relevance of this approach in ridge regression and in Poisson regression. Each iteration of the algorithm consists of two steps, the E- or expectation step, and the M- or maximization step.

Suppose that we have an initial set of parameter estimates $\boldsymbol{\theta}_0$. The E-step consists of evaluating the conditional expectation of the log-likelihood for all the data, where the expectation is taken with regard to the distribution of the 'missing' data, conditional on $\boldsymbol{\theta}_0$ and on the observed data. In other words, if the data are decomposed into observed and missing values as $\mathbf{x} = (\mathbf{x}_{\text{obs}}, \mathbf{x}_{\text{miss}})$, and if $f(\mathbf{x}_{\text{miss}} \mid \mathbf{x}_{\text{obs}}, \boldsymbol{\theta}_0)$ is the conditional p.d.f. of the missing data, given $\mathbf{x}_{\text{obs}}$ and $\boldsymbol{\theta}_0$, then we calculate

$$\int l(\boldsymbol{\theta}; \mathbf{x}) f(\mathbf{x}_{\text{miss}} \mid \mathbf{x}_{\text{obs}}, \boldsymbol{\theta}_0)\, d\mathbf{x}_{\text{miss}}.$$

When the probability distribution of the data belongs to the k-parameter exponential family (see Section 2.6), the log-likelihood is a linear function of the k minimal sufficient statistics, and the E-step reduces to finding the conditional expectations of these sufficient statistics, and then substituting them in the complete log-likelihood function (Exercise 3.5). When the log-likelihood is a linear function of the missing data themselves, the E-step simplifies still further. It then consists of replacing the missing values by their conditional expectations in the complete likelihood function.

The M-step is straightforward. The 'expected log-likelihood' formed in the E-step is maximized with respect to $\boldsymbol{\theta}$, giving a new estimate $\boldsymbol{\theta}_1$. We then return to the E-step, with $\boldsymbol{\theta}_1$ replacing $\boldsymbol{\theta}_0$, and cycle through E- and M-steps until convergence is achieved.

Example 3.2 (*continued*)

In this example we have exponentially distributed random variables $X_1, X_2, \ldots, X_n$, with censoring at time T. The values of $X_1, X_2, \ldots, X_m$ are observed, but all we know about $X_{m+1}, X_{m+2}, \ldots, X_n$ is that they exceed T. We saw earlier that there is an explicit expression for the MLE

$$\hat{\theta} = \frac{m}{(n-m)T + \Sigma_{i=1}^{m} x_i},$$

so we do not need an iterative method. However, the example provides a useful illustration of how the EM-algorithm works.

Suppose that $n = 5$, $m = 3$, $T = 10$, $x_1 = 1.5$, $x_2 = 2.5$, $x_3 = 6.0$. Then

$$\hat{\theta} = \frac{3}{20 + 10} = 0.100.$$

Now treat x_4, x_5 as missing data and use the EM-algorithm to find $\hat{\theta}$, using as an initial estimate θ_0, the MLE based only on x_1, x_2, x_3:

$$\theta_0 = \frac{m}{\Sigma_{i=1}^{m} x_i} = \frac{3}{10} = 0.300.$$

Because the complete log-likelihood is a linear function of x_4 and x_5 (Exercise 3.6), the E-step involves finding $\mathrm{E}[X_i \mid x_1, x_2, x_3, \theta]$, $i = 4, 5$, and substituting these expected values in the likelihood function. For an exponential random variable, X, with parameter θ we have that $\mathrm{E}[X \mid X \geqslant T] = T + 1/\theta$, (Exercise 3.4) so the E-step gives

$$\mathrm{E}[X_i \mid X_i \geqslant 10, \theta = 0.3] = 10 + \frac{10}{3} = \frac{40}{3}, \qquad i = 4, 5.$$

Treating these as genuine observations x_4, x_5, the MLE of θ is

$$\frac{n}{\Sigma_{i=1}^{n} x_i} = \frac{5}{110/3} = 0.136.$$

This gives the new estimate θ_1 and completes the first M-step. Repeating the procedure we have

$$\mathrm{E}[X_i \mid X_i \geqslant 10, \theta = 0.136] = 52/3$$

$$\theta_2 = 15/134 = 0.112$$

$$\mathrm{E}[X_i \mid X_i \geqslant 10, \theta = 0.112] = 284/15$$

$$\theta_3 = 75/718 = 0.104$$

$$\mathrm{E}[X_i \mid X_i \geqslant 10, \theta = 0.104] = 1468/75$$

$$\theta_4 = 3750/3686 = 0.102,$$

and so on.

So far we have implicitly assumed that MLEs are always found by solving $\partial l/\partial\theta_j = 0$, $j = 1, 2, \ldots, k$. The following example shows that this is not always true. The example also reinforces the desirability, noted earlier in the chapter, of plotting the likelihood function. Such a plot will help to avoid trying to solve $\partial l/\partial\theta = 0$ (or $\partial L/\partial\theta = 0$) when no solution exists.

Example 3.4

Suppose that $X_1, X_2, \ldots, X_n$ form a random sample from the uniform distribution, $U[0, \theta]$, so the likelihood function is

$$L(\theta; \mathbf{x}) = \begin{cases} \theta^{-n}, & 0 \leqslant x_{(1)} \leqslant x_{(2)} \leqslant \cdots \leqslant x_{(n)} \leqslant \theta \\ 0, & \text{elsewhere} \end{cases}$$

It is clear that trying to solve $\partial L/\partial\theta = 0$ or $\partial l/\partial\theta = 0$ will not work in this example because L is a monotone function of θ in the range for which L is non-zero. L is maximized by taking θ as small as possible. We have that $\theta \geqslant x_{(n)}$, so the MLE of θ is $X_{(n)}$.

3.2.3 *Properties of MLEs*

In this section we look at how MLEs behave with respect to the properties defined in Chapter 2, namely unbiasedness, consistency, efficiency and sufficiency. First, however, we examine the property of invariance, already mentioned in the proof of Lemma 3.1.

Invariance

Lemma 3.2

Suppose that $\boldsymbol{\theta}$ and $\boldsymbol{\phi}$ represent two alternative parameterizations for some probability distribution, and that $\boldsymbol{\phi}$ is a (1–1) function of $\boldsymbol{\theta}$, so that we can write $\boldsymbol{\phi} = \mathbf{g}(\boldsymbol{\theta})$, $\boldsymbol{\theta} = \mathbf{h}(\boldsymbol{\phi})$ for appropriate functions $\mathbf{g}(.)$, $\mathbf{h}(.)$. If $\hat{\boldsymbol{\theta}}$ is the MLE of $\boldsymbol{\theta}$, then the MLE of $\boldsymbol{\phi}$ is $\mathbf{g}(\hat{\boldsymbol{\theta}})$.

Proof

This follows fairly directly from the definition of MLEs, for suppose that the value of $\boldsymbol{\phi}$ which maximizes L corresponds to $\tilde{\boldsymbol{\theta}} \neq \hat{\boldsymbol{\theta}}$, i.e. $L(\mathbf{g}(\tilde{\boldsymbol{\theta}}); \mathbf{x}) > L(\mathbf{g}(\hat{\boldsymbol{\theta}}); \mathbf{x})$. Taking the inverse function $h(.)$ we have that $L(\tilde{\boldsymbol{\theta}}; \mathbf{x}) > L(\hat{\boldsymbol{\theta}}; \mathbf{x})$, so $\hat{\boldsymbol{\theta}}$ is not the MLE. □

The situation when $\boldsymbol{\phi}$ is not a (1–1) function of $\boldsymbol{\theta}$ is more complicated, with more than one possible approach to finding a MLE for $\boldsymbol{\phi}$. Pal and Berry (1992) give sufficient conditions under which two other approaches give the same estimator $\boldsymbol{\phi}(\hat{\boldsymbol{\theta}})$.

Example 3.5

Consider $X_1, X_2, \ldots, X_n \sim N(\mu, \sigma^2)$, μ, σ^2 both unknown.

Then the log likelihood is

$$l = -\frac{n}{2}\ln(2\pi\sigma^2) - \frac{1}{2\sigma^2}\sum_{i=1}^{n}(x_i - \mu)^2.$$

$$\frac{\partial l}{\partial \mu} = \frac{1}{\sigma^2}\sum_{i=1}^{n}(x_i - \mu), \qquad \frac{\partial l}{\partial \sigma^2} = -\frac{n}{2\sigma^2} + \frac{1}{2\sigma^4}\sum_{i=1}^{n}(x_i - \mu)^2.$$

Equating these two derivatives to zero and solving gives $\hat{\mu} = \bar{x}$ from the first equation, and substituting in the second we have

$$\hat{\sigma}^2 = \frac{1}{n}\sum_{i=1}^{n}(x_i - \bar{x})^2.$$

It can be verified that this solution does indeed provide a maximum, so the MLEs for μ, σ^2 are $\bar{X}$ and $(1/n)\,\Sigma_{i=1}^{n}(X_i - \bar{X})^2$ respectively.

Because we wanted the MLE for σ^2 we differentiated with respect to σ^2. However, we could equally well have differentiated with respect to σ, found the MLE, $\hat{\sigma}$, for σ, and, using the invariance property, squared $\hat{\sigma}$ to obtain the same estimator as before for σ^2; because σ is restricted to non-negative values, σ^2 is a (1–1) function of σ.

In Table 2.2 we saw that the natural parameters for the normal distribution are $1/2\sigma^2$ and μ/σ^2, with corresponding natural sufficient statistics

$$T_1 = \sum_{i=1}^{n} X_i^2, \qquad T_2 = \sum_{i=1}^{n} X_i.$$

It is easy to show that

$$E[T_1] = n[\sigma^2 + \mu^2].$$

$$E[T_2] = n\mu.$$

Using Lemma 3.1, and the invariance property, the MLEs of μ, σ^2 can be found by solving

$$\sum_{i=1}^{n} x_i = n\mu$$

$$\sum_{i=1}^{n} x_i^2 = n[\sigma^2 + \mu^2],$$

leading to the same solutions as before.

Unbiasedness

It is clear from our examples that MLEs can be unbiased or biased.

Example 3.1 (*continued*)

$\bar{X}$ is a biased estimator for θ, although, if we use the parameterization

$$f(x;\,\phi) = \begin{cases} \dfrac{1}{\phi}\,\mathrm{e}^{-x/\phi}, & x > 0 \\ 0, & \text{elsewhere} \end{cases}$$

then the MLE $\bar{X}$ is an unbiased estimator for ϕ.

Example 3.3 (*continued*)

$X_{(n)}$ is a biased estimator for θ – see Exercise 2.3.

Example 3.5 (*continued*)

$\bar{X}$ is an unbiased estimator for μ, but $\hat{\sigma}^2$ is a biased estimator for σ^2.

Note that in all cases in these examples, when the MLE is a biased estimator its bias tends to 0 as $n \to \infty$.

Consistency

Subject to fairly weak regularity conditions, ML estimators are consistent – see, for example, Cox and Hinkley (1974, p. 288).

Efficiency

Since ML estimators may be biased we can only talk in general about *asymptotic* efficiency.

Lemma 3.3

Given the usual regularity conditions, it follows that if $\hat{\theta}$ is the MLE for θ, then asymptotically $\hat{\theta} \sim N(\theta, I_\theta^{-1})$.

Proof

No proof will be given here – see, for example, Cox and Hinkley (1974, p. 294) for an outline. □

This result shows that $\hat{\theta}$ is asymptotically efficient, but it also includes asymptotic normality. Estimators which are consistent, asymptotically efficient, and asymptotically normally distributed are sometimes known as *best asymptotically normal* (BAN).

Lemma 3.3, as well as the earlier results for unbiasedness and consistency, can be generalized simply to the multiparameter case, i.e. if $\hat{\boldsymbol{\theta}}$ is the MLE for $\boldsymbol{\theta}$ then $\hat{\boldsymbol{\theta}} \sim N(\boldsymbol{\theta}, I_\theta^{-1})$, where I_θ is the information matrix.

Lemma 3.3, and its multiparameter extension, is asymptotic, but there is also an important result regarding the efficiency of MLEs which holds for any sample size, namely:

Lemma 3.4

Suppose that there exists an unbiased estimator, $\tilde{\theta}$, which attains the Cramér–Rao bound. Suppose further that $\hat{\theta}$, the MLE for θ, is a solution to $\partial l/\partial\theta = 0$. Then $\tilde{\theta} = \hat{\theta}$.

Proof

From Lemma 2.2, $\tilde{\theta}$ must satisfy $\partial l/\partial\theta = I_\theta(\tilde{\theta} - \theta)$.

But setting $\partial l/\partial\theta = 0$, and solving will give the MLE for θ. Since $I_\theta > 0$ (in all but exceptional circumstances), this gives $\tilde{\theta} - \hat{\theta} = 0$ as the only solution, i.e. $\tilde{\theta} \equiv \hat{\theta}$. □

Thus if the Cramér–Rao lower bound is attained, it is generally the MLE which attains it, i.e. $\hat{\theta}$ is efficient. If the bound is unattainable, but the usual regularity conditions are satisfied, then the MLE is asymptotically efficient. These results effectively ensure efficiency or asymptotic efficiency for members of the exponential family.

Example 3.1 (*continued*)

In the exponential distribution with parameterization

$$f(x;\theta) = \begin{cases} \theta\, \mathrm{e}^{-\theta x}, & x > 0 \\ 0, & \text{elsewhere} \end{cases}$$

the MLE $\hat{\theta} = 1/\bar{X}$ is an asymptotically efficient estimator for θ, but it is not efficient for finite sample sizes. However, if we use the parameterization with $\phi = 1/\theta$, and MLE $\hat{\phi} = \bar{X}$, then $\hat{\phi}$ is an efficient estimator for ϕ for all sample sizes.

Example 3.5 (*continued*)

For the normal distribution $N(\mu, \sigma^2)$, the MLE $\hat{\mu}$ is an efficient estimator for μ, whereas $\hat{\sigma}^2$ is asymptotically efficient for σ^2, but not efficient for finite sample sizes.

If the usual regularity conditions are not satisfied, then the MLE may or may not be efficient, and need not even be asymptotically efficient, as is shown by the following examples.

Example 3.4 (*continued*)

The MLE, $X_{(n)}$, for θ in the uniform distribution, $U[0, \theta]$, is asymptotically efficient, although not efficient for finite sample sizes. We should, of course, remember that because the Cramér–Rao inequality is not applicable here, efficiency is not defined with respect to this bound. We showed, however, in Example 2.14, that $(n+1)X_{(n)}/n$ is an MVUE, from which it follows that $X_{(n)}$ must be asymptotically efficient.

Example 3.6

Suppose that $X_1, X_2, \ldots, X_n$ are a random sample from $U[k\theta, (k+1)\theta]$ where $\theta > 0$, i.e. the distribution with p.d.f.

$$f(x;\theta) = \begin{cases} \dfrac{1}{\theta}, & k\theta \leqslant x \leqslant (k+1)\theta \text{ for some constant } k \\ 0, & \text{elsewhere} \end{cases},$$

Then

$$L(\theta; \mathbf{x}) = \begin{cases} \theta^{-n}, & \text{for } k\theta \leqslant x_{(1)} \leqslant \cdots \leqslant x_{(n)} \leqslant (k+1)\theta \\ 0, & \text{elsewhere} \end{cases}$$

The likelihood is maximized by taking $(k+1)\theta = X_{(n)}$, so the MLE is $\hat{\theta} = X_{(n)}/(k+1)$.

However, because θ is present in both the upper and lower ends of the range of values of X, we might suspect that $X_{(1)}$ also provides information about θ. In fact $(X_{(1)}, X_{(n)})$ is minimal sufficient for θ, which confirms our suspicions. A possible estimator for θ based on $X_{(1)}$ is $X_{(1)}/k = \tilde{\theta}$ and a possible estimator based on both $X_{(1)}$ and $X_{(n)}$ is

$$\theta^* = a\hat{\theta} + (1-a)\tilde{\theta}, \qquad \text{for some constant} \quad 0 \leqslant a \leqslant 1.$$

The estimator θ^*, like $\hat{\theta}$ and $\tilde{\theta}$, is asymptotically unbiased; we now consider its variance. By symmetry, $X_{(1)}$ and $X_{(n)}$ have the same variance, V, say, where $V = n\theta^2/(n+1)^2(n+2)$. Then $\text{Var}[\hat{\theta}] = V/(k+1)^2$ and $\text{Var}[\tilde{\theta}] = V/k^2$ so

$$\text{Var}[\theta^*] \approx \frac{a^2 V}{(k+1)^2} + \frac{(1-a)^2 V}{k^2}.$$

The equality is approximate because there is a covariance term as well. However, intuitively, $X_{(1)}$ and $X_{(n)}$ will be very loosely correlated for large n, because there are so many other Xs in between. It turns out that this covariance term is proportional to n^{-3}, whereas V is proportional to n^{-2}, and because we are looking at asymptotic efficiency, we can ignore the covariance term. The value of a which minimizes $\text{Var}[\theta^*]$ is

$$a = (k+1)^2/[k^2 + (k+1)^2]$$

and substituting this value in the expression for $\text{Var}[\theta^*]$ gives

$$\text{Var}[\theta^*] = \frac{V}{[k^2 + (k+1)^2]} < \frac{V}{(k+1)^2} = \text{Var}[\hat{\theta}].$$

This inequality holds however large n is taken, so that $\hat{\theta}$ is not asymptotically efficient.

Sufficiency

Lemma 3.5

If $\hat{\boldsymbol{\theta}}$ is the unique MLE for a parameter $\boldsymbol{\theta}$, then $\hat{\boldsymbol{\theta}}$ must be a function of the minimal sufficient statistic for $\boldsymbol{\theta}$.

Proof

Consider the factorization criterion, Theorem 2.1.

$L(\boldsymbol{\theta}; \mathbf{x}) = K_1[\mathbf{t} \mid \boldsymbol{\theta}]K_2[\mathbf{x}]$ where $\mathbf{t}$ is (minimal) sufficient. Then maximizing $L(\boldsymbol{\theta}; \mathbf{x})$ with respect to $\boldsymbol{\theta}$ requires K_1 to be maximized; we can forget about K_2 since it does not depend on $\boldsymbol{\theta}$. However K_1 depends on $\mathbf{x}$ only through $\mathbf{t}$, so the value of $\boldsymbol{\theta}$ which maximises K_1 will depend only on $\mathbf{t}$, i.e. $\hat{\boldsymbol{\theta}}$ is a function of $\mathbf{T}$. □

Note that Lemma 3.5 does not mean that $\hat{\boldsymbol{\theta}}$ is necessarily sufficient, although it often is. For illustration, consider the examples given so far in this chapter. In Example 3.6, $\hat{\theta}$ is not sufficient, but in all the other examples the MLE is sufficient.

In cases where $\hat{\boldsymbol{\theta}}$ is sufficient, it must be minimal sufficient. This follows because Lemma 3.5 implies that $\hat{\boldsymbol{\theta}}$ defines a partition of the sample space at least as coarse as that given by the minimal sufficient statistic.

Lemma 3.5 is useful in finding MVUEs, even when $\hat{\boldsymbol{\theta}}$ is not sufficient. If we find a function of $\hat{\boldsymbol{\theta}}$ which is unbiased, this in turn will be a function of $\mathbf{T}$, and if $\mathbf{T}$ is complete we have found an MVUE. Notice that in Example 3.6 it is possible to find an unbiased estimator for θ which is based on $X_{(n)}$, and hence is a function of the minimal sufficient statistic, $\mathbf{T}$. However, $\mathbf{T}$ is not complete (there is more than one unbiased estimator based on $\mathbf{T}$), and the estimator is not an MVUE.

3.2.4 *Problems with MLEs*

In the previous chapter, dominated by discussion of MVUEs, we finished by noting some problems. MLEs, although widely used and often very useful, also have problems. Indeed, unless one adopts a very narrow view of what is required in statistical inference, it turns out that all proposed approaches and techniques have some difficulties associated with them. When the usual regularity conditions hold, the properties of MLEs are generally good, but this is not necessarily the case for non-regular situations. Cheng and Traylor (1995) give a good discussion of the problems that can arise in non-regular cases, and how such problems may be overcome.

We have already seen an example where the MLE was neither sufficient nor asymptotically efficient. With enough ingenuity we can construct examples where MLEs are inconsistent (Korn, 1990), where they do not exist (Bickel and Doksum, 1977, Problem 3.3.12), or where they are not unique (Bickel and Doksum, 1977, Problem 3.3.4). Non-uniqueness in turn opens up the possibility that an MLE need not be a function of the minimal sufficient statistic (Levy, 1985). At a more practical level, in real-world examples the likelihood function may take a complex form, especially when several parameters are being estimated. This can lead to problems actually finding the MLE because of multiple maxima (Gates, 1993), or flat areas on the likelihood surface – see Figure 3.1.

Another type of problem arises when the MLE occurs at the boundary of the region of allowable variables. This will often ruin the 'nice' asymptotic properties of the MLE. A thorough discussion of an example where boundary estimation occurs is given by Catchpole and Morgan (1994).

Having noted possible problems with MLEs, we end with the reassurance that serious difficulties are the exception, not the rule. There are good reasons why MLEs are the estimators of choice in many circumstances, for it is relatively unusual for their properties to be completely unsatisfactory.

3.3 Modifications and extensions of maximum likelihood estimation

Straightforward maximum likelihood estimation, as described in Section 3.2, cannot always be implemented, or there may be circumstances in which it can be improved upon in some sense. For these reasons, a number of modifications and extensions to the basic definition of likelihood, and to the method of maximum likelihood, have been proposed. One class of modifications arises when $\boldsymbol{\theta}$ consists of several parameters, but only a subset of these parameters is of interest. This case is dealt with in Section 3.3.1. Section 3.3.2 briefly describes a number of other modifications to the likelihood function and to maximum likelihood estimation. The idea of *quasi-likelihood*, which arises mainly in the context of generalized linear models, is left until Chapter 10.

3.3.1 Inference in the presence of nuisance parameters

Suppose that in our usual situation of a random sample of observations $X_1, X_2, \ldots, X_n$ from a distribution with p.d.f. $f(x; \boldsymbol{\theta})$, the vector of parameters $\boldsymbol{\theta}$ is partitioned into $\boldsymbol{\theta}_1$ and $\boldsymbol{\theta}_2$, where $\boldsymbol{\theta}_1$ contains parameters of interest, and $\boldsymbol{\theta}_2$ consists of other unknown parameters which are not of direct interest (*nuisance parameters*).

Definition 3.2

Suppose that $\mathbf{T} = (\mathbf{T}_1, \mathbf{T}_2)$ is a minimal sufficient statistic for the vector of parameters $\boldsymbol{\theta} = (\boldsymbol{\theta}_1, \boldsymbol{\theta}_2)$, and that

(a) the p.d.f. of $\mathbf{T}_2$ depends on $\boldsymbol{\theta}_2$, but not on $\boldsymbol{\theta}_1$;
(b) the conditional p.d.f. of $\mathbf{T}_1$ given $\mathbf{T}_2 = \mathbf{t}_2$ depends on $\boldsymbol{\theta}_1$ but not on $\boldsymbol{\theta}_2$, for all values of $\mathbf{t}_2$.

Then $\mathbf{T}_2$ is an *ancillary statistic* for $\boldsymbol{\theta}_1$, and $\mathbf{T}_1$ is *conditionally sufficient* for $\boldsymbol{\theta}_1$ in the presence of $\boldsymbol{\theta}_2$. The concept of ancillary statistics is sometimes useful in making inferences about $\boldsymbol{\theta}_1$ but there are a number of variations on the definition of ancillarity. For example, Cox and Hinkley (1974, p. 35) refer to the definition above as ancillarity in the extended sense, having also provided a more restrictive definition (p. 32).

Definition 3.3

Consider again a vector of parameters $\boldsymbol{\theta} = (\boldsymbol{\theta}_1, \boldsymbol{\theta}_2)$, with likelihood function $L(\boldsymbol{\theta}_1, \boldsymbol{\theta}_2; \mathbf{x})$. Suppose that $\hat{\boldsymbol{\theta}}_{2.1}$ is the MLE of $\boldsymbol{\theta}_2$ for a given value of $\boldsymbol{\theta}_1$. Then the *profile likelihood* for $\boldsymbol{\theta}_1$ is $L(\boldsymbol{\theta}_1, \hat{\boldsymbol{\theta}}_{2.1}; \mathbf{x})$.

This profile likelihood is a modification of the 'ordinary' likelihood which can be used to make inferences about $\boldsymbol{\theta}_1$ in the presence of the nuisance parameters $\boldsymbol{\theta}_2$. The idea is to use the profile likelihood in exactly the same way as the 'ordinary' likelihood is used in making inferences when there is no nuisance parameter. In the case of point estimation, maximizing the profile likelihood with respect to $\boldsymbol{\theta}_1$ leads to the same estimator $\hat{\boldsymbol{\theta}}_1$ as is obtained by maximizing the likelihood simultaneously with respect to $\boldsymbol{\theta}_1$ and $\boldsymbol{\theta}_2$. However, this equivalence does not carry over to hypothesis testing and interval estimation as we shall see in later chapters.

Example 3.5 (*continued*)

We have $X_1, X_2, \ldots, X_n \sim N(\mu, \sigma^2)$, μ, σ^2 both unknown. The likelihood function is

$$L(\mu, \sigma^2; \mathbf{x}) = (2\pi\sigma^2)^{-n/2} \exp\left\{-\frac{1}{2\sigma^2}\sum_{i=1}^{n}(x_i - \mu)^2\right\}.$$

Given μ, the MLE of σ^2 is $(1/n)\ \Sigma_{i=1}^{n}\ (x_i - \mu)^2$, and given σ^2, the MLE of μ is $\bar{x}$. Hence the profile likelihood for μ is

$$\left[\frac{2\pi e}{n}\sum_{i=1}^{n}(x_i - \mu)^2\right]^{-n/2}$$

and the profile likelihood for σ^2 is

$$(2\pi\sigma^2)^{-n/2} \exp\left\{-\frac{1}{2\sigma^2}\sum_{i=1}^{n}(x_i - \bar{x})^2\right\}.$$

Although there is no great virtue in doing so in this example, it is common practice and is frequently useful to work with the profile log-likelihood rather than the profile likelihood.

A number of modifications and adjustments to the profile likelihood have been suggested – see, for example, Cox and Reid (1987) and the discussion which follows. As with profile likelihood itself, these have greater relevance to hypothesis testing and interval estimation than to point estimation and we shall discuss them further in Section 4.6.

Profile likelihood is one of several modifications of the likelihood, collectively known as *pseudo-likelihoods*, which are based on a derived likelihood for a subset of parameters. Also included in this class of pseudo-likelihoods are *marginal likelihood*, *conditional likelihood* and *partial likelihood*.

Suppose that $\mathbf{x}$, or some transformation of $\mathbf{x}$, is partitioned into $\mathbf{u}$ and $\mathbf{v}$, and that $\boldsymbol{\theta} = (\boldsymbol{\theta}_1, \boldsymbol{\theta}_2)$, where $\boldsymbol{\theta}_1$ consists of the parameters of interest, and $\boldsymbol{\theta}_2$ is made up of nuisance parameters. Then the joint density for $\mathbf{x}$, which is the likelihood function, can be written as $f(\mathbf{u}, \mathbf{v}; \boldsymbol{\theta}_1, \boldsymbol{\theta}_2)$. Suppose that $f(\mathbf{u}, \mathbf{v}; \boldsymbol{\theta}_1, \boldsymbol{\theta}_2)$ can be factorized as

$$f(\mathbf{u}; \boldsymbol{\theta}_1) f(\mathbf{v} \mid \mathbf{u}; \boldsymbol{\theta}_1, \boldsymbol{\theta}_2),$$

where the first term depends only on $\boldsymbol{\theta}_1$. If $f(\mathbf{v} \mid \mathbf{u})$ does not depend on $\boldsymbol{\theta}_1$, it seems reasonable to base inferences about $\boldsymbol{\theta}_1$ on $f(\mathbf{u}; \boldsymbol{\theta}_1)$, the *marginal likelihood*. Even when $f(\mathbf{v} \mid \mathbf{u})$ depends on both $\boldsymbol{\theta}_1$, and $\boldsymbol{\theta}_2$, it may still be desirable to base inferences on $f(\mathbf{u}; \boldsymbol{\theta}_1)$ for the sake of simplicity, provided that the information loss in ignoring $f(\mathbf{v} \mid \mathbf{u})$ is small.

Alternatively, if there is a factorization

$$f(\mathbf{u} \mid \mathbf{v}; \boldsymbol{\theta}_1) f(\mathbf{v}; \boldsymbol{\theta}_1, \boldsymbol{\theta}_2)$$

where again the first term depends only on $\boldsymbol{\theta}_1$, it now makes sense to base our inference on $f(\mathbf{u} \mid \mathbf{v}; \boldsymbol{\theta}_1)$, the *conditional likelihood*. In this case, if $(\mathbf{u}, \mathbf{v})$ are minimal sufficient for $(\boldsymbol{\theta}_1, \boldsymbol{\theta}_2)$, and the distribution of $\mathbf{v}$ does not depend on $\boldsymbol{\theta}_1$, it follows that $\mathbf{v}$ is an ancillary statistic for $\boldsymbol{\theta}_1$.

Example 3.7

Suppose that X_1, X_2 are independent Poisson random variables with means θ_2 and $\theta_1\theta_2$ respectively, and let $U = X_2$, $V = X_1 + X_2$. We have

$$f(x_1, x_2; \theta_1, \theta_2) = \frac{\theta_2^{x_1} (\theta_1\theta_2)^{x_2} \, \mathrm{e}^{-[\theta_2 + \theta_1\theta_2]}}{x_1! x_2!}$$

Transforming to U, V gives

$$f(u, v; \theta_1, \theta_2) = \frac{\theta_1^u \theta_2^v \, \mathrm{e}^{-\theta_2(1+\theta_1)}}{(v-u)! u!}$$

$$= \left[\binom{v}{u} \left(\frac{\theta_1}{1+\theta_1}\right)^u \left(1 - \frac{\theta_1}{1+\theta_1}\right)^{v-u}\right] \left[\frac{[\theta_2(1+\theta_1)]^v \, \mathrm{e}^{-\theta_2(1+\theta_1)}}{v!}\right].$$

Inspection of this last expression shows that the conditional distribution of U, given $V = v$, is a binomial distribution with parameters v and $\theta_1/(1+\theta_1)$, which does not depend on θ_2. Inference about θ_1 can therefore be based on the conditional binomial likelihood.

Another type of pseudo-likelihood is the *partial likelihood*. Here it is assumed that the data consist of a sequence of pairs of random variables $X_1, \mathrm{Y}_1; \mathrm{X}_2, \mathrm{Y}_2; \ldots; \mathrm{X}_m, \mathrm{Y}_m$. Let

$$X^{(j)} = (X_1, X_2, \ldots, X_j), \quad Y^{(j)} = (Y_1, Y_2, \ldots, Y_j), \qquad j = 1, 2, \ldots, m.$$

Then the likelihood function can be written

$$L(\boldsymbol{\theta}; \mathbf{x}, \mathbf{y}) = \prod_{j=1}^{m} f(x_j \mid x^{(j-1)}, y^{(j-1)}; \boldsymbol{\theta}) \prod_{j=1}^{m} f(y_j \mid x^{(j)}, y^{(j-1)}; \boldsymbol{\theta}).$$

The second product is the partial likelihood based on $(Y_1, Y_2, \ldots, Y_m)$. Unlike marginal and conditional likelihoods, the partial likelihood does not arise as an 'ordinary' likelihood in a derived experiment. Nevertheless, both marginal and conditional likelihoods arise as special cases of partial likelihood. Discussion of these relationships, and various other properties of partial likelihood, is given by Cox (1975). Partial likelihood is again a useful device for inference in some situations where there are nuisance parameters. Perhaps the best-known example is the proportional hazards model (see Section 8.3) where partial likelihood is used to eliminate the baseline hazard function $\lambda_0(t)$.

A further possible approach is to use an *integrated likelihood*. In this case the likelihood function of interest for $\boldsymbol{\theta}_1$ is

$$\int_{\boldsymbol{\theta}_2} L(\boldsymbol{\theta}_1, \boldsymbol{\theta}_2; \mathbf{x})\pi(\boldsymbol{\theta}_2)\,\mathrm{d}\boldsymbol{\theta}_2$$

where $\pi(\boldsymbol{\theta}_2)$ is a (prior) distribution for the nuisance parameter $\boldsymbol{\theta}_2$. The presence of $\pi(\boldsymbol{\theta}_2)$ means that Bayesian ideas underlie this approach, so further discussion is deferred until Section 7.5.

3.3.2 *Other modifications of the likelihood function*

Grouped likelihood

Cheng and Iles (1987) express the view that the likelihood function for a continuous random variable X takes the general form

$$L(\boldsymbol{\theta}; \mathbf{x}) = \prod_{i=1}^{n} \Pr[x_i \leqslant X \leqslant x_i + h_i \,|\, \boldsymbol{\theta}]$$

$$= \prod_{i=1}^{n} \int_{x_i}^{x_i+h_i} f(x; \boldsymbol{\theta})\,\mathrm{d}x$$

which is a *grouped likelihood*. When regularity conditions hold, we can let the h_i tend to zero, giving rise to the more usual form

$$L(\boldsymbol{\theta}; \mathbf{x}) = \prod_{i=1}^{n} f(x_i; \boldsymbol{\theta}).$$

When regularity conditions do not hold, however, there may be problems with the conventional likelihood function; for example, it may be unbounded. Such problems can sometimes be resolved by using the more general, grouped, definition of likelihood. In other cases it is more appropriate to modify the likelihood so that it is a product of densities for some observations, multiplied by integrals from the grouped form for the other observations – see Cheng and Iles (1987), Cheng and Traylor (1995).

Penalized likelihood

In some circumstances the log-likelihood may be modified to incorporate a secondary criterion, so that the quantity to be maximized becomes

$$l_p(\boldsymbol{\theta}; \mathbf{x}) = l(\boldsymbol{\theta}; \mathbf{x}) - \alpha R(\boldsymbol{\theta})$$

where $R(\boldsymbol{\theta})$ is some penalty function and α is a fixed parameter which governs the relative importance of $l(\boldsymbol{\theta}; \mathbf{x})$ and $R(\boldsymbol{\theta})$ in the penalized likelihood $l_p(\boldsymbol{\theta}; \mathbf{x})$. The most usual context for penalized likelihood is when a density function or a smooth regression curve is being fitted through a set of points. In this case the secondary criterion reflects a desire to have a smooth curve or density, so that $R(\boldsymbol{\theta})$ is a measure of roughness, and α is a so-called smoothing parameter. An example of penalized likelihood estimation in the context of density estimation is given by Good and Gaskins (1980). A penalty function can be used to represent criteria other than 'roughness' – for example, Green (1990) demonstrates its use to represent prior information. In this case we can regard $\exp\{-\alpha R(\boldsymbol{\theta})\}$ as proportional to a so-called *prior distribution* for $\boldsymbol{\theta}$, so that maximizing $l_p(\boldsymbol{\theta}; \mathbf{x})$ is equivalent to finding the maximum of the *posterior distribution* for $\boldsymbol{\theta}$ (see Section 6.1). In both the decision theory and Bayesian approaches to inference (Chapters 6, 7) the posterior distribution plays a central role. As noted in Section 3.2.2, Green (1990) shows how to use a modification of the EM algorithm to find maximum penalized likelihood estimators.

Restricted maximum likelihood

The term *restricted maximum likelihood criterion* can be used simply to denote MLEs under some restrictions or constraints – see, for example, Lütkepohl (1991, Section 5.2.3). The effect of constraints on the method of scoring is discussed by Osborne (1992). However, restricted maximum likelihood is often abbreviated to REML, and is then used in a more specific sense, which we now describe.

Consider a linear model which slightly extends that of Example 2.12 and Section 3.4.2, as follows:

$$\mathbf{Y} = \mathbf{X}\boldsymbol{\beta} + \mathbf{Z}\mathbf{b} + \boldsymbol{\varepsilon}$$

where $\boldsymbol{\beta}$ is a vector of p parameters which we wish to estimate, often treatment effects, with $\mathbf{b}$ a vector of so-called unobservable random effects. Both $\boldsymbol{\beta}$ and $\mathbf{b}$ are unknown vectors, but whereas the elements of $\boldsymbol{\beta}$ are assumed to be fixed, those of $\mathbf{b}$ are considered to be random. $\mathbf{X}$ is an $(n \times p)$ matrix of known constants, but $\mathbf{Z}$ may be a function of some unknown parameters $\boldsymbol{\theta}$. It is assumed that $\mathbf{b}$ and $\boldsymbol{\varepsilon}$ both have zero means, but that the covariance matrices of $\mathbf{b}$ and $\boldsymbol{\varepsilon}$ may each depend on the parameter $\boldsymbol{\theta}$. This model is expressed in a rather general form and looks fairly complicated, with potentially many parameters. However, in practice the number of parameters may be quite small. For example, in an incomplete block design, $\mathbf{b}$ will represent random block effects with $\boldsymbol{\varepsilon}$ an error term. We would typically assume that $\mathbf{Z}$ is fixed, and that the covariance matrices for $\mathbf{b}$, $\boldsymbol{\varepsilon}$ are $\gamma\sigma^2\mathbf{I}$ and $\sigma^2\mathbf{I}$, respectively, so that $\boldsymbol{\theta}$ consists of only two parameters γ and σ^2. ML can be used to estimate $\boldsymbol{\beta}$ and $\boldsymbol{\theta}$, but there are some disadvantages, especially in the context of estimating variance components, as discussed by Harville (1977).

An alternative procedure is obtained by transforming $\mathbf{Y}$ into ($\mathbf{SY}$, $\mathbf{QY}$) where $\mathbf{S}$ is of rank p, $\mathbf{Q}$ is of rank $(n-p)$, and the columns of $\mathbf{SY}$ are uncorrelated with those of $\mathbf{QY}$. It then follows that the likelihood based on $\mathbf{Y}$ is the product of likelihoods based on $\mathbf{SY}$ and $\mathbf{QY}$. If $\mathbf{S}$ is chosen to be $\mathbf{S} = \mathbf{I} - \mathbf{X}(\mathbf{X}^T\mathbf{X})^{-1}\mathbf{X}^T$, where $\mathbf{I}$ is an identity matrix, it follows that $\mathbf{SY}$ represents residuals from the fitted model $\hat{\mathbf{Y}} = \mathbf{X}\hat{\boldsymbol{\beta}}$, where $\hat{\boldsymbol{\beta}} = (\mathbf{X}^T\mathbf{X})^{-1}\mathbf{X}^T\mathbf{Y}$. REML, which can then stand for 'residual' as well as 'restricted' ML, uses the likelihood based on $\mathbf{SY}$ to estimate $\boldsymbol{\theta}$. It can be argued that in estimating $\boldsymbol{\theta}$ it is better to first remove the effect of estimating $\boldsymbol{\beta}$ and this is what is done here. By an appropriate choice of $\mathbf{Q}$, we can also estimate $\boldsymbol{\beta}$ using only the likelihood based on $\mathbf{QY}$. Further discussion of REML can be found in Harville (1977) and Patterson and Thompson (1971).

3.4 Other methods of estimation

3.4.1 The method of moments

This method is interesting because of its simplicity. Its properties are somewhat variable – often estimators produced by the method have good properties, but this is by no means guaranteed, as shown below in a fairly simple example (Example 3.4). Its chief attraction is therefore its ease of use. It can, for example, sometimes be used to find a sensible initial estimate in an iterative method for finding MLEs.

In order to explain the way that the method operates we need to digress and define moments. For our random variable X with p.d.f. $f(x; \theta)$ we define the *ith population moment about the origin* as $E[X^i]$. The *ith (population) moment about the mean* is $E[(X-\mu)^i]$, where $\mu = E[X]$. Given a random sample, $x_1, x_2, \ldots, x_n$ of observations on X, the *ith sample moment about the origin* is $(1/n)\Sigma_{j=1}^n x_j^i$, and the *ith sample moment about the mean* is $(1/n)\Sigma_{j=1}^n (x_j - \bar{x})^i$.

In the method of moments we express the parameters of interest as functions of the population moments of our distribution. We then use as our estimates the corresponding functions of the sample moments.

Example 3.5 (*continued*)

$X_1, X_2, \ldots, X_n \sim N(\mu, \sigma^2)$. $\bar{X}$ is the estimator for μ, and

$$\hat{\sigma}^2 = \frac{1}{n} \sum_{i=1}^{n} (X_i - \bar{X})^2$$

which are the same in both cases as the MLEs found earlier. Note, however, that the same moment estimators would be valid for the mean and variance of any other distribution, whereas the MLEs may well change when the distribution changes.

Example 3.4 (*continued*)

$X_1, X_2, \ldots, X_n \sim U[0, \theta]$. $E[X] = \theta/2$, so $2\bar{X}$ is the method-of-moments estimator for θ. Now it is easy to see that $2\bar{X}$ does not have very good properties as an estimator of θ. It is unbiased and consistent, but its asymptotic efficiency is zero. This follows because its variance is proportional to n^{-1}, whereas that of the MLE is proportional to n^{-2}. Furthermore, if it happens that $2\bar{X} < X_{(n)}$, we know for certain that $2\bar{X}$ will underestimate θ, but we cannot take this knowledge into account in any way. One further comment regarding this example is that we can find more than one function of the moments which is equal to θ. For example, $\sigma^2 = \theta^2/12$, so $[12\hat{\sigma}^2]^{1/2}$ is a possible estimator for θ. Such non-uniqueness is common, and in such situations the convention is to choose the estimator involving the lowest order moment(s), here $2\bar{X}$. This convention will typically optimize the properties of the estimators, because usually estimators of low-order moments have better properties than those of higher-order moments for the same distribution.

As noted above, method-of-moments estimators have rather variable properties. Usually they will be inferior to MLEs, but there are exceptions. For example, Hosking and Wallis (1987) discuss a problem involving the generalized Pareto distribution. MLEs are asymptotically efficient, but for samples as large as 500, moment estimators are superior to MLEs in terms of efficiency. In the same article, a modification of the method-of-moments is described, which does even better. This modification leads to so-called *probability-weighted moment estimators.*

3.4.2 *The method of least squares*

The method of least squares is a well-known method of estimation, although its use is mainly restricted to linear models and various generalizations of such models.

The general linear model can be written (see Example 2.12) as $E(\mathbf{Y}) = \mathbf{X}\boldsymbol{\beta}$, where $\mathbf{Y} = (Y_1, Y_2, \ldots, Y_n)^T$ is a vector of n random variables, $\mathbf{X}$ is an $(n \times p)$ matrix of fixed constants, and $\boldsymbol{\beta}$ is a vector of unknown parameters. The model can alternatively be written

$$\mathbf{Y} = \mathbf{X}\boldsymbol{\beta} + \boldsymbol{\varepsilon},$$

where $\boldsymbol{\varepsilon} = (\varepsilon_1, \varepsilon_2, \ldots, \varepsilon_n)^T$ and the random variables $\varepsilon_1, \varepsilon_2, \ldots, \varepsilon_n$ are independent, each with mean zero and (usually unknown) variance σ^2. The method of least squares then estimates $\boldsymbol{\beta}$ by minimizing

$$\sum_{i=1}^{n} \varepsilon_i^2 = \boldsymbol{\varepsilon}^T\boldsymbol{\varepsilon} = (\mathbf{Y} - \mathbf{X}\boldsymbol{\beta})^T(\mathbf{Y} - \mathbf{X}\boldsymbol{\beta}).$$

A considerable body of literature exists on the properties of least squares estimators in the general linear model – see, for example, Silvey (1970, Chapter 3). We simply note here that the least squares (LS) estimator of $\boldsymbol{\beta}$ in the model above is $\hat{\boldsymbol{\beta}} = (\mathbf{X}^T\mathbf{X})^{-1}\mathbf{X}^T\mathbf{y}$, and that if $\varepsilon_1, \varepsilon_2, \ldots, \varepsilon_n$ are assumed to be normally distributed, then the LS estimate is identical to the MLE (Exercise 3.27).

The basic least squares technique can be modified in various ways, and we now discuss briefly some of these modifications.

Weighted least squares

The basic LS technique makes the assumption that all the ε_i have the same variance and are independent. If this is not the case, it is intuitively reasonable to give less weight in our optimization criterion to those observations which are more variable. Hence, if $\text{var}(\varepsilon_i) = W_i\sigma^2$, where $W_1, W_2, \ldots, W_n$ are known, *weighted LS* minimizes $\Sigma_{i=1}^n \varepsilon_i^2/W_i$.

More generally, if there are also correlations between the ε_i, and the covariance matrix of $\boldsymbol{\varepsilon}$ is $\sigma^2\mathbf{W}$, where $\mathbf{W}$ is known, then weighted LS minimizes $\boldsymbol{\varepsilon}^T\mathbf{W}^{-1}\boldsymbol{\varepsilon}$. This leads to an estimator

$$\hat{\boldsymbol{\beta}} = (\mathbf{X}^T\mathbf{W}^{-1}\mathbf{X})^{-1}\mathbf{X}^T\mathbf{W}^{-1}\mathbf{Y}.$$

When the ε_i have unequal variances, and are correlated, weighted LS estimators have advantages with respect to efficiency, when compared to the ordinary LS estimator, Silvey (1970, Section 3.6).

Conditional and unconditional least squares

The idea of conditional and unconditional least squares arises in the context of time-series modelling. Consider a time series $X_1, X_2, \ldots, X_n$, which follows a first-order autoregressive model so that

$$(X_t - \mu) = \phi(X_{t-1} - \mu) + \varepsilon_t$$

where $\mu = \text{E}[X_t]$, $t = 1, 2, \ldots, n$, ϕ is an unknown parameter which is to be estimated, and $\varepsilon_1, \varepsilon_2, \ldots, \varepsilon_n$ are independent random variables, each with mean zero and variance σ^2. A least squares approach finds ϕ which minimizes

$$\sum_{t=1}^{n} \varepsilon_t^2 = \varepsilon_1^2 + \sum_{t=2}^{n} \varepsilon_t^2.$$

The reason for separating ε_1^2 from the sum of squares is that, because of the time series nature of the data, it takes a different form from subsequent ε_t^2. We have

$$\sum_{t=2}^{n} \varepsilon_t^2 = \sum_{t=2}^{n} [(X_t - \mu) - \phi[(X_{t-1} - \mu)]^2$$

which can easily be minimized with respect to ϕ, giving a *conditional LS estimator* for ϕ. For an unconditional LS estimator we add a term $(1 - \phi^2)\ (X_1 - \mu)^2$, corresponding to ε_1^2, to the quantity to be minimized (Cryer, 1986, Section 7.3). If the ε_t are assumed to be normally distributed, the log-likelihood function is (see Exercise 3.28)

$$l(\phi,\sigma^2;\,\mathbf{x}) = \text{constant} - \frac{n}{2}\ln(\sigma^2) + \tfrac{1}{2}\ln(1-\phi^2) - \frac{(1-\phi^2)}{2\sigma^2}(X_1-\mu)^2$$

$$-\frac{1}{2\sigma^2}\sum_{t=2}^{n}[(X_t-\mu)-\phi(X_{t-1}-\mu)]^2.$$

The MLE maximizes the whole of this expression, the unconditional LS estimator maximizes the last two terms in the expression, while the conditional LS estimator maximizes the last term only.

Iteratively reweighted least squares

Sometimes methods of estimation which do not directly use a least squares criterion lead to an *iteratively reweighted least squares algorithm.* This arises, for example, in robust estimation (Section 8.4.1) and in generalized linear models (Section 10.3). In iteratively reweighted least squares each iteration finds weighted least squares estimates for a given set of weights. These estimates are then used to construct a new set of weights which are used in the next iteration.

3.4.3 *The method of minimum* χ^2

For this method the sample space for X is divided into k classes, and $p_i(\boldsymbol{\theta})$ is defined as

$$p_i(\boldsymbol{\theta}) = \int_{x\,\in\,i\text{th class}} f(x;\,\boldsymbol{\theta})\,\mathrm{d}x$$

the probability of falling in the ith class, $i = 1, 2, \ldots, k$. As indicated, the $p_i(\boldsymbol{\theta})$ depend on the value of $\boldsymbol{\theta}$. Also, $\hat{p}_i$ is the proportion of the observed sample falling in the ith class, $i = 1, 2, \ldots, k$. Then the minimum χ^2 estimate of $\boldsymbol{\theta}$ is the value of $\boldsymbol{\theta}$ which minimizes

$$\sum_{i=1}^{k}\frac{(\hat{p}_i - p_i(\boldsymbol{\theta}))^2}{p_i(\boldsymbol{\theta})}$$

i.e. $\boldsymbol{\theta}$ makes the χ^2 goodness-of-fit test statistic as small as possible. A modified version of this (Stuart and Ord, 1991, pp. 732–3), for which calculations may be simpler because only the numerator involves $\boldsymbol{\theta}$, is to find $\boldsymbol{\theta}$ which minimizes

$$\sum_{i=1}^{k}\frac{(\hat{p}_i - p_i(\boldsymbol{\theta}))^2}{\hat{p}_i}.$$

3.5 Discussion

In this chapter we have discussed possible methods for finding estimators. We have concentrated on maximum likelihood estimation and related methods because of their widespread use in statistical inference. We have seen that this popularity is justified by the general good properties of MLEs. However, like other inference procedures, there are circumstances where MLEs are less than ideal.

The final section of the chapter briefly describes three alternative methods of estimation which may be of use when MLEs are not satisfactory, or for finding initial values in an iterative ML scheme. All these alternative methods remain within the classical framework of inference, based on sampling distributions. Other methods outside this framework will be discussed in subsequent chapters. In particular, estimators based on decision theory and on the Bayesian approach are introduced in Chapter 6 and 7. Robust estimation and estimation based on computationally intensive methods are described in Chapters 8 and 9 respectively.

Exercises

3.1 Find the maximum likelihood estimator for θ in the following cases, where in each case we have available a random sample $X_1, X_2, \ldots, X_n$ from the relevant distribution.

(a) The geometric distribution with probability function $\theta(1-\theta)^{x-1}$, $x = 1, 2, \ldots$.

(b) The uniform distribution on the interval $(-\theta/2, \theta/2)$.

(c) The gamma distribution with parameters 2 and θ, i.e. with p.d.f.

$$f(x \mid \theta) = \frac{1}{\theta^2}\, x\, \mathrm{e}^{-x/\theta}, \qquad x > 0.$$

(d) The Poisson distribution with mean θ.

3.2 Consider the situation in Exercise 3.1(c) when there are n_1 observations, $X_1, X_2, \ldots, X_{n_1}$, whose values are given, and are all $\leqslant a$, for some constant $a > 0$. In addition it is known that $n_2 = (n - n_1)$ observations have values $\geqslant a$, but their exact values are unknown. Write down the likelihood function in this case, and find an equation whose solution will give the maximum likelihood estimator for θ.

3.3 Three independent binomial experiments are conducted with n_1, n_2, n_3 trials and x_1, x_2, x_3 are the respective numbers of successes observed.

(a) Suppose that the probability of success, p, is the same in each trial. Find the maximum likelihood estimator of p, based on all three trials.

(b) Now suppose that the probability of success varies between trials, and is α, $\alpha + \beta$, α, respectively. Find maximum likelihood estimators of α and β, based on all three trials.

3.4 The random variable X has an exponential distribution with p.d.f. $f(x; \theta) = \theta\, \mathrm{e}^{-\theta x}$, $x > 0$, For a fixed positive constant T, show that $\mathrm{E}[X \mid X \geqslant T] = T + 1/\theta$.

3.5 Suppose that $X_1, X_2, \ldots, X_n$ form a random sample from a distribution belonging to the k-parameter exponential family. Show that the E-step in the EM algorithm reduces to finding the conditional expectations of the k minimal sufficient statistics, and then substituting these expected values into the complete log-likelihood function.

3.6 Using Exercise 3.5, or otherwise, show that the complete log-likelihood in Example 3.2 is a linear function of the unknown values $x_{m+1}, x_{m+2}, \ldots, x_n$.

3.7 Let $X_1, X_2, \ldots, X_n$ be a random sample from the distribution having p.d.f.

$$f(x;\boldsymbol{\theta}) = \begin{cases} \dfrac{1}{\theta_2}\,\mathrm{e}^{-(x-\theta_1)/\theta_2}, & x \geqslant \theta_1,\ \theta_2 > 0 \\ 0, & \text{elsewhere.} \end{cases}$$

Find the maximum likelihood estimators of θ_1 and θ_2.

3.8 Observations are made on independent random variables X_{ij} where $X_{ij} \sim N(\mu_i, \theta), j = 1, 2;\ i = 1, \ldots, n$. Write down the likelihood function for $\theta, \mu_1, \mu_2, \ldots, \mu_n$. Obtain the maximum likelihood estimators of the parameters $\mu_1, \ldots, \mu_n$ and show that the maximum likelihood estimator $\hat{\theta}$ of θ is $(1/4n)\Sigma Z_i^2$, where $Z_i = X_{i1} - X_{i2}$. Obtain the expectation and variance of $\hat{\theta}$; hence or otherwise prove that, as $n \to \infty$, $\hat{\theta}$ is not a consistent estimator for θ.

3.9 The random variables $X_1, X_2, \ldots, X_m, X_{m+1}, \ldots, X_n$ are independently, normally distributed with unknown mean μ and variance 1. After $X_1, X_2, \ldots, X_m$ have been observed it is decided to record only the signs of $X_{m+1}, \ldots, X_n$. Obtain the equation satisfied by the maximum likelihood estimator $\hat{\mu}$ of μ.

3.10 A random sample of n observations is taken on a random variable X which has a Poisson distribution with mean θ. Suppose that $\phi = \theta^2$. Find the maximum likelihood estimator, $\hat{\phi}$, for ϕ, and show that $\hat{\phi}$ is a biased, but consistent, estimator. (Note that $\mathrm{E}[X^2] = \theta^4 + 6\theta^3 + 7\theta^2 + \theta$.)

3.11 Let $X_1, X_2, \ldots, X_n$ be a random sample from the distribution having p.d.f.

$$f(x \mid \theta) = \begin{cases} \dfrac{2x}{\theta^2}, & 0 < x < \theta \\ 0 & \text{otherwise.} \end{cases}$$

Determine the maximum likelihood estimator of the median of this distribution and show that this estimator is sufficient. Is it also minimal sufficient?

3.12 Let $X_1, X_2, \ldots, X_n$ be a random sample from the uniform distribution $U[\mu - \sqrt{3}\sigma, \mu + \sqrt{3}\sigma]$. Find the maximum likelihood estimators of μ and σ.

3.13 A random sample of size n is obtained from a bivariate normal distribution with density function

$$f(x, y) = (2\pi)^{-1}(1-\rho^2)^{-1/2}\sigma^{-2}$$
$$\times \exp\left[-\frac{1}{2(1-\rho^2)\sigma^2}\{(x-\mu_1)^2 - 2\rho(x-\mu_1)(y-\mu_2) + (y-\mu_2)^2\}\right]$$
$$-\infty < x, y < \infty.$$

Write down the likelihood function and show that its logarithm may be written

$$l = \text{constant} - n\ln(\sigma^2) - \tfrac{1}{2}n\ln(1-\rho^2)$$
$$- \tfrac{1}{2}n\sigma^{-2}(1-\rho^2)^{-1}[(s_1^2 + s_2^2 - 2\rho s_{12})$$
$$+ (\bar{x}-\mu_1)^2 - 2\rho(\bar{x}-\mu_1)(\bar{y}-\mu_2) + (\bar{y}-\mu_2)^2]$$

where

$$\bar{x} = \frac{1}{n}\sum_{i=1}^{n} x_i,\ \bar{y} = \frac{1}{n}\sum_{i=1}^{n} y_i, \qquad s_1^2 = \frac{1}{n}\sum_{i=1}^{n} (x_i - \bar{x})^2$$

$$s_2^2 = \frac{1}{n}\sum_{i=1}^{n} (y_i - \bar{y})^2, \qquad s_{12} = \frac{1}{n}\sum_{i=1}^{n} (x_i - \bar{x})(y_i - \bar{y}).$$

Hence obtain the maximum likelihood estimators of μ_1, μ_2, σ^2 and ρ.

3.14 A sample of n independent observations is taken on a random variable **x** having a logarithmic series distribution,

$$P(X = x) = \frac{-\theta^x}{x\ \ln(1-\theta)}, \quad x = 1, 2, \ldots$$

where θ is an unknown parameter in the range (0, 1). Show that the maximum likelihood estimater $\hat{\theta}$ of θ satisfies the equation

$$\hat{\theta} + \bar{x}(1 - \hat{\theta})\ln(1 - \hat{\theta}) = 0$$

where $\bar{x}$ is the sample mean, and find the asymptotic distribution of $\hat{\theta}$.

3.15 (a) The gamma distribution

$$f(x; \alpha, \beta) = \beta^\alpha x^{\alpha-1}\, e^{-\beta x}/\Gamma(\alpha), \qquad x > 0$$

has mean α/β and variance α/β^2. If α is known find the maximum likelihood estimator of β and its asymptotic variance.

Find the maximum likelihood estimator of $1/\beta$. What is its mean and variance? Does this estimator attain the minimum variance bound (MVB)? Does the maximum likelihood estimator for β attain the MVB?

(b) Verify that the maximum likelihood estimator of $1/\beta$ obtained in part (a) is sufficient. Is the estimator of β also sufficient?

3.16 Independent observations $x_1, x_2, \ldots, x_n$ are available on a random variable which is normally distributed with mean and standard deviation both equal to μ.

Find the maximum likelihood estimator $\hat{\mu}$ of μ. Show that the asymptotic variance of $\hat{\mu}$ is $\mu^2/3n$.

Compare this with the variance of the maximum likelihood estimator which would have been obtained for the mean if the functional relationship between the mean and variance had been ignored. Comment on your results.

3.17 A closed population of animals contains θ males and θ females. A random sample is taken in which each animal, independently of all others, has a known probability p of being caught. Altogether m males and f females are caught. Write down the likelihood function $L(\theta)$ for θ.

Let $t = \max(m, f)$, $s = \min(m, f)$. By considering the ratio of $L(\theta + 1)$ to $L(\theta)$, or otherwise, show that if

$$\{(t + 1)(1 - p)\}^2 < (t - s + 1)$$

then the maximum likelihood estimator $\hat{\theta}$ of θ is t, and that if the condition above is not satisfied then $\hat{\theta}$ is the largest integer smaller than the larger root of the equation

$$(2p - p^2)\theta^2 - (m + f)\theta + mf = 0.$$

3.18 The number of particles emitted by a radioactive source in unit time has a Poisson distribution with mean ρ. The strength of the source decreases as time goes by, and on days 0, 1, 2, ..., n it is assumed that ρ is $\alpha, \alpha\beta, \alpha\beta^2, \ldots, \alpha\beta^n$ respectively, where α, β are unknown parameters. Independent counts of particles $x_0, x_1, x_2, \ldots, x_n$ are obtained over unit time on days 0, 1, 2, ..., n. Show that

$$\left(\sum_{i=0}^{n} x_i, \sum_{i=0}^{n} ix_i\right)$$

is minimal sufficient for (α, β), and find equations whose solution will give $\hat{\alpha}$, $\hat{\beta}$, the maximum likelihood estimators for α and β.
Write down approximate expressions for the variances of $\hat{\alpha}$, $\hat{\beta}$.

3.19 In families where one parent has a rare hereditary disease the probability that a particular child inherits the disease is p, where $0 < p < 1$. Show that, in a family of fixed size k, the probability of at least one abnormal (diseased) child is $1 - (1 - p)^k$. In a survey, only families of size k with at least one abnormal child are sampled. In all, n such families are observed independently and there are r_i abnormal children in the ith family ($i = 1, 2, \ldots, n$). Show that $\hat{p}$, the maximum likelihood estimator for p, satisfies the equation

$$nk\hat{p} = [1 - (1 - \hat{p})^k] \sum_{i=1}^{n} r_i.$$

3.20 Individuals are given a measurable stimulus, to which they may or may not respond. When the stimulus is x, the probability that an individual responds is $p(x)$ where $p(x)$ and x are related by

$$p(x) = e^{\alpha+\beta x}/\{1 + e^{\alpha+\beta x}\}.$$

Here α and β ($\beta > 0$) are fixed and unknown constants. Thus as x increases from $-\infty$ to ∞, $p(x)$ increases from 0 to 1. Experimental results are available for k groups of individuals as follows. Each of the n_i individuals in the group was given stimulus x_i, and r_i individuals responded out of these n_i ($i = 1, 2, \ldots, k$).

Find two equations whose solution gives the maximum likelihood estimators $\hat{\alpha}$ and $\hat{\beta}$ for the parameters α and β, and describe briefly how you would set about solving for $\hat{\alpha}$ and $\hat{\beta}$ when numerical values for $n_1, n_2, \ldots, n_k; r_1, r_2, \ldots, r_k$ are given.

3.21 A cosmetic company is considering the marketing of a new product for men and wishes to estimate the proportion, θ, of males in a certain age group that would buy the product. Because a direct question may cause embarrassment, a so-called randomized response procedure is used to disguise the interviewee's actual willingness to buy the product.

Each person interviewed throws a fair die, and instead of giving the interviewer his true response 'Yes (will buy the product)' or 'No (will not buy it)', he gives a coded response A, B or C, as indicated by the table below. The interviewer does not see the score on the die.

		Die score					
		1	2	3	4	5	6
True	Yes	C	C	C	A	B	A
Response	No	C	A	A	B	A	B

In a random sample of 1000 men, the numbers of A, B, C responses were 440, 310 and 250 respectively. If each man in the sample has the same probability, θ, of having the response 'Yes', show that the log-likelihood for θ is

$$440 \ln(3 - \theta) + 310 \ln(2 - \theta) + 250 \ln(1 + 2\theta) + \text{constant}$$

and obtain the maximum likelihood estimate of θ.

3.22 Suppose that the number of eggs laid by a particular parasite is a random variable N, where N has a Poisson distribution with mean μ. Each egg, independently of all other eggs, has a probability p of hatching. Given that M denotes the number of hatched eggs, determine the joint distribution of (N, M). Given that $(n_1, m_1), (n_2, m_2), \ldots, (n_s, m_s)$ is a random sample of size s, determine $\hat{\boldsymbol{\theta}}$, the maximum likelihood estimator of $\boldsymbol{\theta}' = (\mu, p)$ and find the asymptotic distribution of $\hat{\boldsymbol{\theta}}$.

3.23 Find the method-of-moments estimator of θ in (a), (b), (c), (d) of Exercise 3.1.

3.24 Let $X_1, X_2, \ldots, X_n$ be a random sample from the uniform distribution $U[\theta - \frac{1}{2}, \theta + \frac{1}{2}]$. Show that the maximum likelihood estimator of θ is any value $\hat{\theta}$ in the interval $[\max(X_i) - \frac{1}{2}, \min(X_i) + \frac{1}{2}]$. What is the method-of-moments estimator.

3.25 Let $X_1, X_2, \ldots, X_n$ be a random sample from the density

$$f(x) = e^{-(x-\theta)}, \qquad x > \theta.$$

(a) Show that the maximum likelihood estimator $\hat{\theta}$ of θ is the minimum of $X_1, \ldots, X_n$.

(b) By finding the density function of $\hat{\theta}$ show that $\hat{\theta}$ is a consistent, but biased estimator of θ with $E(\hat{\theta}) = \theta + 1/n$. Suggest an unbiased and consistent estimator and find its variance.

(c) Compare the sampling properties of the maximum likelihood estimator with that obtained by the method of moments. Is it appropriate to compare the variances of these estimators with that suggested by the Cramér–Rao inequality?

3.26 Consider a random sample of size n from the distribution with p.d.f.

$$f(x; \theta) = \begin{cases} \theta x^{\theta - 1}, & 0 < x < 1 \quad \theta > 0 \\ 0, & \text{elsewhere.} \end{cases}$$

Find

(a) the maximum likelihood estimator for θ;

(b) an estimator based on the method of moments.

3.27 For the general linear model described in Section 3.4.2 show that the maximum likelihood estimator for $\boldsymbol{\beta}$ is identical to the least squares estimator.

3.28 Consider a time series $X_1, X_2, \ldots, X_n$ which follows a first order autoregressive model so that

$$(X_t - \mu) = \phi(X_{t-1} - \mu) + \varepsilon_t$$

where $\mu = E[X_t]$, $t = 1, 2, \ldots, n$, ϕ is an unknown parameter which is to be estimated, and $\varepsilon_1, \varepsilon_2, \ldots, \varepsilon_n$ are independent random variables, each with mean zero and variance σ^2. Show that the conditional distribution of X_t, given $X_{t-1}, \ldots, X_1$ for $t = 2, \ldots, n$, is $N[\phi(X_{t-1} - \mu), \sigma^2]$, and that the marginal distribution of X_1 is $N[\mu, \sigma^2/(1 - \phi^2)]$. Hence, show that the log-likelihood function can be written

$$l(\phi, \sigma^2; \mathbf{x}) = \text{const.} - \frac{n}{2}\ln(\sigma^2) + \tfrac{1}{2}\ln(1-\phi^2) - \frac{(1-\phi^2)}{2\sigma^2}(X_1-\mu)^2$$

$$-\frac{1}{2\sigma^2}\sum_{t=2}^{n}[(X_t-\mu)-\phi(X_{t-1}-\mu)]^2.$$

3.29 The minimum χ^2 estimate of a scalar parameter θ minimizes

$$\sum_{i=1}^{k}\frac{(\hat{p}_i - p_i(\theta))^2}{p_i(\theta)},$$

and the modified minimum χ^2 estimate minimizes

$$\sum_{i=1}^{k}\frac{(\hat{p}_i - p_i(\theta))^2}{\hat{p}_i},$$

(see Section 3.4.3). Show that the two estimates can be obtained by solving the equations

$$\sum_{i=1}^{k}\left(\frac{\hat{p}_i}{p_i(\theta)}\right)^2\frac{\partial p_i(\theta)}{\partial\theta} = 0$$

and

$$\sum_{i=1}^{k}\left(\frac{p_i(\theta)}{\hat{p}_i}\right)\frac{\partial p_i(\theta)}{\partial\theta} = 0$$

respectively. (Recall that $\Sigma_{i=1}^{k}\hat{p}_i = \Sigma_{i=1}^{k}p_i(\theta) = 1$.)

3.30 Using Exercise 3.29, or otherwise, find equations whose solutions give the minimum χ^2 estimate and the modified minimum χ^2 estimate respectively for the mean θ of a Poisson distribution. Assume that a random sample of n observations is available, and that the classes used are all the distinct values of the random variable which have been observed.

Chapter Four

Hypothesis testing

4.1 Introduction and some basic definitions

The most usual formulation of a hypothesis-testing problem is that we have to decide between two hypotheses regarding one or more parameters $\boldsymbol{\theta}$. The two hypotheses are the *null hypothesis*, which we denote by $H_0\colon \boldsymbol{\theta} \in \omega$, and the *alternative hypothesis*, which we denote by $H_1\colon \boldsymbol{\theta} \in \Omega - \omega$. Thus Ω is the set of all possible values for $\boldsymbol{\theta}$, called the *parameter space* and ω is some subset of Ω.

To make our choice between H_0 and H_1 we have a random sample $X_1, X_2, \ldots, X_n$ from a distribution with p.d.f. $f(x; \boldsymbol{\theta})$. We choose some subset C of possible values of $X_1, X_2, \ldots, X_n$ and reject H_0 if and only if $\mathbf{X} \in C$. Usually C is defined in terms of (extreme) values of some statistic $T(\mathbf{X})$. T is the *test statistic* and C is the *critical region* or *rejection region*; $\bar{C}$, the complement of C, is the *acceptance region.*

A hypothesis is *simple* if it specifies a single value for $\boldsymbol{\theta}$ (i.e. ω or $\Omega - \omega$ contain only one point); otherwise it is *composite.*

It seems likely that it will be easier to deal with simple hypotheses than with composite ones, so we start in the next section by looking at the simplest case of all when H_0, H_1 are both simple.

4.2 Simple null and alternative hypotheses – the Neyman–Pearson approach

If both H_0 and H_1 are simple, they can be written

$$H_0\colon \boldsymbol{\theta} = \boldsymbol{\theta}_0, \qquad H_1\colon \boldsymbol{\theta} = \boldsymbol{\theta}_1$$

for some values $\boldsymbol{\theta}_0$, $\boldsymbol{\theta}_1$ of $\boldsymbol{\theta}$. When we decide which of H_0, H_1 to choose there are two ways in which a mistake can be made.

Definition 4.1

Rejection of H_0 when it is true is called a *Type I error*; acceptance of H_0 when it is false is a *Type II error*. The probabilities of making a Type I error or Type II error are denoted by α, β respectively.

Definition 4.2

The probability of Type I error, α, in a test of a hypothesis is called the *size* or *significance level* of the test. The complement of the probability of Type II error, $\eta = 1 - \beta$, is the *power* of the test.

Ideally, we should like a critical region C which makes both α and β small, but it is clear from their definitions that, for fixed sample size, if one is decreased then the other increases. We have $\alpha = \Pr[\mathbf{X} \in C \mid H_0]$, $\beta = \Pr[\mathbf{X} \in \bar{C} \mid H_1]$. To decrease α we need to remove points from C, and hence add them to $\bar{C}$. This, in turn, will usually lead to an increase in β.

The *classical* approach is to fix α at some pre-determined standard level (usually $\alpha = 0.05, 0.01, 0.10$ or 0.001 is chosen for convenience) and then look for a test which minimizes β. If the smallest β possible for our fixed α is unacceptably large we have three options: increase α, increase the sample size, or abandon the test.

Note the lack of symmetry between α and β, and between H_0 and H_1 in this approach. The value of α is fixed at a desired level, whereas there is less emphasis on the actual value of β. A related point is that rejection of H_0 implies a degree of disbelief in H_0, but 'acceptance' of H_0 simply means that there is little evidence against H_0 and does not rule out other possible hypotheses. 'Failure to reject' is a better term than 'acceptance'. This lack of symmetry contrasts with some other approaches to hypothesis testing, for example the decision theory approach – see Section 6.3.

Definition 4.3

A test which minimizes β for fixed α is called a *most powerful* or *best* test of size α. Having defined a best test we need to know how to find such tests. Fortunately there is a surprisingly simple result which allows us to do so.

Lemma 4.1 (The Neyman–Pearson Lemma)

Suppose that $X_1, X_2, \ldots, X_n$ form a random sample from a distribution with p.d.f. $f(\mathbf{x}; \boldsymbol{\theta})$, and we wish to test H_0: $\boldsymbol{\theta} = \boldsymbol{\theta}_0$ vs. H_1: $\boldsymbol{\theta} = \boldsymbol{\theta}_1$. If $L(\boldsymbol{\theta}; \mathbf{x})$ is the likelihood function, then the best test of size α of H_0 vs. H_1 has a critical region of the form

$$\frac{L(\boldsymbol{\theta}_1; \mathbf{x})}{L(\boldsymbol{\theta}_0; \mathbf{x})} \geqslant A$$

for some non-negative constant, A.

Proof

The proof which follows is for continuous X with 'well-behaved' $f(\mathbf{x}; \boldsymbol{\theta})$ – we could replace integrals by summations for a proof for discrete X. An adaptation for multivariate $\mathbf{X}$ is also straightforward.

Let C be the critical region of size α given by the lemma, and let C^* be the critical region of *any other test* with size α. With obvious notation, we shall show that $\beta^* \geqslant \beta$, which proves the lemma.

$$\beta^* - \beta = \Pr[\mathbf{X} \in \bar{C}^* \mid \boldsymbol{\theta} = \boldsymbol{\theta}_1] - \Pr[\mathbf{X} \in \bar{C} \mid \boldsymbol{\theta} = \boldsymbol{\theta}_1]$$

$$= \int_{\mathbf{x} \in \bar{C}^*} L(\boldsymbol{\theta}_1; \mathbf{x})d\mathbf{x} - \int_{\mathbf{x} \in \bar{C}} L(\boldsymbol{\theta}_1; \mathbf{x})d\mathbf{x}$$

$$= \left[\int_{\bar{C}^* \cap C} L(\boldsymbol{\theta}_1; \mathbf{x})d\mathbf{x} + \int_{\bar{C}^* \cap \bar{C}} L(\boldsymbol{\theta}_1; \mathbf{x})d\mathbf{x} \right]$$

$$- \left[\int_{\bar{C} \cap C^*} L(\boldsymbol{\theta}_1; \mathbf{x})d\mathbf{x} + \int_{\bar{C} \cap \bar{C}^*} L(\boldsymbol{\theta}_1; \mathbf{x})d\mathbf{x} \right]$$

$$= \int_{\bar{C}^* \cap C} L(\boldsymbol{\theta}_1; \mathbf{x})d\mathbf{x} - \int_{\bar{C} \cap C^*} L(\boldsymbol{\theta}_1; \mathbf{x})d\mathbf{x}.$$

Now in C, and hence in $\bar{C}^* \cap C$, we have $L(\boldsymbol{\theta}_1; \mathbf{x}) \geqslant AL(\boldsymbol{\theta}_0; \mathbf{x})$.
Also, in $\bar{C}$, and hence in $\bar{C} \cap C^*$, we have $L(\boldsymbol{\theta}_1; \mathbf{x}) < AL(\boldsymbol{\theta}_0; \mathbf{x})$.
Hence

$$\beta^* - \beta \geqslant A\left[\int_{\bar{C}^* \cap C} L(\boldsymbol{\theta}_0; \mathbf{x})d\mathbf{x} - \int_{\bar{C} \cap C^*} L(\boldsymbol{\theta}_0; \mathbf{x})d\mathbf{x} \right].$$

Adding and subtracting $\int_{\bar{C} \cap C^*} L(\boldsymbol{\theta}_0; \mathbf{x})d\mathbf{x}$ on the right-hand side of this inequality gives

$$\beta^* - \beta \geqslant A\left[\int_C L(\boldsymbol{\theta}_0; \mathbf{x})d\mathbf{x} - \int_{C^*} L(\boldsymbol{\theta}_0; \mathbf{x})d\mathbf{x} \right] = A(\alpha - \alpha) = 0, \quad \text{as required.} \ \square$$

$\beta^* > \beta$ will hold in all but very exceptional circumstances, because for equality between β^* and β we need

$$\int_{\bar{C} \cap C^*} L(\boldsymbol{\theta}_0; \mathbf{x})d\mathbf{x} = 0$$

and

$$L(\boldsymbol{\theta}_1; \mathbf{x}) = AL(\boldsymbol{\theta}_0; \mathbf{x}) \qquad \text{throughout} \quad \bar{C}^* \cap C.$$

It is interesting to note the crucial rôle played by the likelihood function in deriving best tests. The tests proposed by the lemma are sometimes known as likelihood ratio (LR) tests.

Example 4.1

$X_1, X_2, \ldots, X_n \sim N(\mu, \sigma^2)$, σ^2 known. H_0: $\mu = \mu_0$, H_1: $\mu = \mu_1$, with $\mu_1 > \mu_0$.

Intuitively we would expect that a test with critical region consisting of large values of $\bar{X}$ would be suitable; we now verify that, as often happens, the best test *is* the one suggested by intuition.

$$L(\mu; \mathbf{x}) = (2\pi\sigma^2)^{-n/2} \exp\left\{ -\frac{1}{2\sigma^2} \sum_{i=1}^{n} (x_i - \mu)^2 \right\},$$

so

$$\frac{L(\mu_1;\ \mathbf{x})}{L(\mu_0;\ \mathbf{x})} = \exp\left\{-\frac{1}{2\sigma^2}\sum_{i=1}^{n}[(x_i-\mu_1)^2-(x_i-\mu_0)^2]\right\},$$

$$= \exp\left\{-\frac{1}{2\sigma^2}\sum_{i=1}^{n}[2x_i(\mu_0-\mu_1)+(\mu_1^2-\mu_0^2)]\right\}.$$

The best test of H_0 vs. H_1 has a critical region of the form

$$\frac{L(\mu_1;\ \mathbf{x})}{L(\mu_0;\ \mathbf{x})} \geqslant A \qquad \text{or equivalently} \qquad \ln\left[\frac{L(\mu_1;\ \mathbf{x})}{L(\mu_0;\ \mathbf{x})}\right] \geqslant \ln(A),$$

for some constant $A \geqslant 0$

i.e.

$$-\sum_{i=1}^{n}[2x_i(\mu_0-\mu_1)+(\mu_1^2-\mu_0^2)] \geqslant 2\sigma^2\ln(A)$$

or

$$\bar{x} \geqslant \frac{\sigma^2\ln(A)}{n(\mu_1-\mu_0)} + \tfrac{1}{2}(\mu_0+\mu_1) = B$$

say, where B is a constant.

To find the appropriate value of B for a test of given size α, we ignore the complicated expression above and just use our knowledge that under H_0, $\bar{X} \sim N(\mu_0, \sigma^2/n)$, so that $\Pr[\bar{X} \geqslant B] = \alpha$ will be achieved if $B = \mu_0 + z_\alpha\sigma/\sqrt{n}$. Here z_α is the standard notation for a critical value of the standard normal distribution, i.e. if $Z \sim N(0, 1)$, then z_α is such that $\Pr[Z \geqslant z_\alpha] = \alpha$.

Note that if μ_1 had been less than μ_0, then the direction of the inequality above would have changed when we divided through by $(\mu_1 - \mu_0)$, so the critical region would have been of the form $\bar{X} \leqslant$ constant, again corresponding with what intuition might have suggested.

Example 4.2

Consider n trials of a binomial experiment, with probability of success, p:

$$H_0\colon p = p_0, \quad H_1\colon p = p_1 \qquad \text{with} \quad p_1 > p_0$$

$$L(p;\ x) = \binom{n}{x} p^x(1-p)^{n-x}$$

where x is the number of successes observed. Going through similar steps to those in Example 4.1, we find a critical region of the form

$$\ln\left[\frac{L(p_1;\ \mathbf{x})}{L(p_0;\ \mathbf{x})}\right] = x\ln(p_1/p_0) + (n-x)\ln[(1-p_1)/(1-p_0)] \geqslant \ln(A)$$

which reduces to

$$x \geqslant \frac{\ln(A) - n \ln[(1-p_1)/(1-p_0)]}{\ln[p_1(1-p_0)/p_0(1-p_1)]} = B$$

say, where B is a constant.

Here we divided both sides of the inequality by $\ln[p_1(1-p_0)/p_0(1-p_1)]$. This is positive since $p_1 > p_0$; if p_1 had been negative, the divisor would have been negative, with a critical region of the form $x \leqslant$ constant.

At first sight it appears that A, and hence B, can take any value, but since X is discrete and can only take the values $0, 1, 2, \ldots, n$ it is pointless looking at values of B other than $0, 1, 2, \ldots, (n+1)$. Table 4.1 gives the values of α and β for all possible distinct values of B for a specific example, namely for the case where $p_0 = 0.5$, $p_1 = 0.8$ and $n = 10$.

A typical entry for α is

$$\Pr[X \geqslant B \mid p = 0.5] = \sum_{x=B}^{10} \binom{10}{x} (0.5)^x (0.5)^{10-x}$$

and a typical entry for β is

$$\Pr[X < B \mid p = 0.8] = \sum_{x=0}^{B-1} \binom{10}{x} (0.8)^x (0.2)^{10-x}.$$

A problem which arises here, and in general with discrete examples, is that we are restricted to a small number of possible values of α, typically not the standard ones of 0.05, 0.01, etc., unless we use a rather artificial procedure. Suppose that here we want $\alpha = 0.05$; then we have three possible options. First, we can get as close as possible to $\alpha = 0.05$ using the values in Table 4.1 by rejecting H_0 if $x \geqslant 8$. This gives $\alpha = 0.055$, whereas we may feel that $\alpha = 0.05$ is the maximum we can tolerate as our probability of Type I error. This implies that we should reject H_0 only if $x \geqslant 9$, leading to $\alpha = 0.011$, far smaller than we require. A third alternative is to use a so-called *randomized* critical region. Here we reject H_0 if $x \geqslant 9$, accept H_0 if $x \leqslant 7$ and use random number tables or some other randomizing device to reject H_0 with probability 0.886 if $x = 8$. This gives a test of size $0.011 + 0.886\,(0.055-0.011) = 0.050$. Although this is a theoretically neat way of achieving the required value of α, the approach has little practical appeal.

Table 4.1 Probabilities of Type I and Type II errors for a binomial example

B	0	1	2	3	4	5	6	7	8	9	10	11
α	1.000	0.999	0.989	0.945	0.828	0.623	0.377	0.172	0.055	0.011	0.001	0.000
β	0.000	<0.001	<0.001	<0.001	0.001	0.006	0.033	0.121	0.322	0.624	0.893	1.000

Note that in both of the above examples, the best test depends on the observations only through the minimal sufficient statistic for the parameter of interest. This is a general result, for suppose that **T** is minimal sufficient for $\boldsymbol{\theta}$. Then the test statistic for the best test is given by the likelihood ratio

$$\frac{L(\boldsymbol{\theta}_1;\ \mathbf{x})}{L(\boldsymbol{\theta}_0;\ \mathbf{x})} = \frac{K_1[\mathbf{t}\,|\,\boldsymbol{\theta}_1]K_2[\mathbf{x}]}{K_1[\mathbf{t}\,|\,\boldsymbol{\theta}_0]K_2[\mathbf{x}]},$$

by the factorization criterion (Theorem 2.1). The factor K_2 cancels, and what is left depends on **X** only through **T**.

4.2.1 *Likelihood ratio (LR) tests for exponential families*

Suppose that $X_1, X_2, \ldots, X_n$ are a random sample from a distribution belonging to the k-parameter exponential family with p.d.f $f(x; \boldsymbol{\theta}) = \exp\{\Sigma_{j=1}^{k} A_j(\boldsymbol{\theta})B_j(x) + C(x) + D(\boldsymbol{\theta})\}$. The LR which is used in the best test of $H_0\colon \boldsymbol{\theta} = \boldsymbol{\theta}_0$, vs. $H_1\colon \boldsymbol{\theta} = \boldsymbol{\theta}_1$ takes the form

$$\frac{\exp\left\{\sum_{j=1}^{k}\sum_{i=1}^{n} A_j(\boldsymbol{\theta}_1)B_j(x_i) + \sum_{i=1}^{n} C(x_i) + D(\boldsymbol{\theta}_1)\right\}}{\exp\left\{\sum_{j=1}^{k}\sum_{i=1}^{n} A_j(\boldsymbol{\theta}_0)B_j(x_i) + \sum_{i=1}^{n} C(x_i) + D(\boldsymbol{\theta}_0)\right\}}$$

$$= \exp\left\{\sum_{j=1}^{k}\sum_{i=1}^{n} [A_j(\boldsymbol{\theta}_1) - A_j(\boldsymbol{\theta}_0)]B_j(x_i) + [D(\boldsymbol{\theta}_1) - D(\boldsymbol{\theta}_0)]\right\}.$$

We reject H_0 when this quantity exceeds a suitably chosen constant.

In the special case where $k = 1$, the LR reduces to

$$\exp\left\{[A(\theta_1) - A(\theta_0)] \sum_{i=1}^{n} B_j(x_i) + [D(\theta_1) - D(\theta_0)]\right\}$$

$$= \exp\{(\phi_1 - \phi_0)t + [D(\theta_1) - D(\theta_0)]\}$$

using the natural parameterization – see Definition 2.8. If $(\phi_1 - \phi_0) > 0$, then the critical region for the best test consists of large values of the minimal sufficient statistic T. For $(\phi_1 - \phi_0) < 0$, the critical region contains small values of T.

Examples 4.1, 4.2 give illustrations of this result. In Example 4.1, we had $T = \bar{X}$, and $\phi = \mu/\sigma^2$, so $(\phi_1 - \phi_0) > 0$ is equivalent to $(\mu_1 - \mu_0) > 0$, in which case we reject H_0 for large values of $\bar{X}$. In Example 4.2, the minimal sufficient statistic is X, and $\phi = \ln[p/(1-p)]$. Thus $(\phi_1 - \phi_0) > 0$ is equivalent to $(p_1 - p_0) > 0$, and for this case we reject H_0 for large values of X.

4.3 Pure significance tests

In the previous section, and indeed in the introduction to this chapter, the framework for hypothesis testing included *two* specific hypotheses, H_0 and H_1. We now look briefly at the situation where we specify only a null hypothesis, H_0, and examine how to decide whether or not a given set of data is consistent with H_0. No explicit alternative hypothesis H_1 is specified, perhaps because we are very vague about our alternatives, and cannot easily parameterize them. We then have a so-called *pure significance* test. For an interesting discussion, from a historical perspective, of the present approach, and comparisons with that of the previous section, see Lehmann (1993).

What is generally done in pure significance testing is to look at the distribution of a test statistic, T, under H_0, and define the *p-value* for the test as

$$\Pr[T \text{ at least as extreme as the value observed} \mid H_0].$$

H_0 will be accepted (rejected) if the p-value is large (small) enough, or we may just quote the p-value and leave things there. The apparent absence of H_1 is somewhat illusory. We need some idea of the likely departures from H_0 before we can decide what is meant by 'T at least as extreme as the value observed' in the definition of the p-value. Furthermore, if we wish to compare different test statistics this will typically involve power comparisons, which are impossible without some idea of alternatives.

Not only have p-values been much used, but they have also been heavily criticized. For example, papers by Casella and Berger (1987) and Berger and Sellke (1987), and the accompanying discussion, strongly suggest that p-values may not be reconcilable with the actual evidence from an experiment. The basic problem is that what we are interested in when assessing the compatibility of a data set with H_0 is $\Pr[H_0 \mid T = t]$. The p-value is $\Pr[T$ at least as extreme as $t \mid H_0]$, which can be quite different.

Example 4.3

Consider again the specific binomial example at the end of Example 4.2, where we have $n = 10$ and H_0: $p = \frac{1}{2}$. Suppose we observe $X = 8$; then the p-value could be defined as $\Pr[X \geqslant 8 \mid H_0] = 0.055$ (i.e. a *one-sided* p-value). This is identical to the size of the test in which we reject for values of $X \geqslant 8$.

Of course, this definition of the p-value makes the implicit assumption that deviations from H_0 must be in the direction $p > \frac{1}{2}$. If values of $p < \frac{1}{2}$ were also entertained (a two-sided alternative to $p = \frac{1}{2}$), then 'X at least as extreme as 8' would have to include small, as well as large, values of X. In the current example, with a symmetric null distribution, the one-sided p-value is simply doubled for a two-sided alternative. However, for a non-symmetric null distribution there are a number of possible different strategies for computing two-sided p-values. Davis (1986) discusses some of these possibilities in the context of (2×2) contingency tables.

We can also use this example to illustrate the difference between the p-value and $\Pr[H_0 \mid T = t]$. For this we need an alternative hypothesis; suppose that we have H_1: $p = 0.98$, i.e. *the only possible values for p are* 0.50 *or* 0.98, which, in itself, may well be considered to be unrealistic.

Then $\Pr[X \geqslant 8 \mid H_0] = 0.0557$ and $\Pr[X \leqslant 8 \mid H_1] = 0.0162$. So, although the p-value is small, a corresponding value based on H_1 is even smaller. Alternatively, if we assume *a priori* that H_0 and H_1 are equally plausible, then

$$\Pr[X = 8 \mid H_0] = 0.0439, \qquad P[X = 8 \mid H_1] = 0.0153$$

leads to

$$\Pr[H_0 \mid X = 8] = \frac{0.0439}{0.0592} = 0.742.$$

4.4 Composite hypotheses – uniformly most powerful tests

For a scalar θ, the most usual composite set-up is to have a simple null hypothesis $H_0: \theta = \theta_0$, and an alternative which is either *one-sided* ($H_1: \theta < \theta_0$ or $H_1: \theta > \theta_0$) or *two-sided* ($H_1: \theta \neq \theta_0$). However, we start by considering the general two-hypothesis framework with

$$H_0: \boldsymbol{\theta} \in \omega, \qquad H_1: \boldsymbol{\theta} \in \Omega - \omega$$

and generalize some of the ideas developed for simple H_0, H_1.

Definition 4.4

The *power function* is defined as $\eta(\boldsymbol{\theta}) = P\{\mathbf{X} \in C \mid \boldsymbol{\theta}\}$, and the *size* of the test is now defined to be $\sup_{\boldsymbol{\theta} \in \omega} \eta(\boldsymbol{\theta}) = \alpha$.

When H_0 is simple, this definition of size reduces to the previous one (Definition 4.2), and when H_1 is simple, then the power function evaluated when $\boldsymbol{\theta} \in \Omega - \omega$ is simply the power defined earlier (Definition 4.2). For composite H_0 a distinction is drawn by some authors between *size* and *significance level*. The latter is defined as a value α such that $\eta(\boldsymbol{\theta}) \leqslant \alpha$ for all $\boldsymbol{\theta} \in \omega$.

Next we introduce an idea which extends the concept of most powerful tests to the composite hypothesis case.

Definition 4.5

Suppose we have a test of size α; then it is *uniformly most powerful* (*UMP*) of size α if its power function $\eta(\boldsymbol{\theta})$ is such that $\eta(\boldsymbol{\theta}) \geqslant \eta^*(\boldsymbol{\theta})$ for all $\boldsymbol{\theta} \in \Omega - \omega$, where $\eta^*(\boldsymbol{\theta})$ is the power function of *any other* size-α test.

Lemma 4.2

If H_0 is simple but H_1 is composite, suppose that we find a best test, from the Neyman–Pearson lemma, for H_0 vs. $H_1': \boldsymbol{\theta} = \boldsymbol{\theta}_1$ where $\boldsymbol{\theta}_1$ is some value of $\boldsymbol{\theta}$ contained in $\Omega - \omega$. If the best test takes the same form for all $\boldsymbol{\theta}_1 \in \Omega - \omega$, then this test is UMP.

Proof

The power of the best test for H_0 vs. H_1' is $\eta(\boldsymbol{\theta}_1)$ and this is no less than $\eta^*(\boldsymbol{\theta}_1)$ for any other size-α test of H_0 vs. H_1'. Since this holds for all $\boldsymbol{\theta}_1 \in \Omega - \omega$, it follows that we have a UMP test.

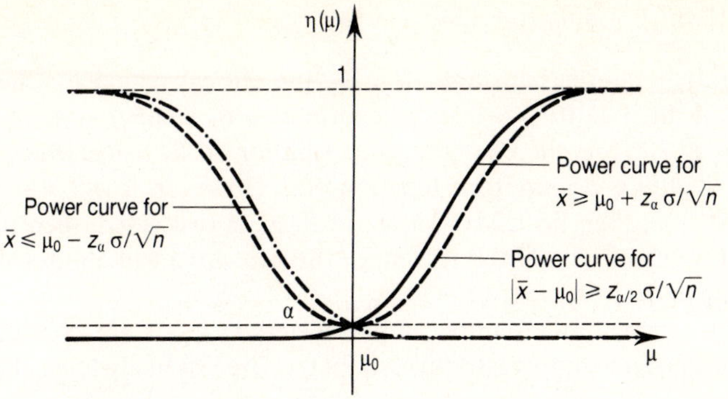

Fig. 4.1 Power functions for three tests of H_0: $\mu = \mu_0$ vs. H_0: $\mu \neq \mu_0$.

Example 4.4

$X_1, X_2, \ldots, X_n \sim N(\mu, \sigma^2)$, σ^2 known. We wish to test

$$H_0: \mu = \mu_0 \text{ vs. } H_1: \mu > \mu_0.$$

We have seen in Example 4.1 that, if we pick *any* $\mu_1 > \mu_0$, then the best test of H_0 vs. H_1': $\mu = \mu_1$ has a critical region of the form $\bar{X} \geqslant B$, where $B = \mu_0 + z_\alpha \sigma/\sqrt{n}$ for a test of size α. Thus this test is UMP for H_0 vs. H_1.

Example 4.5

Consider the same situation as in Example 4.4, except that H_1 now becomes H_1: $\mu \neq \mu_0$. Then the best test of H_0 vs. H_1': $\mu = \mu_1$ is not the same for all $\mu_1 \in \Omega - \omega$.

If $\mu_1 < \mu_0$ the critical region has the form $\bar{X} \leqslant B_1$, whereas if $\mu_1 > \mu_0$ the critical region has the form $\bar{X} \geqslant B_2$, where B_1, B_2 are appropriately chosen constants. An obvious possible critical region for a sensible test is $|\bar{X} - \mu_0| \geqslant B_3$, where B_3 is again a constant chosen to give size α, but as can be seen by plotting its power function (see Figure 4.1) it is not UMP; it is, naturally, bettered by $\bar{X} \geqslant B_2$ for $\mu_1 > \mu_0$, and by $\bar{X} \leqslant B_1$ for $\mu_1 < \mu_0$. In fact, since neither of these two tests can be bettered for $\mu_1 > \mu_0$, $\mu_1 < \mu_0$ respectively, there is no UMP test.

Examples 4.4 and 4.5 display typical behaviour for UMP tests. Often a UMP test will exist, and is easily found, for one-sided H_1, but for two-sided H_1 there is usually no UMP test. We illustrate this for members of exponential families of distributions.

4.4.1 *UMP tests for exponential families*

Suppose that we have a random sample, $X_1, X_2, \ldots, X_n$, from a distribution belonging to the one-parameter exponential family, and that we wish to test H_0: $\theta = \theta_0$ vs. H_1: $\theta > \theta_0$. If θ_1 is such that $\theta_1 > \theta_0$, and we find the best test of H_0 vs. H_1': $\theta = \theta_1$, then the LR used in this test takes the form (see Section 4.2.1)

$$\exp\{[A(\theta_1) - A(\theta_0)]t + [D(\theta_1) - D(\theta_0)]\}.$$

For appropriately chosen constants, B_1, B_2, the critical region for the test has the form $T \geqslant B_1$ if $A(\theta_1) - A(\theta_0) > 0$ and the form $T \leqslant B_2$ if $A(\theta_1) - A(\theta_0) < 0$. A UMP test will exist if $A(\theta_1) - A(\theta_0)$ has the same sign for all $\theta_1 > \theta_0$, which will happen, for example, if $A(\theta)$ is a monotone function of θ. However, if $A(\theta_1) - A(\theta_0)$ changes sign within $\Omega - \omega$, then no UMP will exist. If H_1 is two-sided, then it is extremely unlikely that we can find a UMP test, since $A(\theta_1) - A(\theta_0)$ will almost always change sign as we move from $\theta_1 > \theta_0$ to $\theta_1 < \theta_0$.

Clearly, if we increase the number of parameters to $k > 1$, the conditions required for a UMP test become more restrictive, so that we are less likely to be able to find one.

4.4.2 *UMP tests within restricted subsets of tests*

The idea of UMP tests is an attractive one. Unfortunately, we have seen that, unlike best tests for simple H_0 and simple H_1, UMP tests may often not exist. The question thus arises of what to do when no UMP exists? One strategy is to design a general procedure for constructing tests; such a procedure should be widely applicable, relatively straightforward to implement, and produce tests whose properties are generally good. We look at a number of general hypothesis testing procedures later in the chapter, but first we examine an alternative strategy.

The idea is to define some desirable property of tests and restrict attention to tests satisfying this property.

In this section we discuss briefly three possible ways of restricting the tests we examine. We then look for a UMP test within this subset of tests.

Unbiased tests

Definition 4.6

Suppose that, as before, we wish to test H_0: $\boldsymbol{\theta} \in \omega$ vs. H_1: $\boldsymbol{\theta} \in \Omega - \omega$.

A test of size α is said to be *unbiased* if $\eta(\boldsymbol{\theta}) \geqslant \alpha$ for all $\boldsymbol{\theta} \in \Omega - \omega$, i.e. the power function is never smaller for a point in $\Omega - \omega$ than it is for any point in ω. This is intuitively a reasonable requirement. If it does not hold, then there are points in H_1 for which we are less likely to reject H_0 than for any value in H_0, which seems undesirable.

We can see that in Example 4.5 neither of the one-sided tests is unbiased. However, the two-sided test *is* unbiased and in fact it is uniformly most powerful unbiased, i.e. UMP among the restricted set of unbiased tests. We shall not prove this here – see, for example, Lehmann (1986, Section 5.2). Indeed, we shall skim very quickly over unbiasedness, and the topics of the next two subsections. To do otherwise would get us much deeper into theoretical arguments than we wish to go in this text. Lehmann (1986) can be consulted as a suitable reference for this theory.

Similar tests

Definition 4.7

With the same notation as before, a test of size α is *similar* if $\eta(\boldsymbol{\theta}) = \alpha$ for all $\boldsymbol{\theta} \in \omega$. The term 'similar' is an abbreviation of 'similar to the sample space'. The (trivial) test whose critical region consists of the whole sample space has this property.

The definition is more restrictive than that for unbiasedness, and it is not immediately obvious why this restriction might be useful. In fact, the concept of similar tests has its main use in situations where the null hypothesis is composite, but only because one or more nuisance parameters is unspecified under H_0.

Example 4.6

Suppose that $X_1, X_2, \ldots, X_n \sim N(\mu, \sigma^2)$ and we wish to test H_0: $\mu = \mu_0$, σ^2 unknown vs. H_1: $\mu > \mu_0$, σ^2 unknown. H_0 would be simple, were it not for the presence of the nuisance parameter σ^2, whose value is not of direct interest, but which has a direct influence on the exact distribution of the X_i. It can be shown that in this example the usual t-test is uniformly most powerful similar. Once again, we shall stop short of giving details of the theory to justify this result – see, for example, Stuart and Ord (1991, pp. 833–4). However, in outline what happens is that we can condition on the statistic $V = \Sigma_{i=1}^{n} (X_i - \mu_0)^2$. If we condition on the value of V, the best test for H_0 vs. H_1': $\mu = \mu_1$, σ^2 unknown, where $\mu_1 > \mu_0$, has the same critical region for all values of μ_1. This test, which is equivalent to the t-test, is thus UMP, conditional on the value of V. But because the critical region has the same value of $\eta(\boldsymbol{\theta})$ for all values of V, the test is similar.

Invariant tests

It is difficult to give a concise definition of invariant tests, but the basic idea is as follows.

Suppose that our hypotheses are unaffected by certain transformations of the data; then look only at tests which are unchanged by the same transformations. For example, with H_0: $\mu = 0$ vs. H_1: $\mu \neq 0$, the hypotheses are not affected by changing the sign of all observations. We therefore look only at critical regions which are unchanged if all observations have their signs changed, for example regions based on $|\bar{X}|$. In Example 4.5, the test with critical region $|\bar{X} - \mu_1| \geqslant B_3$ is invariant, and in fact is uniformly most powerful invariant. Lehmann (1986, Chapter 6) gives a thorough discussion of the theory of invariance, including some connections with unbiasedness (Section 6.6).

4.4.3 *Locally most powerful tests*

In this section we adopt a somewhat different strategy to that of the previous section. Rather than restrict ourselves to a search for *uniformly* most powerful tests within subsets of all possible tests, we look for *locally* most powerful (LMP) tests, i.e. tests which are most powerful for alternatives close to H_0. This is easiest to do for simple H_0, so we restrict attention to this case; we also restrict ourselves to a single parameter, and consider $H_0: \theta = \theta_0$ vs. $H_1: \theta = \theta_0 + \delta$ for some small δ ($\delta > 0$).

Now the best test of H_0 vs. H_1 has a critical region of the form

$$\frac{L(\theta_0 + \delta;\, \mathbf{x})}{L(\theta_0;\, \mathbf{x})} \geqslant A$$

or, in terms of log likelihood,

$$l(\theta_0 + \delta;\, \mathbf{x}) - l(\theta_0;\, \mathbf{x}) \geqslant \ln(A)$$

for an appropriately chosen constant A. Using a Taylor expansion, and assuming standard regularity conditions, we have that, for small δ,

$$l(\theta_0 + \delta;\, \mathbf{x}) \approx l(\theta_0;\, \mathbf{x}) + \delta \left(\frac{\partial l}{\partial \theta}\right)_{\theta = \theta_0}$$

so a locally most powerful test has a critical region of the form

$$\left(\frac{\partial l}{\partial \theta}\right)_{\theta = \theta_0} \geqslant B \text{ for some constant } B.$$

This test is based on the first derivative of the log-likelihood (the score) and hence is an (efficient) *score test*; such tests will be discussed further in Section 7 of this chapter. Under the usual regularity conditions, we know from Chapter 2 that

$$\mathrm{E}\left[\frac{\partial l}{\partial \theta}\right] = 0 \quad \text{and} \quad \mathrm{var}\left[\frac{\partial l}{\partial \theta}\right] = I_\theta.$$

Furthermore, for large n, $\partial l/\partial \theta$ is approximately normally distributed so that, in order to achieve a size α test we need $B = z_\alpha I_\theta^{1/2}$.

Example 4.7

Suppose that $X_1, X_2, \ldots, X_n$ are a random sample from the one-parameter Cauchy distribution with p.d.f.

$$f(x;\, \theta) = [\pi(1 + (x - \theta)^2)]^{-1}.$$

It can be shown that

$$\left(\frac{\partial l}{\partial \theta}\right) = 2 \sum_{i=1}^{n} \frac{(X_i - \theta)}{[1 + (X_i - \theta)^2]} \quad \text{and} \quad I_\theta = n/2.$$

Asymptotically, the locally most powerful test therefore rejects $H_0: \theta = \theta_0$ in favour of $H_1: \theta > \theta_0$ if and only if

$$2 \sum_{i=1}^{n} \frac{(X_i - \theta_0)}{[1 + (X_i - \theta_0)^2]} \geqslant z_\alpha \left[\frac{n}{2}\right]^{1/2}.$$

Note that, as with UMP tests, the idea of locally most powerful tests is usually only useful if the alternative hypothesis is one-sided, or in one specific direction if $\boldsymbol{\theta}$ is a vector. This is because different directions away from H_0 will generally lead to different LMP tests. In Example 4.7, for instance, if H_1 becomes $H_1: \theta_0 - \delta$, then the direction of the inequality for the critical region is reversed.

4.5 Further properties of tests of hypothesis

So far in this chapter we have looked at 'optimal' tests, according to various definitions. Best tests, UMP tests, UMP unbiased tests, LMP tests, and so on, all require definitions of some desirable properties of tests. In the present section we define two more desirable properties.

Definition 4.8

Suppose that a test depends on the sample size n, and that we can write the power function as $\eta_n(\boldsymbol{\theta})$. A test is said to be *consistent* if $\text{Lim}_{n\to\infty}\,\eta_n(\boldsymbol{\theta}) = 1$ for all $\boldsymbol{\theta} \in \Omega - \omega$.

As with consistency in point estimation, this is an asymptotic property which we would usually like a test to possess, and, as in point estimation, most 'common-sense' procedures do possess it. For example, tests given by the Neyman–Pearson lemma are consistent (Bickel and Doksum, 1977, Theorem 6.6.1).

Definition 4.9

Suppose that we have two possible tests of H_0 vs. H_1 where H_0, H_1 are both simple. If n_1, n_2 are the minimum possible sample sizes for tests 1 and 2 for which we can achieve a size α and power $\geqslant \eta$, then the *relative efficiency* of test 1 compared with test 2 is n_2/n_1. If test 2 is the best test, then this ratio is the *efficiency* of test 1.

These definitions depend on η (and α); a possible once-and-for-all measure (corresponding to $\eta \to 1$ for fixed α) is the *asymptotic efficiency or asymptotic relative efficiency*, in which we look at the limit of n_2/n_1 as $n_1, n_2 \to \infty$.

We can extend the ideas of efficiency to composite H_1, but there are complications such as taking $H_1 \to H_0$, and we shall omit the details – see, for example, Stuart and Ord, (1991, pp. 947–949). The concept of asymptotic relative efficiency is particularly useful in non-parametric inference (Gibbons, 1985, Chapter 14). Here we wish to compare optimal parametric tests with non-parametric tests which typically use less information but at the same time are not tied by the restrictive assumptions underlying parametric tests. Further discussion of non-parametric inference is given in Chapter 8.

4.6 Maximum likelihood ratio tests (MLRTs)

In point estimation we looked at possible desirable properties of estimators (Chapter 2), and then (Chapter 3) at general methods of estimation. We follow the same scheme with hypothesis testing; most of the chapter so far has been concerned with looking at desirable properties, but in the next two sections we discuss general methods for constructing tests. The first class of tests which we consider is that of maximum likelihood ratio tests (MLRTs).

Consider, as usual, $X_1, X_2, \ldots, X_n$, a random sample from a distribution with p.d.f. $f(x; \boldsymbol{\theta})$, and suppose that we wish to test

$$H_0: \boldsymbol{\theta} \in \omega \quad \text{vs.} \quad H_1: \boldsymbol{\theta} \in \Omega - \omega.$$

Then define

$$\lambda = \left\{ \underset{\boldsymbol{\theta} \in \omega}{\text{Max}}\, L(\boldsymbol{\theta}; \mathbf{x}) \Big/ \underset{\boldsymbol{\theta} \in \Omega}{\text{Max}}\, L(\boldsymbol{\theta}; x) \right\}$$

and construct a test with critical region of the form $\lambda \leqslant A$ for some constant A. This gives an MLRT. Such tests are often known simply as likelihood ratio tests, but we add the word 'maximum' to distinguish them from the tests produced in the simple vs. simple case by the Neyman–Pearson lemma. The alternative terminology 'generalized likelihood ratio tests' is also occasionally encountered in the literature.

The form of this test is intuitively plausible; we have $0 \leqslant \lambda \leqslant 1$, and λ should be close to 1 if $\boldsymbol{\theta} \in \omega$, but further from 1 (smaller) if $\boldsymbol{\theta} \notin \omega$. Its respectability is intuitively enhanced by its connection with the ubiquitous ML method of estimation. Also when H_0, H_1 are both simple, the MLRT will reduce to the best test given by the Neyman–Pearson lemma, with the slight complication that values of α between $(1-p)$ and 1 are not achievable, where

$$p = \Pr[L(\boldsymbol{\theta}_0; \mathbf{x}) \geqslant L(\boldsymbol{\theta}_1; \mathbf{x}) \,|\, H_0].$$

This follows because

$$\alpha = \Pr[\lambda \leqslant A \,|\, H_0]$$

where $\lambda = \text{Min}\{1, L(\boldsymbol{\theta}_0; \mathbf{x})/L(\boldsymbol{\theta}_1; \mathbf{x})\}$, and $\alpha = 1$ when $A \geqslant 1$, but $\alpha \leqslant (1-p)$ when $A < 1$.

It is a little confusing that the critical region consists of *small* values of λ, whereas tests given by the Neyman–Pearson lemma use a test statistic which is related to the *reciprocal* of λ, and reject H_0 for *large* values of the statistic.

Some authors (for example, see Silvey, 1970, p. 109) try to standardize the two situations, but it is more usual to look at λ. Like the test statistic given by the Neyman–Pearson lemma, and like ML estimators, λ depends on $\mathbf{x}$ only through the minimal sufficient statistic for $\boldsymbol{\theta}$. Thus it is no surprise that tests based on λ have good properties. These properties are generally asymptotic – its small sample properties depend on the particular example. Cox and Hinkley (1974), for example, cite the following properties:

(1) λ gives a UMP test, if one exists, for simple H_0 (p. 313);
(2) the MLRT is asymptotically most powerful unbiased (p. 319);
(3) the MLRT is asymptotically similar (p. 323);

(4) the MLRT is asymptotically efficient (p. 337).

Although it is possible to produce examples, such as Example 4.9 below, where MLRTs are not very sensible, they usually give workable tests with reasonable properties.

Example 4.8

Suppose that $X_1, X_2, \ldots, X_n \sim N(\mu, \sigma^2)$, σ^2 unknown, and we wish to test $H_0: \mu = \mu_0$ vs. $\mu \neq \mu_0$.

Note that none of the approaches presented in Section 4.4 will give an optimal test here. For one-sided H_1 and σ^2 known we had a UMP test; for two-sided H_1 and σ^2 known there is a UMP unbiased or UMP invariant test; for one-sided H_1 and σ^2 unknown there is a UMP similar test but we have not considered the two-sided H_1 and σ^2 unknown case. Conversely, the MLRT is equivalent to the optimal test in any of the cases just listed, as well as producing a test in the present case.

The likelihood function is

$$L(\boldsymbol{\theta}; \mathbf{x}) = (2\pi\sigma^2)^{-n/2} \exp\left\{-\frac{1}{2\sigma^2} \sum_{i=1}^{n} (x_i - \mu)^2\right\}.$$

Under H_0, $\mu = \mu_0$, and the MLE for σ^2 is

$$\tilde{\sigma}^2 = \frac{1}{n} \sum_{i=1}^{n} (x_i - \mu_0)^2.$$

Thus

$$\begin{aligned}\operatorname*{Max}_{\boldsymbol{\theta} \in \omega} L(\boldsymbol{\theta}; \mathbf{x}) &= L(\tilde{\boldsymbol{\theta}}; \mathbf{x}) = (2\pi\tilde{\sigma}^2)^{-n/2} \exp\left\{-\frac{1}{2\tilde{\sigma}^2} \sum_{i=1}^{n} (x_i - \mu_0)^2\right\} \\ &= (2\pi\tilde{\sigma}^2)^{-n/2} \exp\left\{-\frac{1}{2\tilde{\sigma}^2}\, n\tilde{\sigma}^2\right\} = (2\pi\tilde{\sigma}^2)^{-n/2} \exp\left\{-\frac{n}{2}\right\}.\end{aligned}$$

Under $H_0 \cup H_1$, the MLEs for μ, σ^2 are

$$\hat{\mu} = \bar{x},\ \hat{\sigma}^2 = \frac{1}{n} \sum_{i=1}^{n} (x_i - \bar{x})^2$$

and

$$\begin{aligned}\operatorname*{Max}_{\boldsymbol{\theta} \in \Omega} L(\boldsymbol{\theta}; \mathbf{x}) &= L(\hat{\boldsymbol{\theta}}; \mathbf{x}) = (2\pi\hat{\sigma}^2)^{-n/2} \exp\left\{-\frac{1}{2\hat{\sigma}^2} \sum_{i=1}^{n} (x_i - \bar{x})^2\right\} \\ &= (2\pi\hat{\sigma}^2)^{-n/2} \exp\left\{-\frac{n}{2}\right\}.\end{aligned}$$

Hence

$$\lambda = \left(\frac{\tilde{\sigma}^2}{\hat{\sigma}^2}\right)^{-n/2}.$$

Now common sense and common practice suggest that we should use a t-test for H_0 vs. H_1, and we now show that using λ is, in fact, equivalent to the t-test.

$$\lambda = \left[\sum_{i=1}^{n} (x_i - \mu_0)^2 \Big/ \sum_{i=1}^{n} (x_i - \bar{x})^2 \right]^{-n/2}$$

$$= \left[\frac{\sum_{i=1}^{n} (x_i - \bar{x})^2 + n(\bar{x} - \mu_0)^2}{\sum_{i=1}^{n} (x_i - \bar{x})^2} \right]^{-n/2}$$

using $[x_i - \mu_0]^2 = [(x_i - \bar{x}) + (\bar{x} - \mu_0)]^2$, and noting that the cross-product disappears,

$$= \left[1 + n(\bar{x} - \mu_0)^2 \Big/ \sum_{i=1}^{n} (x_i - \bar{x})^2 \right]^{-n/2}$$

$$= [1 + t^2/(n-1)]^{-n/2}$$

where

$$t = \frac{(\bar{x} - \mu_0)}{s/\sqrt{n}} \text{ is the usual } t\text{-statistic.}$$

Now the MLRT rejects H_0 for small λ, i.e. large t^2, so the MLRT is equivalent to the well-known two-sided t-test.

4.6.1 *The distribution of the test statistic in MLRTs*

In Example 4.8 we were able to find the exact distribution of t, and hence λ (though it is simpler to stick with t), and hence construct an exact test of any particular size α. Quite often, however, λ does not have a straightforward distribution and neither does any equivalent test statistic. In such cases we are rescued, at least for large n, by an approximate result which holds quite widely. This approximate result is that, under H_0, $-2 \ln(\lambda) \sim \chi^2_d$, where d is the difference in dimensionality of H_0 and $H_0 \cup H_1$, and where the union $H_0 \cup H_1$ of H_0 and H_1 includes all possible values of $\boldsymbol{\theta} \in \Omega$. To be more specific, if p_0 is the number of parameters estimated under H_0, less the number of constraints relating these parameters to each other, and p_1 is similarly defined under $H_0 \cup H_1$, then $d = p_1 - p_0$. Although we do not need the approximation in this example we give an outline justification that it is valid, and show how well it works. Under H_0, we estimate one parameter σ^2, whereas under $H_0 \cup H_1$ we estimate two, μ and σ^2, so $d = 1$, and, approximately, $-2 \ln(\lambda) \sim \chi^2_{(1)}$.

Now

$$-2 \ln(\lambda) = n \ln[1 + t^2/(n-1)] \approx \frac{nt^2}{(n-1)} \approx t^2$$

for large n. But, for large n the t-statistic is approximately distributed as $N(0, 1)$, and its square will have, approximately, a $\chi^2_{(1)}$ distribution, as required by the general result. The upper 5% critical value for $\chi_{(1)}$ is 3.84 compared with exact 5% points for $-2\ln(\lambda)$ of 5.38, 4.50, 4.29, 3.94 for $n = 5, 10, 21, 61$ respectively.

The general proof of $-2\ln(\lambda) \sim \chi^2_d$ is quite lengthy, and we shall not include it here. It constructs Taylor expansions of $\ln(L)$ at the MLEs of $\boldsymbol{\theta}$ under H_0 and $H_0 \cup H_1$ about the true value of $\boldsymbol{\theta}$, then uses a result from multivariate analysis to get χ^2 distributions for the truncated expressions, and finally differences two χ^2 distributions to get the final χ^2 distribution – see, for example, Cox and Hinkley (1974, Section 9.3) or Andersen (1980, Sections 3.6–3.8). This approximate distributional result is of considerable practical importance, but its validity is by no means universal. Discussion of its limitations, together with possible improvements and adjustments, is deferred until some rival test procedures to MLRTs have been introduced in Section 4.7.

4.6.2 *MLRTs for exponential families of distributions*

Suppose that we have a random sample of observations from a distribution belonging to the k-parameter exponential family. Then, with the usual notation,

$$-2\ln(\lambda) = 2\left\{\sum_{j=1}^{k}\sum_{i=1}^{n} B_j(x_i)[A_j(\hat{\boldsymbol{\theta}}) - A_j(\tilde{\boldsymbol{\theta}})] + n[D(\hat{\boldsymbol{\theta}}) - D(\tilde{\boldsymbol{\theta}})]\right\}$$

where $\tilde{\boldsymbol{\theta}}$, $\hat{\boldsymbol{\theta}}$ are MLEs of $\boldsymbol{\theta}$ under H_0 and $H_0 \cup H_1$ respectively. Using the natural parameterization, we have

$$-2\ln(\lambda) = 2\left\{\sum_{j=1}^{k} T_j(\hat{\phi}_j - \tilde{\phi}_j) + n[D(\hat{\boldsymbol{\phi}}) - D(\tilde{\boldsymbol{\phi}})]\right\}.$$

This expression, not surprisingly, is similar in form to that given in Section 4.2.1 for best tests. An example of this result is given in Exercise 4.20.

4.6.3 *A pathological example of an MLRT*

Throughout this book we identify a number of general inference procedures, which are widely used and which usually give satisfactory results. However, for any procedure, with sufficient inventiveness an example can be constructed which highlights shortcomings of the procedure. Such an example is now given for MLRTs.

Example 4.9

An experiment has $N + 1$ possible outcomes $x_0, x_1, \ldots, x_N$, $N > 2$. A null hypothesis, H_0, assigns probabilities to these outcomes as follows:

$$\Pr[X = x_0] = \tfrac{1}{2}, \quad \Pr[X = x_i] = \frac{1}{2N}, \qquad i = 1, 2, \ldots, N.$$

The alternative hypothesis, H_1, assigns probability $(N-1)/N$ to x_0, but leaves the remaining probabilities unspecified. We have a single observation x, on X, and we wish to find the MLRT of H_0 vs. H_1. Let $p_i = \Pr[X = x_i]$, $i = 0, 1, 2, \ldots, N$, and let L_0, L_1 be the likelihoods under H_0 and H_1 respectively. Then, if $L_\Omega = \text{Max}\{L_0, L_1\}$, it follows that the test statistic λ is L_0/L_Ω.

If $X = x_0$, then $L_0 = p_0 = \frac{1}{2}$, and $L_1 = (N-1)/N$. For $N > 2$, we have $L_1 > L_0$, so $L_\Omega = L_1$, and

$$\lambda = L_0/L_1 = N/2(N-1) > \tfrac{1}{2}.$$

If $X = x_i$, $i \neq 0$, $L_0 = 1/2N$, and $L_1 = p_i$. L_1 is clearly maximized when $p_i = 1/N$, because $p_0 + p_i \leqslant 1$, and $p_0 = (N-1)/N$. Thus $L_\Omega = L_1$, and $\lambda = L_0/L_1 = \frac{1}{2}$.

Because λ takes only two distinct values, we only have a choice of three values for the size, α, of a test based on λ. With a test whose critical region is of the form $\lambda \leqslant A$,we have three options:

(1) if $A < \frac{1}{2}$, then $\alpha = \Pr[\lambda \leqslant A \mid H_0] = 0$;
(2) if $A \geqslant N/2(N-1)$, then $\alpha = 1$;
(3) if $\frac{1}{2} \leqslant A < N/2(N-1)$, then $\alpha = \frac{1}{2}$.

Taking the only non-trivial test, corresponding to (c), we find that the power function can be calculated as

$$\eta = \Pr[\lambda \leqslant A \mid H_1] = \Pr[X \neq x_0 \mid H_1] = \frac{1}{N}.$$

This is less than α, so our test is biased.

Of course, Example 4.9 is not a very realistic one, and does not detract too much from the general usefulness of MLRTs.

4.7 Alternatives to and modifications of maximum likelihood ratio tests

MLRTs are widely used, but there are circumstances where other general test procedures may be preferred. Two of these, score tests and Wald tests, are described in Sections 4.7.1 and 4.7.2. In addition, it may sometimes be necessary to modify MLRTs, especially when, as in Section 3.3.1, the vector of parameters, $\boldsymbol{\theta}$, is subdivided into $\boldsymbol{\theta}_1$ and $\boldsymbol{\theta}_2$, consisting respectively of the parameters of interest and nuisance parameters. Such modifications are discussed in Section 4.7.3.

4.7.1 *Score tests*

In Section 4.4.3 score tests arose as locally most powerful tests, but more generally they can be viewed as a class of test procedures. The test statistic is based on $\partial l/\partial\theta$ when θ is a scalar, or the vector

$$\mathbf{u}(\boldsymbol{\theta}) = \left(\frac{\partial l}{\partial \theta_1}, \frac{\partial l}{\partial \theta_2}, \ldots, \frac{\partial l}{\partial \theta_k}\right)'$$

for vector $\boldsymbol{\theta}$. Under the usual regularity conditions, if the null hypothesis is $H_0: \boldsymbol{\theta} = \boldsymbol{\theta}_0$, with $H_1: \boldsymbol{\theta} \in \Omega - \{\boldsymbol{\theta}_0\}$, then the test statistic is written as $\mathbf{u}^{\mathrm{T}}(\boldsymbol{\theta}_0)\mathbf{I}_{\boldsymbol{\theta}_0}^{-1}\mathbf{u}(\boldsymbol{\theta}_0)$. Under H_0 this statistic has asymptotically a χ^2 distribution with k degrees of freedom, and is asymptotically equivalent to the test statistic for the corresponding MLRT, although the two statistics will differ in small samples (Cox and Hinkley, 1974, Section 9.3). The score test is sometimes known as the *Lagrange multiplier test*, especially in the econometrics literature (Buse, 1982). It has the advantage compared to the MLRT that it does not require the MLE of $\boldsymbol{\theta}$ to be found under $H_0 \cup H_1$.

4.7.2 Wald tests

With the same null and alternative hypotheses as in 4.7.1, the test statistic for the *Wald test* is based on the MLE, $\hat{\boldsymbol{\theta}}$, of $\boldsymbol{\theta}$ under $H_0 \cup H_1$. Assuming the usual regularity conditions, the test statistic $(\hat{\boldsymbol{\theta}} - \boldsymbol{\theta}_0)^{\mathrm{T}}\mathbf{I}_{\hat{\boldsymbol{\theta}}}(\hat{\boldsymbol{\theta}} - \boldsymbol{\theta}_0)$ has asymptotically a χ^2 distribution with k degrees of freedom under H_0. As with score tests, the test statistics for Wald tests are asymptotically equivalent to the test statistics for the corresponding MLRTs (Cox and Hinkley, 1974, Section 9.3). The Wald test statistic is sometimes known as the *maximum likelihood test statistic* and also appears in the modified, but asymptotically equivalent, form $(\hat{\boldsymbol{\theta}} - \boldsymbol{\theta}_0)^{\mathrm{T}}\mathbf{I}_{\boldsymbol{\theta}_0}(\hat{\boldsymbol{\theta}} - \boldsymbol{\theta}_0)$.

Example 4.10

In a binomial experiment with n trials and probability of success p, X successes are observed. We wish to test $H_0: p = p_0$ vs. $H_1: p \neq p_0$,

$$l(p; x) = \ln\left[\binom{n}{x}\right] + x\ln(p) + (n-x)\ln(1-p)$$

$$\frac{\partial l}{\partial p} = \frac{x}{p} - \frac{(n-x)}{(1-p)} = \frac{x - np}{p(1-p)}$$

$$I_p = \frac{n}{p(1-p)} \quad \text{and} \quad \hat{p} = \frac{x}{n}.$$

The score test statistic is

$$\left(\frac{x - np_0}{p_0(1-p_0)}\right)^2 \bigg/ \frac{n}{p_0(1-p_0)} = \frac{(x - np_0)^2}{np_0(1-p_0)}$$

a familiar statistic.

The Wald test statistic is

$$(\hat{p} - p_0)^2 \frac{n}{\hat{p}(1-\hat{p})} = \frac{(x - np_0)^2}{n\hat{p}(1-\hat{p})}.$$

which replaces p_0 by $\hat{p}$ in the denominator as compared to the score test statistic. The modified version of the Wald test noted above is identical to the score test statistic in this example.

The MLRT statistic has a completely different form from the other two tests in this example, namely

$$x \ln\left[\frac{\hat{p}(1-p_0)}{p_0(1-\hat{p})}\right] + n \ln\left[\frac{1-\hat{p}}{1-p_0}\right]$$

although it is easy to show, using Taylor expansions, that it is asymptotically equivalent to the score test statistic – see Exercise 4.24.

Although the discussion of score tests and Wald tests has been in terms of simple H_0, they can readily be extended to cases when H_0 is composite (Cox and Hinkley, 1974, Section 9.3.) Various relationships exist between these tests and MLRTs. Buse (1982) gives useful insights into some of these relationships, especially for simple H_0. Engle (1984) gives a thorough discussion of the use of the three classes of test in econometrics. Lütkepohl (1991, Section 12.3.2) discusses a non-trivial example in vector time series where each of the three classes of test is the preferred choice in some part of the problem.

4.7.3 *Modifications to MLRTs*

There is a considerable literature on modifications to MLRTs and, to a lesser extent, score tests and Wald tests. The modifications are made for two main reasons, namely, to improve the basic χ^2 approximation to the null distribution of the test statistic and to take into account the effect of nuisance parameters.

An early modification, originally due to Bartlett, relies on expressing the expectation of the test statistic $-2\ln(\lambda)$ as, approximately, $d(1 + a/n)$, where d represents the degrees of freedom in the usual χ^2 approximations, n is the sample size and a is a known constant. Not only does the statistic $\{-2/[1 + (a/n)]\}\ln(\lambda)$ have an expectation which is closer than that of $-2\ln(\lambda)$ to the expectation of the approximating distribution but the whole distribution of the modified statistic is closer to χ_d^2 (Stuart and Ord, 1991, p. 873). An example of this type of correction in the context of time series modelling is given by Ravishanker *et al.* (1990).

A second class of adjustments to MLRTs arises when there are nuisance parameters and inference is based on a modification of the likelihood function, such as profile likelihood or conditional likelihood – see Section 3.3. As noted at the beginning of the present section, there is a considerable and growing literature in this area, much of it of a technical nature. It would be inappropriate to give details here, but a few references are noted which will allow the interested reader to find out more. Cox and Reid (1987) define a conditional likelihood ratio statistic, and compare its properties to those of a profile likelihood ratio statistic in the case where the nuisance parameters can be made orthogonal to the parameters of interest. Pierce and Peters (1992) examine various approximations in the context of conditional inference. The papers by Pierce and Peters (1992) and Cox and Reid (1987) are both followed by extensive published discussions (14 and 25 contributions respectively), thus providing a good review of opinions on the subject, at the time the papers were written.

The adjustment to the profile likelihood test statistic given by Cox and Reid (1987) assumes *parameter orthogonality*, defined as follows. Suppose that $\boldsymbol{\theta} = (\boldsymbol{\theta}_1, \boldsymbol{\theta}_2)$ where, as in Section 3.3.1, $\boldsymbol{\theta}_1$ contains parameters of interest, and $\boldsymbol{\theta}_2$ consists of nuisance parameters. The sets of parameters $\boldsymbol{\theta}_1$, $\boldsymbol{\theta}_2$ are orthogonal if for all θ_s and θ_t, where θ_s and θ_t are elements of $\boldsymbol{\theta}_1$ and $\boldsymbol{\theta}_2$ respectively, the corresponding element of the information matrix, I_{st}, is zero. An adjustment which does not make this orthogonality assumption is provided by Cox and Reid (1993). Bartlett-type adjustments can be made to likelihood ratio tests based on modified likelihoods, as well as to standard MLRTs. Ghosh and Mukerjee (1994) discuss some problems related to tests based on adjusted conditional likelihood and based on adjusted profile likelihood, from both frequentist and Bayesian points of view – see also DiCiccio and Stern (1994). Adjustments analogous to Bartlett corrections may also be made to score tests and Wald tests – see, for example, Cordeiro *et al.* (1993).

Many of the distributional approximations outlined in this section, as well as the original χ^2 approximation to $-2\ln(\lambda)$, rely for their validity on an assumption of some version of the usual regularity conditions. When these conditions do not hold, the distributional approximations will often break down, although it is sometimes possible to derive alternative distributional results. For example, Stuart and Ord (1991, pp. 876–880) discuss an exact distributional result, $-2\ln(\lambda) \sim \chi^2_{2d}$, which holds in some situations where the range of X depends on $\boldsymbol{\theta}$. Cox and Hinkley (1974, Section 9.3) describe some implications of a null value $\boldsymbol{\theta}_0$ falling on the boundary of Ω.

4.8 Discussion

In this chapter we have discussed the standard frequentist approach to hypothesis testing. Discussion of alternative approaches based on decision theory and Bayesian ideas are deferred until Chapters 6 and 7 respectively. A number of topics, both elementary and advanced, have been omitted entirely.

At the elementary level, some of the more practical aspects of hypothesis testing, whilst indisputably important, do not fall naturally into the framework of this text. These include the choice of sample size to achieve a desired power, and the distinction between practical significance and statistical significance.

At an advanced level, details have been omitted for many of the more complicated techniques, for example in Sections 4.4 and 4.7.3. A further topic for which no details have been given is the comparison of *non-nested* composite hypotheses. Work in this area dates back to Cox (1961). One type of approach to problems of this type is to optimize a criterion such as Akaike's information criterion (AIC) – Akaike (1973). Such criteria are based on the likelihood function, but also include a penalty function, which increases as the number of parameters in the model increases. Optimization of AIC, or similar criteria, therefore involves a trade-off between those models which give a good fit to the data in terms of likelihood, and those models which are parsimonious (have few parameters).

Exercises

4.1 Use the Neyman–Pearson lemma to find the form of the critical region for the best test of H_0 against H_1 when
(a) $X_1, X_2, \ldots, X_n$ are a random sample from a Poisson distribution with mean θ, and $H_0: \theta = \theta_0$, $H_1: \theta = \theta_1$, $\theta_1 > \theta_0$.
(b) $X_1, X_2, \ldots, X_n$ are a random sample from the exponential distribution with p.d.f. $f(x; \theta) = 1/\theta\, e^{-x/\theta}$, $x > 0$, and $H_0: \theta = \theta_0$, $H_1: \theta = \theta_1$, $\theta_1 > \theta_0$.
(c) $X_1, X_2, \ldots, X_n$ are a random sample from the exponential distribution with p.d.f. $f(x; \theta) = \theta e^{-\theta x}$, $x > 0$, and $H_0: \theta = \theta_0$, $H_1: \theta = \theta_1$, $\theta_1 > \theta_0$.
(d) $X_{11}, X_{12}, \ldots, X_{1n_1} \sim N(\mu_1, \sigma_1^2)$, $X_{21}, X_{22}, \ldots, X_{2n_2} \sim N(\mu_2, \sigma_2^2)$, all X_{ij} are independent of each other, σ_1^2, σ_2^2 are known, and $H_0: \mu_2 = \mu_1$, $H_1: \mu_2 = \mu_1 + \delta$ with $\delta > 0$ (μ_1 and δ are both known constants).

4.2 In Exercise 1(d), suppose that $\sigma_1^2 = \sigma_2^2 = \delta = 1$, that $n_1 = n_2 = n$, and that we wish to perform a best test with $\alpha = 0.01$. Find
(a) the power of the test when $n = 10$;
(b) the smallest value for n for which we can achieve a power $\geqslant 0.95$.

4.3 $X_1, X_2, \ldots, X_n$ are a random sample from the gamma distribution with parameters 3 and θ, which has p.d.f.

$$f(x; \theta) = \frac{1}{2\theta^3} x^2 e^{-x/\theta}, \qquad x > 0.$$

(a) Find the best test of $H_0: \theta = \theta_0$ against $H_1: \theta = \theta_1$ where $\theta_1 > \theta_0$.
(b) Use (a) to find the uniformly most powerful test of H_0 against $H_1': \theta > \theta_0$.

4.4 $X_1, X_2, \ldots, X_n$ form a random sample from a uniform distribution on $[0, \theta]$. Find the form of a best test of size α for $H_0: \theta = \theta_0$ against $H_1: \theta = \theta_1$, where $\theta_1 > \theta_0$. Suppose now that the alternative hypothesis is $H_1: \theta > \theta_0$; show that there exists a uniformly most powerful test, and plot the power function of such a test.

4.5 A random sample $X_1, X_2, \ldots, X_n$ is taken from the distribution with probability density function

$$f(x; \theta) = \frac{1}{\sqrt{2\pi\theta x}} \exp\left\{ -\frac{1}{2}\left(\frac{\ln(x)}{\theta}\right)^2 \right\}, \qquad x > 0$$

where $\theta > 0$. Show that there is a uniformly most powerful test of the null hypothesis $H_0: \theta = \theta_0$, against the alternative $H_1: \theta > \theta_0$, and find the form of this test.

4.6 A random sample $X_1, X_2, \ldots, X_n$ is taken from the distribution with probability density function

$$f(x; \theta) = \frac{\theta^\lambda}{\Gamma(\lambda)} x^{\lambda - 1} e^{-\theta x} \qquad x > 0$$

where $\lambda > 0$ is known, but $\theta > 0$ is an unknown parameter.

Show that the critical region for the most powerful α-level test of $H_0: \theta = \theta_0$ against $H_1: \theta = \theta_1$, where $\theta_1 > \theta_0$, is of the form $\Sigma_{i=1}^n X_i \leqslant C$, for some constant C. Show also that when $n = 1$, and $\lambda = 1$, the power of this test is $1 - (1 - \alpha)^{\theta_1/\theta_0}$.

4.7 Suppose that $X_1, X_2, \ldots, X_n$ is a random sample of observations on a random variable X, which takes values only in the range (0, 1). Under the null hypothesis H_0, the distribution of X is uniform on (0, 1), whereas under an alternative hypothesis, H_1, the distribution is the truncated exponential with probability density function

$$f(x; \theta) = \frac{\theta\, e^{\theta x}}{e^{\theta} - 1}, \qquad 0 \leqslant x \leqslant 1, \quad \theta > 0$$

where θ is unknown. Show that there is a uniformly most powerful test of H_0 vs. H_1 and find, approximately, the critical region for such a test when n is large.

4.8 For a random sample of observations from a Poisson distribution, as in Exercise 1(a), consider testing H_0 against the composite alternative $H_1': \theta \neq \theta_0$. Show that no uniformly most powerful test exists for H_0 against H_1'.

An intuitively plausible acceptance region for H_0 against H_1' above is of the form $C_1 < \Sigma_{i=1}^n X_i < C_2$, where C_1, C_2 are suitably chosen integers such that $C_2 > C_1 + 1$. Using the result that $Y = \Sigma_{i=1}^n X_i$ has a Poisson distribution with mean $n\theta$, show that the power function of the test with this acceptance region has the form

$$1 - e^{-n\theta} \sum_{y=C_1+1}^{C_2-1} (n\theta)^y / y!$$

4.9 Suppose that $X_1, X_2, \ldots, X_n$ are a random sample from a $N(\mu, \sigma^2)$ distribution with both parameters unknown. We wish to test the null hypothesis $H_0: \mu = 0, 0 < \sigma^2 < \infty$ against the alternative $H_1: \mu \neq 0, 0 < \sigma^2 < \infty$. Find the form of

(a) a uniformly most powerful similar test,

(b) a uniformly most powerful invariant test.

4.10 Suppose $X_1, X_2, \ldots, X_{n_1}$ are independent binary random variables such that $\Pr[X_i = 1] = \phi$ and $\Pr[X_i = 0] = 1 - \phi$, $i = 1, 2, \ldots, n_1$, and suppose $Y_1, Y_2, \ldots, Y_{n_2}$ are independent binary random variables such that $\Pr[Y_i = 1] = \psi$, and $\Pr[Y_i = 0] = 1 - \psi$, $i = 1, 2, \ldots, n_2$. The following model for these two samples is proposed:

$$\ln\left\{\frac{\phi}{1 - \phi}\right\} = \theta_1 + \theta_2$$

$$\ln\left\{\frac{\psi}{1 - \psi}\right\} = \theta_2.$$

The null hypothesis is $H_0: \theta_1 = 0$ and the alternative is $H_1: \theta_1 > 0$. Derive a uniformly most powerful similar test of H_0 against H_1.

4.11 Suppose $X_1, X_2, \ldots, X_n \sim N(\mu, \sigma^2)$, for $n > 1$, and we wish to test $H_0: \sigma = \sigma_0$ vs. $H_1: \sigma = \sigma_1$, where $\sigma_1 > \sigma_0$ and μ is unknown. Show that the most powerful similar test is based on the statistic $\Sigma_{i=1}^n (X_i - \bar{X})^2$. Is this result also true if $\sigma_1 < \sigma_0$?

4.12 The random variable X has probability density function $f(x)$, the functional form of which is unknown. A random sample of size n is drawn to test the null hypothesis

$$H_0: f(x) = f_0(x)$$

against the alternative

$$H_1: f(x) = f_1(x)$$

The functional forms of f_0 and f_1 are known. They have no unknown parameters and the same domain.

By considering the probability density function

$$\lambda f_0(x) + (1 - \lambda) f_1(x)$$

show that H_0 and H_1 may be expressed parametrically. Hence show that if

$$f_0(x) = (2\pi)^{-1/2}(-\tfrac{1}{2}x^2), \qquad -\infty < x < \infty$$

$$f_1(x) = \tfrac{1}{2}\exp(-|x|), \qquad -\infty < x < \infty$$

then the best critical region for the test of H_0 against H_1 is given by

$$\sum_{i=1}^{n} (|x_i| - 1)^2 \geq k, \text{ for some constant } k.$$

Evaluate the best critical region for $n = 1$ with (a) $k = 1$ and (b) $k = \frac{1}{4}$. In the case (a) is the test unbiased?

4.13 Let $X_1, X_2, \ldots, X_n$ be a random sample from a normal distribution with unknown mean μ and variance σ^2. Show that the maximum likelihood ratio test of the null hypothesis H_0: $\sigma^2 = \sigma_0^2$ against the alternative H_1: $\sigma^2 \neq \sigma_0^2$ has a test statistic

$$\left(\frac{Q}{n}\right)^{n/2} e^{(n-Q)/2} \quad \text{where} \quad Q = \sum_{i=1}^{n} \frac{(X_i - \bar{X})^2}{\sigma_0^2}.$$

Show that this test has a critical region consisting of both large and small values of Q, i.e. H_0 is rejected for $Q \leq C_1$ or $Q \geq C_2$ where $C_1 < C_2$ are appropriately chosen constants. By considering the minimum value of the power function of such a test, derive conditions that C_1 and C_2 must satisfy if the test is to be unbiased.

4.14 A survey of the use of a particular product was conducted in four areas, with a random sample of 200 potential users interviewed in each area. The results were that in the four areas, respectively x_1, x_2, x_3 and x_4 of the 200 interviewees said that they used the product. Construct a maximum likelihood ratio test to test whether the proportion of the population using the product is the same in each area.

Carry out the test, with $\alpha = 0.05$, when $x_1 = 76$, $x_2 = 53$, $x_3 = 59$ and $x_4 = 48$, using the large sample approximation for the distribution of the test statistic.

4.15 Random samples are available from k exponential distributions; the ith distribution has parameter θ_i, and the sample from this distribution is of size n_i. Construct a maximum likelihood ratio test of the null hypothesis $\theta_1 = \theta_2 = \cdots = \theta_k$ against a general alternative, and show how an approximate test may be performed by comparing the value of a suitable statistic with tabulated values of a χ^2 distribution. Show also that, if $k = 2$ and $n_1 = n_2$, an exact test based on an F-distribution is possible.

4.16 Random samples, of size n_1 and n_2 respectively, are drawn from the distributions $N(\mu_1, \sigma_1^2)$ and $N(\mu_2, \sigma_2^2)$, all four parameters being unknown. Suppose that $\sigma_2^2/\sigma_1^2 = c$, and we wish to test the null hypothesis H_0: $c = c_0$, for some specified c_0. Construct a maximum likelihood ratio test for H_0 against the two-sided alternative H_1: $c \neq c_0$ and, in the particular case when $n_1 = n_2$, show that the test may be performed by comparing the observed value of a suitable statistic with a standard tabulation of critical values of the F-distribution.

4.17 Suppose that $X_1, X_2, \ldots, X_n$ form a random sample from a distribution with probability density function

$$f(x;\theta)=\begin{cases} c(\theta)d(x), & a \leqslant x \leqslant b(\theta) \\ 0 & \text{elsewhere,} \end{cases}$$

where $b(\theta)$ is a monotone increasing function of the single parameter θ. Show that the maximum likelihood ratio test (MLRT) statistic for testing H_0: $\theta = \theta_0$ against the two-sided alternative $H_1: \theta \neq \theta_0$ is given by

$$W = -2\ln(\lambda) = -2n\ln\left[\int_a^{X(n)} c(\theta_0)d(x)\,\mathrm{d}x\right]$$

where $X(n) = \max(X_1, X_2, \ldots, X_n)$.

Given that

$$f(x;\theta)=\begin{cases} \dfrac{2x}{\theta^2}, & 0 \leqslant x \leqslant \theta \\ 0 & \text{elsewhere} \end{cases}$$

obtain the MLRT statistic W, as above, and show that when H_0 is true W follows exactly a chi-squared distribution on two degrees of freedom.

4.18 Let $X_1, X_2, \ldots, X_m$ and $Y_1, Y_2, \ldots, Y_n$ be independent random samples from two exponential distributions with unknown parameters λ and μ, i.e. the Xs and Ys have, respectively, probability density functions

$$f(x;\lambda) = \lambda\,\mathrm{e}^{-\lambda x}, \qquad x \geqslant 0$$
$$g(y;\mu) = \mu\,\mathrm{e}^{-\mu y}, \qquad y \geqslant 0.$$

Show that the critical region for the maximum likelihood ratio test of the null hypothesis H_0: $\lambda = \mu$ against the alternative hypothesis $H_1: \lambda \neq \mu$ depends on the data only through the ratio $\bar{y}/\bar{x}$ of the sample means.

Given that $2\lambda\,\Sigma_{i=1}^{m}\,X_i$ and $2\mu\,\Sigma_{j=1}^{n}\,Y_j$ each has a χ^2 distribution, with $2m$ and $2n$ degrees of freedom respectively, derive the distribution of $\bar{Y}/\bar{X}$. Use this distribution to find an unbiased test of H_0 above against the one-sided alternative H_1': $\lambda > \mu$.

4.19 A random sample of n observations $X_1, X_2, \ldots, X_n$ is taken from a Poisson distribution with standard deviation θ (i.e. the usual parameter of the distribution is θ^2). Construct a score test for the null hypothesis H_0: $\theta = \theta_0$ against the alternative $H_1: \theta > \theta_0$, and show that this test is uniformly most powerful.

If $\theta_0 = 1$ and $n = 5$, find a critical region for the above test with size $\alpha = 0.0318$. What is the power of this test when $\theta = 2$?

4.20 Suppose that $X_1, X_2, \ldots, X_n$ form a random sample from $N(\mu, \sigma^2)$ where μ and σ^2 are both unknown. Use the natural parameterization for $N(\mu, \sigma^2)$ to find an expression for $-2\ln(\lambda)$ in a maximum likelihood ratio test of H_0: $\mu = \mu_0$ vs. $H_1: \mu \neq \mu_0$ – see Section 4.6.2.

4.21 Using a random sample of size n from a Poisson distribution with mean θ, it is required to test H_0: $\theta = \theta_0$ against H_1: $\theta \neq \theta_0$. Find the test statistic for

(a) a score test of H_0 vs. H_1;

(b) a Wald test of H_0 vs. H_1.

Compare these statistics with that of the MLRT for the same pair of hypotheses.

4.22 For an exponential distribution with p.d.f. $f(x;\theta) = \theta\,\mathrm{e}^{-\theta x}$, $x > 0$, construct
(a) the score test;
(b) the Wald test
for the hypotheses $H_0\colon \theta = \theta_0$ vs. $H_1\colon \theta \neq \theta_0$.

4.23 Suppose that $X_1, X_2, \ldots, X_n$ form a random sample from a distribution with p.d.f. $f(x;\boldsymbol{\theta})$, where $\boldsymbol{\theta} = (\theta_1, \theta_2)$. Here θ_1 is a parameter of interest and θ_2 is a nuisance parameter, and it is required to test $H_0\colon \theta_1 = \theta_{10}$ vs. $H_1\colon \theta_1 \neq \theta_{10}$. The score test statistic in this case is $u^2(\theta_{10})I_{\tilde{\theta}}^{-1}$ and the Wald test statistic is $(\hat{\theta}_1 - \theta_{10})^2 I_{\hat{\theta}_1|\hat{\theta}_2}$, where $\tilde{\boldsymbol{\theta}}$, $\hat{\boldsymbol{\theta}}$ are the MLEs of $\boldsymbol{\theta}$ under H_0, $H_0 \cup H_1$ respectively,

$$u(\theta_{10}) = \left.\frac{\partial l}{\partial \theta_1}\right|_{\theta_1 = \theta_{10}}, \qquad I_{\tilde{\theta}} = \mathrm{E}\left[\frac{-\partial^2 l}{\partial \theta_1^2}\right]_{\theta = \tilde{\theta}}$$

$$I_{\hat{\theta}_1|\hat{\theta}_2} = I_{11} - I_{12}^2/I_{22}, \qquad \text{and} \qquad I_{ij} = \mathrm{E}\left[\frac{-\partial^2 l}{\partial \theta_i \partial \theta_j}\right]_{\theta = \hat{\theta}}.$$

Find the Wald and score test statistics for the hypotheses in Exercise 4.13.

4.24 Show that the MLRT in Example 4.10 (Section 4.7.2) is asymptotically equivalent to the Wald and score tests.

Chapter Five

Interval estimation

5.1 Introduction

In Chapters 2 and 3 we discussed point estimation at some length. However, as was noted in Chapter 1, a point estimate on its own is of little use – some measure of its precision is also necessary. This leads naturally to the idea of interval estimation, which is of great practical importance. Nevertheless we devote less space in this book to interval estimation than to either point estimation or hypothesis testing. This is because the theory underlying interval estimation is closely related to that already covered for point estimation and hypothesis testing, and the theory is most conveniently developed in these latter contexts. In fact, there is a close relationship between interval estimation and hypothesis testing, so that many of the ideas of hypothesis testing carry over directly to interval estimation; we shall show this later for some of the more useful ideas.

After defining confidence sets in this introductory section, Section 2 discusses various ways of constructing such sets, mainly based on ideas from point estimation and hypothesis testing. Section 3 defines a number of desirable (optimal) properties of confidence sets, and Section 4 describes some problems associated with interval estimation. Some of these problems can be overcome by using a Bayesian approach, which is discussed in detail in Chapter 7.

Suppose that we are interested in a vector of parameters $\boldsymbol{\theta}$. Then we divide the possible values of $\boldsymbol{\theta}$ into a 'plausible' region and a 'less plausible' region, with the plausible region usually constructed to have a predetermined probability of including the true value of $\boldsymbol{\theta}$. Note that the region depends on data and is random, whereas $\boldsymbol{\theta}$ is fixed. More formally we have the following definition.

Definition 5.1

Suppose that $\{f(x;\ \boldsymbol{\theta});\ \boldsymbol{\theta} \in \Omega\}$ defines a family of distributions, indexed by a vector parameter, $\boldsymbol{\theta}$, and that a random sample of observations, denoted by $\mathbf{X}$, is taken from $f(x;\ \boldsymbol{\theta})$ with $\boldsymbol{\theta}$ fixed, but unknown. If $S_{\mathbf{X}}$ is a subset of Ω, depending on $\mathbf{X}$, such that

$$\Pr[\mathbf{X}\colon S_{\mathbf{X}} \supset \boldsymbol{\theta}] = 1 - \alpha$$

then $S_{\mathbf{X}}$ is a *confidence set* for $\boldsymbol{\theta}$ with *confidence coefficient* $1 - \alpha$.

For a given value of $\boldsymbol{\theta}$, $S_{\mathbf{X}}$ has a probability of $1 - \alpha$ of enclosing $\boldsymbol{\theta}$.

When θ is a scalar, the confidence set often takes the form of an interval on the real line, and is thus a *confidence interval.*

Example 5.1

Many types of confidence interval are very familiar. For example, suppose that $X_1, X_2, \ldots, X_n \sim N(\theta, \sigma^2)$, with σ^2 unknown. Then

$$\frac{\bar{X} - \theta}{S/\sqrt{n}} \sim t_{(n-1)} \qquad \text{and} \qquad \Pr\left[\left|\frac{\bar{X} - \theta}{S/\sqrt{n}}\right| \leqslant t_{(n-1);\,\alpha/2}\right] = 1 - \alpha$$

using the notation established in Section 1.3.1. Manipulation of the probability inequality leads to the familiar confidence interval for θ with end points $\bar{X} \pm t_{(n-1);\,\alpha/2} S/\sqrt{n}$.

Example 5.2

Suppose that an estimator for $\boldsymbol{\theta}$, say $\hat{\boldsymbol{\theta}}$, has a multivariate normal distribution, centred at $\boldsymbol{\theta}$, i.e. $\hat{\boldsymbol{\theta}} \sim N(\boldsymbol{\theta}, \Sigma)$, using standard notation. Then $(\hat{\boldsymbol{\theta}} - \boldsymbol{\theta})^{\mathrm{T}}\Sigma^{-1}(\hat{\boldsymbol{\theta}} - \boldsymbol{\theta}) \sim \chi^2_k$, where k is the number of parameters in $\boldsymbol{\theta}$. Thus a confidence set for $\boldsymbol{\theta}$, with confidence coefficient $(1 - \alpha)$, is given by those $\boldsymbol{\theta}$ for which

$$(\hat{\boldsymbol{\theta}} - \boldsymbol{\theta})^{\mathrm{T}}\Sigma^{-1}(\hat{\boldsymbol{\theta}} - \boldsymbol{\theta}) \leqslant \chi^2_{k;\,\alpha}$$

with notation as given in Section 1.3.1. Hence the confidence set is an ellipsoid in k-dimensional space.

5.2 Construction of confidence sets

As noted in Example 5.1, examples of confidence intervals are well known in standard statistical problems. Such intervals are often constructed by first finding a point estimator for the parameter of interest, or setting up a hypothesis test for the parameter, and then finding a corresponding interval. Later in this section we shall look in detail at the connections between interval estimation on the one hand, and point estimation and hypothesis testing on the other, and at how these connections can be used in confidence set construction. First, however, we look at a concept which underlies these approaches, as well as being useful in its own right.

5.2.1 *Pivotal quantities*

Definition 5.2

In its simplest form a *pivotal quantity* is a function $g(\mathbf{x}; \boldsymbol{\theta})$ of $\boldsymbol{\theta}$ and the observations, whose distribution is known and is independent of $\boldsymbol{\theta}$. In addition, if $\boldsymbol{\theta}$ is scalar, some definitions require that g be a monotone function of θ.

Pivotal quantities can be used to construct confidence sets. The basic idea is that the known distribution of the pivotal quantity g can be used to write down a probability statement of the form

$$\Pr[g_1 \leqslant g \leqslant g_2] = 1 - \alpha.$$

For scalar θ, and given the monotonicity condition on g, the inequalities can then be manipulated, leading to

$$\Pr[\theta_1(\mathbf{X}) \leqslant \theta \leqslant \theta_2(\mathbf{X})] = 1 - \alpha.$$

The interval $[\theta_1(\mathbf{X}), \theta_2(\mathbf{X})]$ satisfies the definition of a confidence interval, with confidence coefficient $1 - \alpha$.

For a non-monotonic function g, or for vector $\boldsymbol{\theta}$, the initial probability statement for g and the final confidence set for $\boldsymbol{\theta}$ may take more complicated forms, but the same basic principle is employed.

Example 5.1 (*continued*)

In this case the pivotal quantity is $(\bar{X} - \theta)/(S/\sqrt{n})$, and the probability statement

$$\Pr\left[-t_{(n-1);\,\alpha/2} \leqslant \frac{\bar{X} - \theta}{S/\sqrt{n}} \leqslant t_{(n-1);\,\alpha/2}\right] = 1 - \alpha$$

can be readily manipulated to give

$$\Pr[\bar{X} - t_{(n-1);\,\alpha/2} S/\sqrt{n} \leqslant \theta \leqslant \bar{X} + t_{(n-1);\,\alpha/2} S/\sqrt{n}] = 1 - \alpha.$$

Example 5.3

This example is not a particularly straightforward one, but it is a classic illustration of the use of a pivotal quantity, in a case where monotonicity of g does not hold. Let $X \sim N(\mu_X, 1)$, $Y \sim N(\mu_Y, 1)$, where X, Y are independent, $\theta = \mu_X/\mu_Y$, and we wish to find a confidence interval for θ. The example seems very restrictive, but could easily be extended so that X, Y become means of samples, with their variances unequal but in a known ratio.

Consider $U = X - \theta Y$; then $U \sim N(0, 1 + \theta^2)$, so that if

$$g(X, Y; \theta) = \frac{X - \theta Y}{[1 + \theta^2]^{1/2}}$$

then $g(X, Y; \theta) \sim N(0, 1)$.

$g(X, Y; \theta)$ therefore has a known distribution which is independent of θ, and so is a pivotal function according to Definition 5.2, although it is not a monotone function of θ.

To obtain a confidence set from a pivotal quantity, we make some statement about g which holds with probability $1 - \alpha$, and then manipulate inequalities, etc. to get a statement about θ. For example, here we can say that

$$\Pr\left[\left|\frac{X - \theta Y}{[1 + \theta^2]^{1/2}}\right| \leqslant z_{\alpha/2}\right] = 1 - \alpha$$

so we solve

$$\frac{(x - \theta y)^2}{(1 + \theta^2)} \leqslant z_{\alpha/2}^2$$

to get a confidence set for θ. Dropping the subscript from z^2 for notational convenience, this leads to the quadratic equation

$$\theta^2(y^2 - z^2) - 2xy\theta + (x^2 - z^2) \leqslant 0 \qquad \text{or} \qquad \psi(\theta) \leqslant 0, \text{ say.}$$

Now depending on the values of x and y the equation $\psi(\theta) = 0$ may have two real roots, no real roots, or a repeated root, and our confidence set could consist of the whole real line, a single interval or two disjoint intervals. This latter, rather unsatisfactory, situation can arise for example if $y^2 < z^2 < 2y^2$ and $z^2 < 2x^2$; the even more unsavoury situations where the confidence set is empty or contains only one point cannot occur.

We can sometimes 'pull a pivotal quantity out of the air' as in the example just given, but more usually we find one via point estimation or hypothesis testing.

5.2.2 *Confidence sets derived from point estimators*

Suppose that $\hat{\boldsymbol{\theta}}$, an estimator for $\boldsymbol{\theta}$, has a known distribution. We can often use this sampling distribution to find a confidence set for $\boldsymbol{\theta}$. Frequently a pivotal quantity is used in the derivation although it may not be immediately apparent that this is so.

Example 5.4

$X_1, X_2, \ldots, X_n \sim N(\mu, \sigma^2)$, σ^2 known; find a confidence set (interval) for μ. $\bar{X}$ is the obvious estimator for μ, and $\bar{X} \sim N(\mu, \sigma^2/n)$. If

$$Z = \frac{\bar{X} - \mu}{\sigma/\sqrt{n}}$$

then $Z \sim N(0, 1)$, and Z is a pivotal quantity. Starting with the statement $\Pr[|Z| \leqslant z_{\alpha/2}] = 1 - \alpha$, and substituting for Z, leads to

$$\Pr[-z_{\alpha/2} \leqslant \sqrt{n}(\bar{X} - \mu)/\sigma \leqslant z_{\alpha/2}] = 1 - \alpha$$

which in turn is easily manipulated to become

$$\Pr[\bar{X} - z_{\alpha/2}\sigma/\sqrt{n} \leqslant \mu \leqslant \bar{X} + z_{\alpha/2}\sigma/\sqrt{n}] = 1 - \alpha.$$

Hence, we have derived the usual confidence interval for μ.

Confidence sets based on maximum likelihood estimators (MLEs)

Consider the MLE $\hat{\theta}$ for a single parameter θ of the distribution whose p.d.f. is $f(x; \theta)$. If $f(x; \theta)$ satisfies appropriate regularity conditions, we have, as $n \to \infty$ (and sometimes, as in Example 5.4, for all n), that

$$\hat{\theta} \sim N(\theta, I_\theta^{-1})$$

Hence $(\hat{\theta} - \theta)I_\theta^{1/2} \sim N(0, 1)$, and is (approximately) a pivotal quantity, so using the same manipulations as in Example 5.4, we have a confidence interval for θ of the form $\hat{\theta} \pm I_\theta^{-1/2} z_{\alpha/2}$, with confidence coefficient approximately (and sometimes exactly) equal to $(1 - \alpha)$.

Note, however, that the above expression is not immediately a confidence interval unless, as in Example 5.4, I_θ does not depend on θ. If I_θ does depend on θ then the end points of the given interval also depend on θ, whereas a confidence interval requires end points which can be computed from the data. We can obtain end points which do not depend on θ, albeit at the expense of introducing an extra level of approximation, if we replace I_θ by $I_{\hat{\theta}}$. Alternatively, further manipulation will often lead to a properly defined confidence set, as the following example demonstrates.

Example 5.5

Consider a binomial experiment with n trials and probability of success p. Then $I_p = n/p(1 - p)$ and

$$\Pr\left[\left(\frac{X - np}{np(1 - p)}\right)^2 \leqslant z_{\alpha/2}^2\right] \approx 1 - \alpha.$$

This is equivalent to

$$\Pr[np^2(n + z^2) - np(2X + z^2) + X^2 \leqslant 0] = \Pr[\psi(p) \leqslant 0] \approx 1 - \alpha.$$

The inequality in the probability statement is satisfied for those values of p which lie between the roots, p_1, p_2 of the quadratic equation $\psi(p) = 0$. We therefore have

$$\Pr[p_1 \leqslant p \leqslant p_2] = 1 - \alpha$$

and $[p_1, p_2]$ forms a confidence interval for p, with coefficient approximately $1 - \alpha$, where p_1, p_2 are given by

$$\left\{(2X + z^2) \pm z\left[z^2 + 4X\left(1 - \frac{X}{n}\right)\right]^{1/2}\right\}/\{2(n + z^2)\}$$

The alternative strategy of replacing θ by $\hat{\theta}$ in I_θ, leads to the interval with end points

$$\hat{p} \pm z_{\alpha/2}[\hat{p}(1 - \hat{p})/n]^{1/2}, \qquad \text{where} \quad \hat{p} = X/n.$$

The distributional result for MLEs generalizes to the case of vector $\boldsymbol{\theta}$, where it becomes $\hat{\boldsymbol{\theta}} \sim N(\boldsymbol{\theta}, I_\theta^{-1})$ approximately for large n. It follows that $(\hat{\boldsymbol{\theta}} - \boldsymbol{\theta})^{\mathrm{T}} I_\theta (\hat{\boldsymbol{\theta}} - \boldsymbol{\theta}) \mathrel{\dot{\sim}} \chi_k^2$, and we have ellipsoidal confidence regions for $\boldsymbol{\theta}$ (as in Example 5.2) if I_θ does not depend on $\boldsymbol{\theta}$.

It is, of course, possible to use estimators other than MLEs to construct confidence intervals. For example, a method-of-moments estimator can be used provided that its distribution is known. In Examples 5.4 and 5.5 the MLE and moment estimators are equivalent, so the same interval will result. In other situations, the MLE is often more efficient than a moment estimator, which would usually lead to the interval based on the MLE having the desirable property of a shorter length than that based on the moment estimator.

5.2.3 *Confidence sets derived from hypothesis tests*

Rather than finding confidence sets via point estimation we can look instead at hypothesis testing. There is duality between hypothesis testing and interval estimation as follows.

Lemma 5.1

Suppose that $\bar{C}(\boldsymbol{\theta}_0)$ is the acceptance region for a test of size α of H_0: $\boldsymbol{\theta} = \boldsymbol{\theta}_0$ vs. H_1: $\boldsymbol{\theta} \in \Omega - \{\boldsymbol{\theta}_0\}$; then a confidence set for $\boldsymbol{\theta}$, with confidence coefficient $(1 - \alpha)$, is given by

$$S_{\mathbf{X}} = \{\boldsymbol{\theta}_0 \colon \mathbf{X} \in \bar{C}(\boldsymbol{\theta}_0)\}.$$

Proof

Note that $\Pr[\mathbf{X}\colon S_{\mathbf{X}} \supset \boldsymbol{\theta} \mid \boldsymbol{\theta} = \boldsymbol{\theta}_0] = \Pr[\mathbf{X} \in \bar{C}(\boldsymbol{\theta}_0) \mid \boldsymbol{\theta} = \boldsymbol{\theta}_0] = 1 - \alpha$, by definition of $\bar{C}(\boldsymbol{\theta}_0)$. Thus $S_{\mathbf{X}}$ satisfies Definition 5.1, and so is a confidence set, with confidence coefficient $1 - \alpha$. What the result means is that we can form a confidence set for $\boldsymbol{\theta}$ by collecting together those null values of $\boldsymbol{\theta}$ which would be accepted in a test of a simple H_0 against a general H_1, given the observed data set. When α is the size of the test, the associated confidence coefficient is $1 - \alpha$. □

Example 5.4 (*continued*)

$$X_1, X_2, \ldots, X_n \sim N(\mu, \sigma^2), \qquad \sigma^2 \text{ known}.$$

In a test of H_0: $\mu = \mu_0$ vs. H_1: $\mu \neq \mu_0$ the usual acceptance region has the form

$$\left| \frac{\bar{X} - \mu_0}{\sigma/\sqrt{n}} \right| \leqslant z_{\alpha/2}$$

and

$$\Pr\left[\left| \frac{\bar{X} - \mu_0}{\sigma/\sqrt{n}} \right| \leqslant z_{\alpha/2} \,\middle|\, \mu = \mu_0 \right] = 1 - \alpha.$$

This is equivalent to $\Pr[\bar{X} - z_{\alpha/2}\sigma/\sqrt{n} \leqslant \mu \leqslant \bar{X} + z_{\alpha/2}\sigma/\sqrt{n}] = 1 - \alpha$, so, as we saw before with this example, the interval with end points $\bar{X} \pm z_{\alpha/2}\sigma/\sqrt{n}$ gives a confidence interval for μ with confidence coefficient $1 - \alpha$.

Example 5.6

Let $X_1, X_2, \ldots, X_n \sim N(\mu, \sigma^2)$, μ unknown, and we require a confidence set for σ^2. Consider H_0: $\sigma^2 = \sigma_0^2$ vs. H_1: $\sigma^2 > \sigma_0^2$.

Using obvious notation, the acceptance region for the usual test of H_0 vs. H_1 is

$$\frac{(n-1)S^2}{\sigma_0^2} \leqslant \chi^2_{(n-1);\,\alpha}.$$

This is easily manipulated to give a confidence interval for σ^2 of the form

$$\sigma^2 \geqslant (n-1)S^2/\chi^2_{(n-1);\,\alpha}.$$

Note that again we have used a pivotal quantity, namely $(n-1)S^2/\sigma_0^2$, which under H_0 has a χ^2 distribution with $(n-1)$ degrees of freedom. Note also that if a two-sided confidence interval is required, this is easily obtainable from a two-sided test of hypothesis.

Example 5.5 (*continued*)

Earlier we found approximate confidence intervals for p, the probability of success in a binomial experiment, using the normal approximation to the binomial distribution. These same intervals can also be derived via hypothesis testing, based on the same normal approximation. Alternatively, a confidence interval may be derived based on the discrete binomial distribution itself. For a test of size α of H_0: $p = p_0$ vs. H_1: $p \neq p_0$, if we observe $X = x$ successes then we reject H_0 if and only if $\Pr[X \leqslant x \mid p = p_0] < \alpha/2$ or $\Pr[X \geqslant x \mid p = p_0] < \alpha/2$. Manipulating these statements to find which values of p_0 lead to acceptance of H_0, gives a confidence interval $[p_L, p_U]$, where p_L, p_U satisfy

$$\sum_{i=x}^{n} \binom{n}{i} p_L^i (1-p_L)^{n-i} = \alpha/2$$

and

$$\sum_{i=0}^{x} \binom{n}{i} p_U^i (1-p_U)^{n-i} = \alpha/2.$$

Note that although this interval is based on the 'exact' binomial distribution rather than on a normal approximation, it is not 'exact' in the sense of giving a coverage probability (confidence coefficient) of exactly $1-\alpha$ for all values of p. The discreteness of the random variable means that the coverage probability can only be achieved at a discrete set of values for p, and possibly not at all (Angus and Schafer, 1984). Finding confidence intervals for a parameter based on a discrete test statistic is by no means straightforward, and research continues (Tingley and Li, 1993). An interesting example is given by Ridout (1994).

Confidence sets derived from maximum likelihood ratio tests (MLRTs) and related procedures

Consider the general hypothesis testing problem where $X_1, X_2, \ldots, X_n$ form a random sample from a distribution with p.d.f. $f(x; \boldsymbol{\theta})$, and we test $H_0: \boldsymbol{\theta} = \boldsymbol{\theta}_0$ vs. $H_1: \boldsymbol{\theta} \neq \boldsymbol{\theta}_0$. In Section 4.6 we saw that we can use an MLRT for such a pair of hypotheses. An MLRT has test statistic.

$$-2\ln(\lambda) = 2[l(\hat{\boldsymbol{\theta}}; \mathbf{x}) - l(\boldsymbol{\theta}_0; \mathbf{x})]$$

where $\hat{\boldsymbol{\theta}}$ is the unrestricted MLE for $\boldsymbol{\theta}$ and $l(.;.)$ is the log-likelihood function. We have the asymptotic result that $-2\ln(\lambda) \sim \chi_d^2$, where d is the difference between the number of parameters estimated independently under $H_0 \cup H_1$ and under H_0, which is usually simply the dimension of $\boldsymbol{\theta}$.

The acceptance region for an MLRT based on this approximate distributional result has the form $-2\ln(\lambda) \leqslant \chi^2_{d;\alpha}$, that is

$$2[l(\hat{\boldsymbol{\theta}}; \mathbf{x}) - l(\boldsymbol{\theta}_0; \mathbf{x})] \leqslant \chi^2_{d;\alpha}$$

or

$$l(\boldsymbol{\theta}_0; \mathbf{x}) \geqslant l(\hat{\boldsymbol{\theta}}; \mathbf{x}) - \tfrac{1}{2}\chi^2_{d;\alpha}.$$

Thus, a confidence set for $\boldsymbol{\theta}$, with approximate confidence coefficient $1 - \alpha$, consists of all values for which $l(\boldsymbol{\theta}; \mathbf{x}) \geqslant l(\hat{\boldsymbol{\theta}}; \mathbf{x}) - \frac{1}{2}\chi^2_{d;\alpha}$. In other words, all values of $\boldsymbol{\theta}$ which have a log-likelihood within a threshold $\frac{1}{2}\chi^2_{d;\alpha}$ of the maximum log-likelihood are included in the confidence set – see Figure 5.1 for the case $d = 1$.

Example 5.4 *(continued)*

For illustration we consider the same example once more. In fact, for this example the distributional result $-2\ln(\lambda) \sim \chi_d^2$ is exact, rather than, as is usually the case, approximate. Recall that $X_1, X_2, \ldots, X_n \sim N(\mu, \sigma^2)$, σ^2 known, and we require a confidence interval for μ. The log-likelihood is

$$l(\mu; \mathbf{x}) = \frac{n}{2}\ln(2\pi\sigma^2) - \frac{1}{2\sigma^2}\sum_{i=1}^{n}(x_i - \mu)^2$$

and

$$l(\hat{\mu}; \mathbf{x}) - l(\mu_0; \mathbf{x}) = -\frac{1}{2\sigma^2}\left[\sum_{i=1}^{n}(x_i - \bar{x})^2 - \sum_{i=1}^{n}(x_i - \mu_0)^2\right].$$

Using the fact that

$$\sum_{i=1}^{n}(x_i - \mu_0)^2 = \sum_{i=1}^{n}(x_i - \bar{x})^2 + n(\bar{x} - \mu_0)^2$$

the confidence interval based on

$$l(\mu; \mathbf{x}) \geqslant l(\hat{\mu}; \mathbf{x}) - \tfrac{1}{2}\chi^2_{1;\alpha}$$

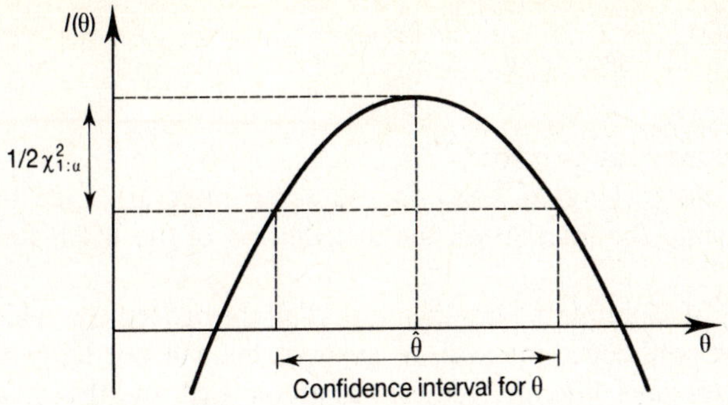

Fig. 5.1 Confidence set for a single parameter based on the MLRT.

becomes

$$\frac{n(\bar{x}-\mu)^2}{2\sigma^2} \leqslant \tfrac{1}{2}\chi^2_{1;\alpha}$$

which is equivalent to

$$\left|\frac{(\bar{x}-\mu)}{\sigma/\sqrt{n}}\right| \leqslant z_{\alpha/2}$$

leading to the same region as previously found.

In Section 4.7, we outlined a number of alternatives to MLRTs. Virtually all of these testing procedures have counterparts in interval estimation, and many of the references given in Section 4.7 are also relevant to interval estimation.

Score tests and Wald tests are easily converted into confidence interval procedures, as we now illustrate with a simple example

Example 5.4 (*continued*)

$$\frac{\partial l}{\partial \mu} = \frac{n}{\sigma^2}(\bar{x}-\mu)$$

and

$$I_\mu = -E\left[\frac{\partial^2 l}{\partial \mu^2}\right] = \frac{n}{\sigma^2}$$

so the score statistic is

$$u^{\mathrm{T}}(\mu_0)I_{\mu_0}^{-1}u(\mu_0) = \frac{n(\bar{x}-\mu_0)^2}{\sigma^2}$$

and the confidence interval based on the result $u^{\mathrm{T}}(\mu_0)I_{\mu_0}^{-1}u(\mu_0) \sim \chi^2_1$ consists of those values of μ for which

$$\frac{n(\bar{x}-\mu)^2}{\sigma^2} \leqslant \chi^2_{1;\alpha}.$$

This is the same interval as before.

Using the Wald test again leads to the same interval, since the procedure is equivalent to basing the interval on the distribution of the MLE for μ, as described in Section 5.2.2.

The simplicity of Example 5.4 has meant that all the procedures which we illustrated lead to the same confidence interval. In general this will not happen – for example, the test procedures gave different results in Example 4.10, and this difference, although asymptotically negligible, would carry over to the corresponding confidence intervals.

Ridout (1994) provides an interesting comparison of confidence intervals based on the MLRT, the Wald test, and the score test with an 'exact' interval, for a discrete example in the context of dilution assays. Doganaksoy and Schmee (1993) compare confidence intervals based on the MLRT and the Wald test for distributions used in life-data analysis. Their comparisons also include modifications of the MLRT-based interval. One adjustment is based on a Bartlett correction (see Section 4.7.3). Another uses the signed square root of the MLRT statistic, $-2\ \ln(\lambda)$, and approximates its distribution by the standard normal distribution, rather than using the χ^2 approximation to the distribution of $-2\ \ln(\lambda)$ itself. This alternative statistic for inference about a single parameter has been widely discussed in the literature – see, for example, Pierce and Peters (1992) and DiCiccio and Martin (1993), who both describe modifications to the statistic which improve the distributional approximation to the test statistic, and hence the coverage probabilities of the corresponding intervals.

The numerous adjustments to the MLRT and other tests, outlined in Section 4.7, which are appropriate to situations in which there are nuisance parameters, are also relevant to the corresponding interval estimation procedures. For example, Cox and Snell (1989, Example 2.2, p. 37) give an illustration of confidence intervals based on profile likelihood. They consider a two-parameter example (one parameter of interest, one nuisance parameter) and comparisons are made with corresponding intervals based on the joint likelihood of both the parameters. There are differences, but these are small. A later example (Example 2.4, p. 48) in Cox and Snell (1989) compares confidence intervals based on unconditional and conditional likelihoods.

5.3 Optimal properties of confidence sets

In Chapters 2 and 4 we looked at desirable properties for point estimators and for tests of hypothesis respectively; we now do the same for confidence sets. Because of the correspondence between tests of hypothesis and confidence sets (acceptance regions can be used to define confidence sets, and vice versa), the properties that we defined for tests can be carried over to confidence sets. We shall demonstrate this carry-over for two of the properties.

Definition 5.3

Suppose that $S_{\mathbf{X}}$, $S'_{\mathbf{X}}$ are both confidence sets for $\boldsymbol{\theta}$, each with confidence coefficient $(1-\alpha)$. Then $S_{\mathbf{X}}$ is said to be *more selective* than $S'_{\mathbf{X}}$ if and only if

$$\Pr[S_{\mathbf{X}} \supset \boldsymbol{\theta}_2 \mid \boldsymbol{\theta} = \boldsymbol{\theta}_1] \leqslant \Pr[S'_{\mathbf{X}} \supset \boldsymbol{\theta}_2 \mid \boldsymbol{\theta} = \boldsymbol{\theta}_1]$$

for any $\boldsymbol{\theta}_1, \boldsymbol{\theta}_2 \in \Omega$, $\boldsymbol{\theta}_1 \neq \boldsymbol{\theta}_2$

i.e. $S_{\mathbf{X}}$, $S'_{\mathbf{X}}$ have the same probability of covering the true value of $\boldsymbol{\theta}$, but $S_{\mathbf{X}}$ has a probability which is never greater than that for $S'_{\mathbf{X}}$ of covering any false value. Note that the terminology here is by no means universal. Stuart and Ord, (1991, p. 753) use the phrase 'more selective' in preference to the alternative 'shorter', which could be misinterpreted. Bickel and Doksum (1977, Section 6.3) use the term 'more accurate' for an equivalent idea.

Ideally, we should like to find a most selective/shortest/uniformly most accurate confidence set, i.e. one which is more selective than any other set with the same confidence coefficient. It may not always be possible to find such a set; in fact we have the following result.

Lemma 5.2

A UMP test of H_0: $\boldsymbol{\theta} = \boldsymbol{\theta}_0$ vs. H_1: $\boldsymbol{\theta} \in \Omega - \{\boldsymbol{\theta}_0\}$ exists if and only if a most selective confidence set exists for $\boldsymbol{\theta}$.

Proof

Suppose that a UMP test of size α exists for H_0 vs. H_1 and that it has a critical region C_1. The corresponding confidence set, with confidence coefficient $(1-\alpha)$, is $S_{1\mathbf{X}}$.

Let $S_{2\mathbf{X}}$ be *any other* confidence set with confidence coefficient $(1-\alpha)$, with the corresponding test of size α having critical region C_2.

Then for any $\boldsymbol{\theta}_1 \in \Omega - \{\boldsymbol{\theta}_0\}$,

$$\Pr[S_{i\mathbf{X}} \supset \boldsymbol{\theta}_0 \mid \boldsymbol{\theta} = \boldsymbol{\theta}_1] = \Pr\{\mathbf{X} \in \bar{C}_i \mid \boldsymbol{\theta} = \boldsymbol{\theta}_1\} \qquad i = 1, 2,$$

writing $\Pr[S_{i\mathbf{X}} \supset \boldsymbol{\theta}_0 \mid \boldsymbol{\theta} = \boldsymbol{\theta}_1]$ as a briefer version of $\Pr[\mathbf{X}\colon S_{i\mathbf{X}} \supset \boldsymbol{\theta}_0 \mid \boldsymbol{\theta} = \boldsymbol{\theta}_1]$. But

$$\Pr[\mathbf{X} \in C_1 \mid \boldsymbol{\theta} = \boldsymbol{\theta}_1] \geqslant \Pr[\mathbf{X} \in C_2 \mid \boldsymbol{\theta} = \boldsymbol{\theta}_1]$$

because C_1 is UMP, so

$$\Pr\{\mathbf{X} \in \bar{C}_1 \mid \boldsymbol{\theta} = \boldsymbol{\theta}_1\} \leqslant \Pr\{\mathbf{X} \in \bar{C}_2 \mid \boldsymbol{\theta} = \boldsymbol{\theta}_1\}.$$

Hence

$$\Pr\{S_{1\mathbf{X}} \supset \boldsymbol{\theta}_0 \mid \boldsymbol{\theta} = \boldsymbol{\theta}_1\} \leqslant \Pr\{S_{2\mathbf{X}} \supset \boldsymbol{\theta}_0 \mid \boldsymbol{\theta} = \boldsymbol{\theta}_1\}$$

i.e. $S_{1\mathbf{X}}$ is most selective.

The argument above can clearly be reversed to show that the existence of a most selective confidence set implies that the corresponding test is UMP. □

We noted in Section 4.4 that UMP tests for two-sided alternative hypotheses are rare. It follows that two-sided most selective confidence intervals are also unusual.

A second possibly desirable property of confidence sets is unbiasedness.

Definition 5.4

A confidence set, S_X, with confidence coefficient $(1 - \alpha)$, is unbiased if

$$\Pr[S_X \supset \theta_2 \mid \theta = \theta_1] \leqslant 1 - \alpha \qquad \text{for all} \quad \theta_1, \theta_2 \in \Omega, \quad \theta_1 \neq \theta_2$$

i.e. the probability of covering a false value of θ is no greater than the probability of covering the true value, whatever that true value may be. Thus, in a certain sense, the confidence set is 'centred on' the true value. As with our 'most selective' property, there is a direct connection between unbiasedness in the interval estimation context, and a corresponding property in hypothesis testing.

Lemma 5.3

A confidence set S_X is unbiased if and only if the corresponding test is unbiased.

Proof

The proof follows similar steps to those for Lemma 5.2, and will not be given in detail – see Exercise 5.18. □

Other properties corresponding to those discussed in Chapter 4 for tests of hypothesis carry over to interval estimation. For example, confidence regions based on similar tests and invariant tests (Section 4.4.2) will themselves possess the properties of 'similarity' and 'invariance' – Cox and Hinkley (1974, p. 228). Because such properties do not illustrate any new ideas, we do not discuss them further here.

As well as properties dual to those from hypothesis testing, a number of additional properties have been introduced which are specifically relevant to confidence sets alone. For example, if we are dealing with confidence intervals it is desirable to make the physical length of the interval as small as possible. Zacks (1971, Section 10.3) discusses this property, and notes some problems which may arise. Juola (1993) presents a procedure which gives the (physically) shortest interval for a given pivotal quantity.

5.4 Some problems with confidence sets

Confidence intervals, or more generally confidence sets, are widely used in practice, and provide a useful tool. However, as with most techniques, careful examination reveals problems which, although not detracting from their general worth, show that caution is needed in using them.

Perhaps the trickiest aspect of confidence sets is their interpretation. The confidence coefficient, $1 - \alpha$, refers to the proportion of a large number of such sets which will include the true value θ. Any particular confidence set either includes θ or it does not – we cannot say, as we might like to and as many non-specialist users are prone to do, that our individual confidence set has probability $1 - \alpha$ of including θ. For such an interpretation we need to treat $\boldsymbol{\theta}$ as a random variable, and adopt a Bayesian approach – see Chapter 7.

Another problem is illustrated by the following example.

Example 5.7 (from De Groot, 1986, p. 400)

Suppose that X_1, X_2, are independent observations from the uniform distribution on $[\theta - \frac{1}{2}, \theta + \frac{1}{2}]$, and that we require a confidence interval for θ. Consider the interval $[X_{(1)}, X_{(2)}]$, where $X_{(1)} = \text{Min}[X_1, X_2]$ and $X_{(2)} = \text{Max}[X_1, X_2]$.

$$\begin{aligned}\Pr[X_{(1)} \leqslant \theta \leqslant X_{(2)}] &= \Pr[X_1 \leqslant \theta \leqslant X_2] + \Pr[X_2 \leqslant \theta \leqslant X_1] \\ &= 2\Pr[X_1 \leqslant \theta]\Pr[X_2 \geqslant \theta] = 2 \times \tfrac{1}{2} \times \tfrac{1}{2} = \tfrac{1}{2}.\end{aligned}$$

Thus, the interval is a confidence interval for θ with confidence coefficient $\frac{1}{2}$.

But suppose now we realize that $X_{(2)} - X_{(1)} > \frac{1}{2}$. It follows that $[X_{(1)}, X_{(2)}]$ must contain θ, i.e.

$$\Pr[X_{(1)} \leqslant \theta \leqslant X_{(2)} \mid X_{(2)} - X_{(1)} > \tfrac{1}{2}] = 1$$

which seems to contradict the earlier result. The contradiction is due to the interpretation of $1 - \alpha$, which is related to what happens in many repetitions of our experiment. This does not allow us to condition on special features of the data actually observed which might, as here, allow us to sharpen our inference.

Such problems can often be removed by taking a Bayesian approach to interval estimation (see Chapter 7). Bayesian intervals may be equivalent numerically to standard confidence intervals when prior information is vague, but the interpretation is different in the two cases.

Another complication arises with confidence sets or regions if we wish to construct several regions *simultaneously*. If the confidence level is $1 - \alpha$ for each region, then the *overall confidence level* relating to the simultaneous coverage of all parameters of interest will be less than $1 - \alpha$. In some circumstances this overall level may be of more interest than the confidence level for individual parameters. For example in simple linear regression, with the model $\text{E}[Y] = \beta_0 + \beta_1 x$, the usual confidence interval for $\text{E}[Y]$, based on n pairs of data $(x_1, y_1), (x_2, y_2), \ldots, (x_n, y_n)$, has end points

$$\hat{\beta}_0 + \hat{\beta}_1 x \pm t_{n-2;\alpha/2}\, s\left[\frac{1}{n} + \frac{(x - \bar{x})^2}{\sum_{i=1}^{n} (x_i - \bar{x})^2}\right]^{1/2}$$

where $\hat{\beta}_0$, $\hat{\beta}_1$ are least squares estimates of β_0, β_1 (Section 3.4.2), s^2 is the residual variance, and $t_{n-2;\alpha/2}$ is a critical value of the t-distribution with $(n - 2)$ degrees of freedom – Hogg and Tanis (1993, Section 8.5). This interval has probability $(1 - \alpha)$ of including $\text{E}[Y]$ when we consider *one single value of* x. However, if we require a probability $1 - \alpha$ of *simultaneously* including $\text{E}[Y]$ in intervals constructed *for all possible* x, we must replace $t_{n-2;\alpha/2}$ by $[2F_{2,n-2;\alpha}]^{1/2}$ – see Miller (1966, p. 111).

Yet another problem with confidence sets or regions is the lack of exact confidence regions in examples including nuisance parameters. Weerahandi (1993) addresses this problem by introducing the notion of *generalized confidence intervals*, in which the requirements placed on the behaviour of the interval in repetitions of the *same* experiment are relaxed. Weerahandi's (1993) approach also provides an alternative resolution of the problem in Example 5.7.

Finally, we note that other types of interval, such as *tolerance intervals* and *prediction intervals*, each with their own interpretation, may be useful in some circumstances – see Vardeman (1992).

Exercises

5.1 $X_1, X_2, \ldots, X_n$ are a random sample from the exponential distribution with p.d.f. $f(x; \theta) = 1/\theta\, e^{-x/\theta}$, $x > 0$. Using the result that $Y = 2\sum_{i=1}^{n} X_i/\theta$ has a $\chi^2_{(2n)}$ distribution, construct a confidence interval for θ based on the pivotal quantity Y.

5.2 Find a confidence interval for the variance of the exponential distribution given in Exercise 5.1.

5.3 Independent normally distributed random variables $X_1, X_2, \ldots, X_n$ are such that X_k has expectation $f(k)\theta$ and variance $g(k)$, where $f(k)$ and $g(k)$ are known functions of k, $k = 1, \ldots, n$. Find a sufficient statistic for θ, and from it construct a two-sided 95% confidence interval for θ.

If $f(k) = g(k) = k$, $k = 1, 2, \ldots, n$, what is the smallest value of n for which the length of this confidence interval is less than 0.5?

5.4 Use the maximum likelihood estimator for θ (see Exercise 3.26) to find an approximate $100(1-\alpha)$% confidence interval for θ, given a random sample of size n from the distribution with probability density function

$$f(x; \theta) = \theta x^{\theta - 1}, \qquad 0 < x < 1, \quad \theta > 0.$$

5.5 $X_1, X_2, \ldots, X_n$ are a random sample from a normal distribution with known mean μ and unknown variance σ^2. Three possible confidence intervals for σ^2 are:

(a) $$\left(\sum_{i=1}^{n} \frac{(X_i - \bar{X})^2}{a_1}, \sum_{i=1}^{n} \frac{(X_i - \bar{X})^2}{a_2}\right)$$

(b) $$\left(\sum_{i=1}^{n} \frac{(X_i - \mu)^2}{b_1}, \sum_{i=1}^{n} \frac{(X_i - \mu)^2}{b_2}\right)$$

(c) $$\left(\frac{n(\bar{X} - \mu)^2}{c_1}, \frac{n(\bar{X} - \mu)^2}{c_2}\right),$$

where $a_1, a_2, b_1, b_2, c_1, c_2$ are constants.

Find values of these six constants which give confidence coefficient 0.90 for each of the three intervals when $n = 10$ and compare the expected widths of the three intervals in this case.

With $\sigma^2 = 1$, what value of n is required to achieve a 90% confidence interval of width less than 2 in cases (b) and (c) above?

5.6 For a random sample of size n from an exponential distribution with mean θ, find a pivotal quantity based on the approximate large sample distribution of the maximum likelihood estimator for θ. Hence show that

$$\left(\frac{\sqrt{n}\,\bar{X}}{(\sqrt{n}+z_{\alpha/2})}, \frac{\sqrt{n}\,\bar{X}}{(\sqrt{n}-z_{\alpha/2})}\right)$$

provides an approximate confidence interval for θ, with confidence coefficient $1-\alpha$.

5.7 Use the results of Exercise 3.16 to find an approximate confidence interval for μ, when μ is both the mean and standard deviation of a normal distribution.

5.8 For the data given in Exercise 3.21 find an approximate 95% confidence interval for θ. Suppose that you are able to ascertain the true response for a random sample of n men. Estimate how large n must be in order to achieve a 95% confidence interval for θ with the same width as that obtained above.

5.9 A random sample $X_1, X_2, \ldots, X_n$ is obtained from a distribution whose p.d.f. depends upon a scalar parameter θ. Given that $l(\theta; \mathbf{x})$ is the log likelihood function, write down the asymptotic distribution of $\partial l/\partial\theta$, and show how it can be used to provide a pivotal quantity.

Hence, or otherwise, derive an asymptotically valid confidence interval for σ using a sample of size n drawn from the normal distribution with mean zero and standard deviation σ.

5.10 One success occurs in 10 trials of a binomial experiment. By considering the simple null hypothesis H_0: $p = p_0$, and the fact that a confidence interval can be constructed by finding values of p_0 for which H_0 is accepted, find an 'exact' two-sided confidence interval for p, the probability of success, with confidence coefficient 0.95 – see the continuation of Example 5.5 in Section 5.2.3.

5.11 Given that X, a Poisson random variable with mean θ, takes the value x, show that a confidence interval (θ_L, θ_U) has confidence coefficient $1-\alpha$ when θ_L, θ_U satisfy the equations

$$\sum_{i=x}^{\infty} \frac{e^{-\theta_L}\theta_L^i}{i!} = \frac{\alpha}{2}$$

$$\sum_{i=0}^{x} \frac{e^{-\theta_U}\theta_U^i}{i!} = \frac{\alpha}{2}.$$

5.12 Suppose that Y_{2k} has a χ^2-distribution with $2k$ degrees of freedom, and hence p.d.f.

$$f(x) = \frac{x^{k-1}\,e^{-x/2}}{2^k(k-1)!} \qquad x > 0.$$

By integrating $f(x)$ by parts, show that, for any $c > 0$,

$$\Pr[Y_{2k} \geqslant c] = \frac{e^{-c/2}(c/2)^{k-1}}{(k-1)!} + \Pr[Y_{2(k-1)} \geqslant c]$$

where $Y_{2(k-1)} \sim \chi^2_{2(k-1)}$.

Use this result together with an expression for $\Pr[Y_2 \geqslant c]$, where $Y_2 \sim \chi^2_2$, to derive a relationship between the c.d.f.s of the χ^2 and Poisson distributions.

Deduce from this relationship and from Exercise 5.11 that the interval

$$(\tfrac{1}{2}\chi^2_{2x;(1-\alpha/2)}, \tfrac{1}{2}\chi^2_{2x+2;\alpha/2})$$

based on a single observation, x, from a Poisson distribution, gives a confidence interval, with confidence coefficient $1 - \alpha$, for the mean of that distribution.

5.13 The random variable X_1 has a Poisson distribution with parameter θ_1 and, independently, the random variable X_2 has a Poisson distribution with parameter θ_2. By considering the conditional distribution of X_1, given the value of $X_1 + X_2$, show how a test of the null hypothesis that $\theta_1 = k\theta_2$, for fixed k, reduces to testing that the 'success probability' parameter of a binomial distribution is $k/(1 + k)$.

Modifications are made to a machine, which runs continuously (except for breakdowns, which result in a negligible loss of running time), in an attempt to reduce the number of breakdowns which occur. In a one-month period prior to the modifications 15 breakdowns occurred, while in a three-month period after the modifications, 20 breakdowns occurred. Find an approximate 95% confidence interval for the ratio, θ_1/θ_2, of the breakdown rates before and after the modifications were made.

5.14 A random variable, X, having a Poisson distribution, is observed to take the value x. Using the result that, for large θ, $X \stackrel{.}{\sim} N(\theta, \theta)$, obtain a quadratic equation in θ, whose roots give the end points of a confidence interval for the mean, θ, of the distribution.

The following table gives the frequency distribution of the number of breakdowns in a year for 550 army vehicles. Assuming the data arise as a random sample from a Poisson distribution, find a confidence interval with approximate confidence coefficient 95% for the expected number of breakdowns per vehicle per year.

Number of breakdowns	0	1	2	3	4	5
Number of vehicles	295	190	53	5	5	2

5.15 Random samples, each of size n, are drawn from the distributions $N(\mu_1, \sigma_1^2)$ and $N(\mu_2, \sigma_2^2)$, all four parameters being unknown. Use the results of Exercise 4.16 to find a confidence interval for σ_1^2/σ_2^2.

5.16 Given that λ, μ are the means of two exponential distributions as defined in Exercise 4.18, use the results of that Exercise to find a lower confidence limit for λ/μ.

5.17 Let $X_1, X_2, \ldots, X_n$ be a random sample from the uniform distribution on the range $[0, \theta]$. Use a pivotal quantity based on the sufficient statistic $Y = X_{(n)}$, to derive a family of $100(1 - \alpha)$% confidence intervals for θ.

Show that the shortest $100(1 - \alpha)$% confidence interval for θ in this family is of length $Y(\alpha^{-1/n} - 1)$.

5.18 Show that the critical region C provides an unbiased test of $H_0\colon \theta = \theta_0$ if and only if the confidence set for θ based on C is unbiased.

Chapter Six

The decision theory approach to inference

6.1 Introduction

Earlier chapters of this book have adopted what might be termed a classical, or frequency-based approach to inference. In these next two chapters we describe a different approach, based on Bayesian methods and decision theory. This chapter focuses on decision theory – Bayesian concepts are introduced sparingly and discussed more generally in Chapter 7. Other approaches to inference will be discussed in subsequent chapters.

The frequency-based theory of Chapters 2–5 treats parameters as *fixed* but unknown. The essential difference in the Bayesian approach to inference is that parameters are treated as random variables and hence have probability distributions.

Prior information about $\boldsymbol{\theta}$, a parameter or vector of parameters, is quantified by means of a *prior* distribution. This distribution will be denoted by $p(\boldsymbol{\theta})$ and may be a p.d.f. or probability mass function. It reflects our feelings, before collecting the data, about the probabilities of the possible values of $\boldsymbol{\theta}$. The information provided by the data about $\boldsymbol{\theta}$ is contained in the likelihood $L(\boldsymbol{\theta}, \mathbf{x}) = f(\mathbf{x}; \boldsymbol{\theta})$. Bayes' theorem is used to combine it with the information contained in $p(\boldsymbol{\theta})$. Now

$$f(\mathbf{x}; \boldsymbol{\theta})p(\boldsymbol{\theta}) = q(\boldsymbol{\theta}; \mathbf{x})h(\mathbf{x})$$

where $h(\mathbf{x})$ is the marginal distribution of $\mathbf{X}$ and $q(\boldsymbol{\theta}; \mathbf{x})$ is the conditional distribution of $\boldsymbol{\theta}$ given $\mathbf{X} = \mathbf{x}$. Thus

$$\begin{aligned} q(\boldsymbol{\theta}; \mathbf{x}) &= f(\mathbf{x}; \boldsymbol{\theta})p(\boldsymbol{\theta})/h(\mathbf{x}) \qquad \text{(Bayes' theorem)} \\ &= L(\boldsymbol{\theta}; \mathbf{x})p(\boldsymbol{\theta})/h(\mathbf{x}). \end{aligned}$$

$q(\boldsymbol{\theta}; \mathbf{x})$ is referred to as the *posterior* distribution. In theory it contains all available information about $\boldsymbol{\theta}$ and so should be used for making decisions, estimates or inferences.

Decision theory seems an obvious approach in hypothesis testing, where we must decide between a null and alternative hypothesis. We discuss decision theory in this context in Section 6.5. It is perhaps less natural to use decision theory in estimation, but we can nevertheless look at estimation problems in this framework, as we see in

Sections 6.3 and 6.6 respectively for point estimation and interval estimation. Before this, however, we present the main elements of a decision problem.

6.2 Elements of decision theory

The following are the basic features of a decision theory problem.

(1) A number of 'actions' are possible; we must decide which to take.
(2) A number of 'states of nature' are possible; we do not know in general which holds.
(3) The relative desirabilities of the various actions for each state of nature can be quantified.
(4) Prior information may be available regarding the relative probabilities of the various states of nature.
(5) Data may be available which will add to our knowledge of the relative probabilities of the states of nature.

Let $\boldsymbol{\theta}$ denote the 'true state of nature'. Decision theory problems sometimes involve no data, but here we suppose data are available and consist of an observation of the random vector $\mathbf{X}$, whose distribution depends on $\boldsymbol{\theta}$. A *decision procedure* δ specifies which action to take for each possible value of $\mathbf{X}$. Thus if we adopt the procedure δ and $\mathbf{X} = \mathbf{x}$, then we must take the action $\delta(\mathbf{x})$.

Whether $\delta(\mathbf{x})$ is a good choice usually depends on the true state of nature and we suppose we have a *loss function*, $L_s(\boldsymbol{\theta}, \delta(\mathbf{x}))$, which measures the loss from action $\delta(\mathbf{x})$ when $\boldsymbol{\theta}$ holds. Alternatively, we could assume we have a *utility function*, which measures the benefit from action $\delta(\mathbf{x})$ for given $\boldsymbol{\theta}$. However, the choice is of little consequence, as the negative of a utility function is a loss function. We will talk about loss, rather than utility, as this is more usual in the context of inference.

The usefulness of a decision procedure depends on a quantity called the risk function.

Definition 6.1

The *risk function* $R(\boldsymbol{\theta}, \delta)$ is defined as

$$R(\boldsymbol{\theta}, \delta) = \int L_S(\boldsymbol{\theta}, \delta(\mathbf{x}))L(\boldsymbol{\theta}; \mathbf{x})\, \mathrm{d}\mathbf{x}$$

i.e. the expected loss, where expectation is with respect to the joint distribution of the *x*s, which is simply the likelihood.

Sometimes there are decision procedures that can clearly be improved upon and hence are not optimal.

Definition 6.2

A procedure δ_1 is *inadmissible* if there exists another procedure δ_2 such that

$$R(\boldsymbol{\theta}, \delta_1) \geqslant R(\boldsymbol{\theta}, \delta_2) \qquad \text{for all} \quad \boldsymbol{\theta}$$

with strict inequality for some $\boldsymbol{\theta}$, i.e. δ_1 is never better than δ_2 and sometimes worse. A procedure which is not inadmissible is *admissible*.

Clearly, inadmissible estimators need not be considered, because we can always find an estimator which is more desirable with respect to risk.

In subsequent sections we concentrate on two decision-making strategies that lead to the minimax and Bayes decision procedures.

Definition 6.3

The *minimax procedure* is such that $\text{Max}_{\boldsymbol{\theta}}\, R(\boldsymbol{\theta}, \delta)$ is minimized. Such a procedure takes a pessimistic view – it assumes that 'nature is malevolent', and will choose the worst possible value of $\boldsymbol{\theta}$. It also ignores any prior information regarding $\boldsymbol{\theta}$.

There are situations, for example in game theory, where a minimax approach is very desirable and, as we see briefly later, it *is* possible to use it in inference. However, a better strategy is usually to use the Bayes procedure, which uses prior information about $\boldsymbol{\theta}$, quantified by means of a prior distribution, $p(\boldsymbol{\theta})$.

Definition 6.4

A *Bayes procedure* is such that $\int R(\boldsymbol{\theta}, \delta)p(\boldsymbol{\theta})\, \mathrm{d}\boldsymbol{\theta}$ is minimized. The quantity $\int R(\boldsymbol{\theta}, \delta)p(\boldsymbol{\theta})\, \mathrm{d}\boldsymbol{\theta}$ is called the *Bayes risk* of the procedure δ, and is the expected risk, where expectation is with respect to the prior distribution for $\boldsymbol{\theta}$.

From the definition of the posterior distribution, $q(\boldsymbol{\theta}; \mathbf{x})$, we have

$$\begin{aligned}\int R(\boldsymbol{\theta}, \delta)p(\boldsymbol{\theta})\, \mathrm{d}\boldsymbol{\theta} &= \iint L_S(\boldsymbol{\theta}, \delta(\mathbf{x}))L(\boldsymbol{\theta}; \mathbf{x})p(\boldsymbol{\theta})\, \mathrm{d}\mathbf{x}\, \mathrm{d}\boldsymbol{\theta}\\ &= \iint L_S(\boldsymbol{\theta}, \delta(\mathbf{x}))q(\boldsymbol{\theta}; \mathbf{x})h(\mathbf{x})\, \mathrm{d}\mathbf{x}\, \mathrm{d}\boldsymbol{\theta}\\ &= \int h(\mathbf{x})\left[\int L_S(\boldsymbol{\theta}, \delta(\mathbf{x}))q(\boldsymbol{\theta}; \mathbf{x})\, \mathrm{d}\boldsymbol{\theta}\right] \mathrm{d}\mathbf{x}.\end{aligned}$$

Hence for any given value of $\mathbf{x}$, we minimize the integral $\int L_S(\boldsymbol{\theta}, \delta(\mathbf{x}))q(\boldsymbol{\theta}; \mathbf{x})\, \mathrm{d}\boldsymbol{\theta}$, and do not need to consider values of $\mathbf{x}$ other than that actually observed. This exemplifies another difference between the frequency-based and Bayesian approaches to inference. In the former, because our inference is based on sampling distributions, we are effectively taking into account values of $\mathbf{X}$ which might have occurred, but did not in our sample. The Bayesian approach concentrates solely on the value $\mathbf{x}$ which has actually been observed.

The Bayes procedure minimizes $\int L_S(\boldsymbol{\theta}, \delta(\mathbf{x}))q(\boldsymbol{\theta}; \mathbf{x})\, \mathrm{d}\boldsymbol{\theta}$, for the observed value of $\mathbf{x}$. In other words it minimizes expected loss, where expectation is taken with respect to the posterior distribution for $\boldsymbol{\theta}$. Under certain conditions (1) a Bayes procedure is necessarily admissible and (2) every admissible procedure is a Bayes procedure for some prior distribution. In particular, if $\boldsymbol{\theta}$ is discrete and can take only a finite number of values, then (2) holds and if, in addition, $p(\boldsymbol{\theta}) > 0$ for all $\boldsymbol{\theta}$, then (1) holds. Proofs of these results and appropriate conditions for continuous $\boldsymbol{\theta}$ are given in Ferguson (1967, Chapter 2).

A result which links Bayes and minimax procedures is that a Bayes procedure, with constant risk for all $\boldsymbol{\theta}$, is minimax. In fact, the minimax procedure is generally a Bayes procedure for some prior distribution, the so-called *least favourable prior distribution* (Ferguson, 1967, Chapter 2). The reason for this name is that, among all possible prior distributions, it is the prior distribution for which the minimum expected risk is a maximum. Whether or not the minimax procedure is sensible will depend on whether the least favourable prior distribution is plausible as a representation of prior knowledge. If we have *any* prior knowledge at all, it is unlikely to coincide exactly with the least favourable prior, so the Bayes procedure will usually be preferred to the minimax procedure.

6.3 Point estimation

Let us consider how point estimation can be fitted into the decision theory framework; for simplicity we stick to a single parameter θ, but most results are easily generalized to a vector of parameters.

(1) The possible actions are possible estimators.
(2) The possible states of nature correspond to the true value of θ.
(3) Loss is some function of the estimator $\hat{\theta}$, and the true value θ, which measures how good our estimator/action is. We denote it by the *loss function* $L_S(\theta, \hat{\theta})$, which increases for bad $\hat{\theta}$.
(4) Prior information is quantified by means of a prior distribution for θ, $p(\theta)$.
(5) The data are simply our usual random sample $x_1, x_2, \ldots, x_n$, from a distribution with p.d.f, or probability mass function, $f(x; \theta)$.

The loss function should reflect the consequence of estimating θ by $\hat{\theta}$, but the choice is often somewhat arbitrary. We look at three simple loss functions and see what form the Bayes and minimax estimators take. More complicated loss functions can be used if it is believed that they are more appropriate. The three we consider give easily interpreted expressions for the Bayes estimator, but one restriction common to all three is that they are symmetric. This may not always be appropriate; for example, if we are estimating the thickness of ice on a pond, then underestimation is worse (will incur a greater loss) than a corresponding degree of overestimation. Conversely, if future prices in a stock market are being estimated with a view to possible investment, most individuals would consider overestimation to be worse than underestimation.

Our three simple loss functions are:

(1) *Zero–one loss*

$$L(\theta, \hat{\theta}) = \begin{cases} 0, & |\hat{\theta} - \theta| < b \\ a, & |\hat{\theta} - \theta| \geqslant b \end{cases}$$

where $a, b > 0$ are some constants.

(2) *Absolute error loss* $L_S(\theta, \hat{\theta}) = a|\hat{\theta} - \theta|$ where $a > 0$ is some constant.
(3) *Quadratic loss* $L_S(\theta, \hat{\theta}) = a(\hat{\theta} - \theta)^2$ where $a > 0$ is some constant.

In each of (1), (2), (3) we lose no generality in taking $a = 1$, and we shall do so in what follows. All are non-decreasing functions of the absolute error of estimation, $|\hat{\theta} - \theta|$, and so they are symmetric about $\hat{\theta} = \theta$.

Minimax estimators can be difficult to find and do not exist in all problems. An example where a minimax estimator can be found is for the zero–one loss function with $X_i \sim N(\theta, 1)$. Then $\bar{X}$ is minimax for estimating θ – Zacks (1971, Example 6.7). This is not a particularly exciting example, because $\bar{X}$ is the obvious estimator for θ. An example where the minimax estimator is not the obvious one is given below in Example 6.1.

Let us now look at the Bayes estimators for our three loss functions.

Zero–one loss

The quantity to be minimized is

$$\int_{-\infty}^{\infty} q(\theta; \mathbf{x}) L_S(\theta, \hat{\theta})\, \mathrm{d}\theta = \int_{\hat{\theta}+b}^{\infty} q(\theta; \mathbf{x})\, \mathrm{d}\theta + \int_{-\infty}^{\hat{\theta}-b} q(\theta; \mathbf{x})\, \mathrm{d}\theta$$

$$= 1 - \int_{\hat{\theta}-b}^{\hat{\theta}+b} q(\theta; \mathbf{x})\, \mathrm{d}\theta$$

so we want to maximize

$$\int_{\hat{\theta}-b}^{\hat{\theta}+b} q(\theta; \mathbf{x})\, \mathrm{d}\theta.$$

If $q(\theta; \mathbf{x})$ is unimodal, maximization is achieved by choosing $\hat{\theta}$ to be the midpoint of the interval of length $2b$ for which $q(\theta; \mathbf{x})$ has the same value at both ends. To see this, note that the quantity to be maximized is the area under $q(\theta; \mathbf{x})$ between $\hat{\theta} \pm b$. Figure 6.1 shows that if we move $\hat{\theta}$ from its suggested position to the left, the shaded area lost is bigger than the shaded area gained. A similar result occurs if we move $\hat{\theta}$ to the right, so our suggested $\hat{\theta}$ is optimal.

In the special case where $q(\theta; \mathbf{x})$ is symmetric, the optimal $\hat{\theta}$ is equal to the mean, median and mode of the posterior distribution. More generally if we let $b \to 0$, then $\hat{\theta} \to$ the global mode of the posterior distribution, regardless of the symmetry or otherwise of $q(\theta; \mathbf{x})$. The mode does seem to be a fairly sensible estimator, since it generalizes the MLE to take account of prior information – it reduces to the MLE if $p(\theta)$ is 'vague' or 'flat' – see Section 6.4.2 for discussion of the choice of priors.

In the case where $q(\theta; \mathbf{x})$ is multimodal, implementation of zero–one loss can be much trickier, and may lead to non-unique $\hat{\theta}$.

Absolute error loss function

With an absolute error loss function, $\hat{\theta}$ is chosen to minimize

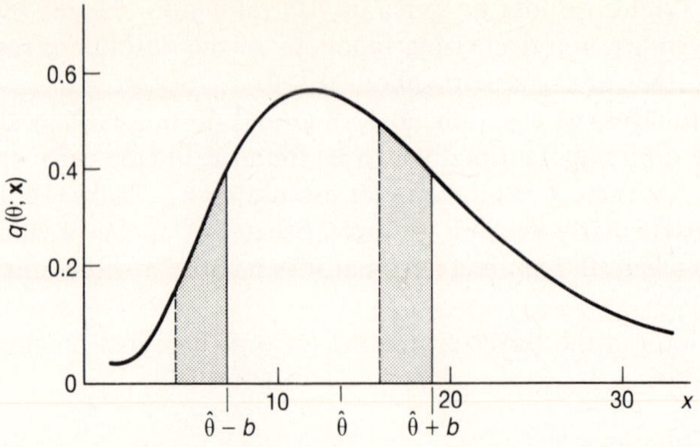

Fig. 6.1 Bayes estimate for zero–one loss.

$$\int |\hat{\theta} - \theta| q(\theta; \mathbf{x}) \, d\theta = \int_{-\infty}^{\hat{\theta}} (\hat{\theta} - \theta) q(\theta; \mathbf{x}) \, d\theta + \int_{\hat{\theta}}^{\infty} (\theta - \hat{\theta}) q(\theta; \mathbf{x}) \, d\theta.$$

This must be differentiated with respect to $\hat{\theta}$ (assuming the derivative exists) and the result equated to 0. A standard result is that

$$\frac{\partial}{\partial y} \int_{a(y)}^{b(y)} g(x, y) \, dx = \frac{\partial b}{\partial y} g(b, y) - \frac{\partial a}{\partial y} g(a, y) + \int_{a(y)}^{b(y)} \frac{\partial g(x, y)}{\partial y} \, dx$$

so $\hat{\theta}$ must satisfy

$$\int_{-\infty}^{\hat{\theta}} q(\theta; \mathbf{x}) \, d\theta - \int_{\hat{\theta}}^{\infty} q(\theta; \mathbf{x}) \, d\theta = 0.$$

Thus the optimal $\hat{\theta}$ is the median of the posterior distribution.

Although we have assumed that the expected loss can be differentiated with respect to $\hat{\theta}$, the result holds more generally. If X is a random variable and d is a constant, then $E(|X - d|)$ is minimized when d is the median of the distribution of X.

Quadratic loss function

With a quadratic loss function, the best estimator is the mean of the posterior distribution. To show this, consider the expected loss

$$\begin{aligned} E_{\theta|\mathbf{x}}[(\hat{\theta} - \theta)^2] &= E[\{(\hat{\theta} - \bar{\theta}) - (\theta - \bar{\theta})\}^2] \\ &= E[(\hat{\theta} - \bar{\theta})^2] - 2(\hat{\theta} - \bar{\theta})E[\theta - \bar{\theta}] + E[(\theta - \bar{\theta})^2] \end{aligned}$$

where $\bar{\theta}$ is the posterior mean of θ.

The second term equals 0, as $E[\theta - \bar{\theta}] = 0$, and the third term does not involve $\hat{\theta}$. Hence the expected loss is minimized when $\hat{\theta} = \bar{\theta}$, the required result. This also shows that the minimum expected loss is $E[(\theta - \bar{\theta})^2]$, the posterior variance of θ.

Example 6.1

Suppose that X is the number of successes in a binomial experiment with n trials and probability of success θ. Find an estimator for θ based on X. As a prior distribution for θ we use a Beta distribution with parameters α, β, and p.d.f.

$$p(\theta) = \begin{cases} \dfrac{\Gamma(\alpha+\beta)}{\Gamma(\alpha)\Gamma(\beta)}\, \theta^{\alpha-1}\,(1-\theta)^{\beta-1}, & 0 \leqslant \theta \leqslant 1 \\ 0, & \text{elsewhere.} \end{cases}$$

The reason for this choice will become clear when we discuss conjugate priors in Section 6.4.2. For appropriate choices of α, β we can model a wide range of unimodal, flat, or U-shaped prior distributions on the interval [0, 1]. This is illustrated in Figure 6.2.

The mean of a Beta distribution with parameters α, β is $\alpha/(\alpha+\beta)$, and, if $\alpha > 1$ and $\beta > 1$, the distribution is unimodal with $(\alpha-1)/(\alpha+\beta-2)$ as its mode (Winkler 1972, p. 151). The median does not have a closed form, but tables are available for a wide range of values of α and β. See, for example, Winkler (1972, Table 5).

It turns out that the posterior distribution for θ is another Beta distribution with parameters $\alpha + x$, $\beta + n - x$; see Example 6.4. From this it follows that with zero–one loss and $b \to 0$, the Bayes estimator is $(\alpha + x - 1)/(\alpha + \beta + n - 2)$, assuming the posterior distribution is unimodal. For a quadratic loss function, the Bayes estimator is $(\alpha + x)/(\alpha + \beta + n)$, and for an absolute error loss function the Bayes estimator is the median of the Beta distribution with parameters $\alpha + x$, $\beta + n - x$, which can be found from tables.

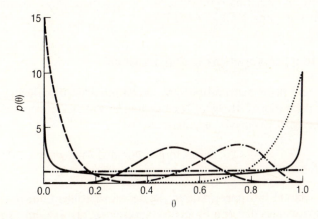

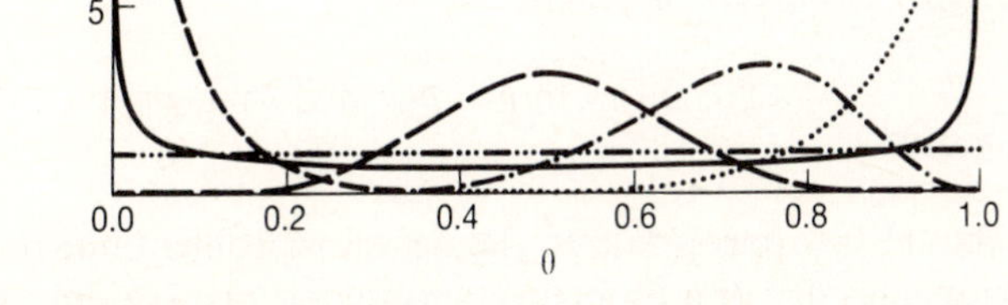

Fig. 6.2 Six beta distributions.

To derive the minimax estimator with quadratic loss, we use the result noted earlier that a Bayes estimator with constant risk is minimax. The risk function is

$$\begin{aligned}
E_{X|\theta}[(\hat{\theta} - \theta)^2] &= [\text{Bias}(\hat{\theta})]^2 + \text{var}[\hat{\theta}] \\
&= \left[\theta - E\left(\frac{\alpha + X}{\alpha + \beta + n}\right)\right]^2 + \text{var}\left[\frac{\alpha + X}{\alpha + \beta + n}\right] \\
&= \left[\theta - \left(\frac{\alpha + n\theta}{\alpha + \beta + n}\right)\right]^2 + \frac{n\theta(1 - \theta)}{(\alpha + \beta + n)^2} \\
&= \frac{[\theta(\alpha + \beta) - \alpha]^2 + n\theta(1 - \theta)}{[\alpha + \beta + n]^2}.
\end{aligned}$$

For this risk function to be constant over all θ, the coefficients of θ and θ^2 in the numerator must both be zero. To achieve this, we need $\alpha = \beta = \sqrt{n}/2$, so that the minimax estimator, using quadratic loss, is

$$(x + \tfrac{1}{2}\sqrt{n})/(n + \sqrt{n}).$$

The least favourable prior in the present case is symmetric with mean $\frac{1}{2}$, and variance $[4(\sqrt{n} + 1)]^{-1}$.

As with other decision procedures it is desirable that an estimator should be admissible, but there are surprising cases where an apparently reasonable estimator is inadmissible. For example, with a quadratic loss function, the sample mean is an inadmissible estimator of the mean of a multivariate normal distribution with three or more variables (James and Stein, 1961). The concept of admissibility also reveals examples in which MLEs and MVUEs may be flawed. The multivariate normal mean example just cited shows that an MLE may be inadmissible, and Muirhead (1985) gives an example of an inadmissible MVUE. Of course, it can be argued that a particular choice of loss function is the reason for inadmissibility in these cases, but provided that the loss function is sensible, examples such as these indicate again that caution is needed in using MLEs or MVUEs.

6.4 Loss functions and prior distributions

Loss functions and prior distributions have been introduced earlier in this chapter and we discuss them more fully here. They are important elements of decision theory but they may be hard to specify in practice.

6.4.1 *Loss functions*

A loss function should be appropriate to the decision problem under consideration. If all potential outcomes are of a monetary nature and involve only small amounts of money, then it is usually reasonable to equate loss to 'financial loss'. However, outcomes can also involve customer goodwill, risk to health, quality of life, and so forth. Then choosing a loss function is difficult and subjective judgement may be needed to quantify the relative desirabilities of different outcomes.

Monetary loss can be unsuitable as a loss function if large sums are involved. To illustrate, suppose a company could take a gamble that is equally likely to result in a profit of £5 million or a loss of £3 million. Then taking the gamble gives an expected gain of

$$0.5 \times £5 \text{ million} - 0.5 \times £3 \text{ million} = £1 \text{ million}.$$

However, if the company is small, a loss of £3 million may force it into liquidation, so it could well be in the company's best interests to refuse the gamble despite the expected financial gain. If so, then financial loss could not be used as a loss function in this example, because decision theory methods require every risk (expected loss) to be preferable to any larger risk.

To state the properties required of a loss function, suppose A, B and C are three potential outcomes. (By an 'outcome' we mean the combination of a state of nature and an action.) Then $L_S(.)$ is a loss function if the following hold.

(1) $L_S(A) \leqslant L_S(B)$ if A is preferred to B.
(2) $L_S(A)$ is bounded.
(3) If $L_S(A) = \lambda L_S(\mathrm{B}) + (1 - \lambda)L_S(C)$, $0 \leqslant \lambda \leqslant 1$, then the decision maker is indifferent between (a) outcome A occurring for certain, and (b) a gamble in which exactly one of B and C occurs, B occurring with probability λ and C with probability $(1 - \lambda)$.

The properties ensure both that a small risk (expected loss) is preferable to a large risk and that minimizing the Bayes (expected) risk is a reasonable criterion for choosing a decision procedure.

The properties may be exploited to construct a loss function if the number of different possible outcomes is small. First, the worst and best possible outcomes are identified. Denote these by B and C and define $L_S(B) = 1$ and $L_S(C) = 0$. Let A be some other outcome. Then the decision maker must assess the value λ such that he or she is indifferent between A occurring for certain and the gamble defined in (3). For property (3) to hold, $L_S(A)$ must be set equal to λ. The procedure is repeated with A representing each outcome in turn.

The selection of a loss function can be difficult and its choice is often made for reasons of mathematical convenience, without any serious consideration of the particular decision problem of current interest. Also, in theoretical work there may be various types of decision problem of interest, rather than one specific problem, and then the choice of loss function may be somewhat arbitrary. For point estimation, Christensen and Hoffman (1985) argue that the posterior mean of $q(\theta; \mathbf{x})$ has advantages as an estimator if we have no real 'feel' for how to quantify a loss function. This implicitly means adoption of quadratic loss. However, the same authors also note that others have put forward reasons for preferring the posterior median or mode, and hence, implicitly, absolute error loss or zero–one loss.

6.4.2 Prior distributions

In earlier chapters, probability has been interpreted in terms of relative frequency. If an event A is one possible result of an experiment that could be repeated many times, then $P(A)$ is axiomatically defined as the relative frequency with which A occurs. However, probability is also useful for describing personal opinion about an unknown state of nature, or about a possible future event, even when the relevant situation cannot be repeated.

For example, a person might assess the probability as 0.2 that 'the first manned space-rocket to Mars will be launched before the year 2020'. This assessment of probability is clearly meaningful, even though it relates to a one-off unrepeatable event. If the person were to select other events which, in the person's opinion, each had a probability of 0.2 of occurring then, obviously, the person should consider each of these events equally likely to occur and should expect 20% of them to occur.

The prior distribution, $p(\boldsymbol{\theta})$, quantifies information about $\boldsymbol{\theta}$ prior to any (further) data being gathered. Sometimes $p(\boldsymbol{\theta})$ can be constructed on the basis of past data. For example, if a quality inspection program has been running for some time, the distribution of the number of defectives in past batches can be used as the prior distribution for the number of defectives in a future batch. More commonly, however, a prior distribution must be based on an expert's experience and personal judgement. Then the expert must be questioned about his or her opinion in such a way that the answers determine a prior distribution (cf. Example 6.2, below).

Conjugate prior distributions

When θ is a scalar, the decision maker (or an expert) could specify its prior distribution by subjectively assessing the values of quantiles. Based on background knowledge and any available past data, the decision maker chooses θ_p so that $\Pr[\theta \leqslant \theta_p]$ is thought to equal p. This is done for an increasingly fine grid of values of p. For example, we could start with $\theta_{0.5}$, then $\theta_{0.25}, \theta_{0.75}$, then $\theta_{0.1}, \theta_{0.2}, \ldots, \theta_{0.9}$, and so on. However, this method becomes impractical if $\boldsymbol{\theta}$ is a vector of several parameters, especially if the parameters are not independent in the prior distribution.

A more convenient way to proceed is to choose, if at all possible, a *conjugate prior distribution* for θ. This is a prior distribution which, when combined with the likelihood function, produces a posterior distribution which is in the same family as the prior. For example, we saw in Example 6.1 that if θ is the probability of success in a binomial experiment, and the prior distribution for θ is a Beta distribution, then the posterior distribution is also a Beta distribution. If we find a conjugate prior distribution which adequately fits our prior beliefs regarding θ we should use it, because it will usually simplify calculations considerably. The task of quantifying prior knowledge then amounts to specifying the parameters of the prior distribution. The following example illustrates the assessment of parameters for a beta distribution.

Example 6.2

A medical researcher was questioned about θ, the proportion of asthma sufferers who would be helped by a new drug. She was first asked about the median, and thought the probability that θ exceeded 0.3 was equal to the probability that θ was below 0.3, so $\theta_{0.5} = 0.3$. She then assessed $\theta_{0.25}$ and $\theta_{0.75}$ as 0.2 and 0.45, respectively. That is, she considered it equally likely that θ lay in each of the intervals [(0, 0.2), (0.2, 0.3), (0.3, 0.45) and (0.45, 1)]. From tables giving fractiles of beta distributions (Winkler, 1972, Table 5), the researcher's opinion might reasonably be represented by the beta distribution with $\alpha = 2$ and $\beta = 4$, for which $\theta_{0.25} = 0.194$, $\theta_{0.5} = 0.314$ and $\theta_{0.75} = 0.454$.

As in the above example, median and quartile assessments can often be used to determine a prior distribution that is parameterized by two unknown scalars. However, assessing prior distributions that involve unknown vector and/or matrix parameters is typically a difficult task and suitable procedures for quantifying prior opinion vary with the form of the prior distribution. Also conjugate priors should not always be used. For example, the conjugate prior when making inferences about the mean of a normal distribution is another normal distribution – see Example 6.3. If our prior knowledge corresponds to a skew distribution, we could not use a conjugate prior in this case.

As indicated at the start of the chapter, the posterior distribution is given by $q(\theta; \mathbf{x}) = L(\theta; \mathbf{x})p(\theta)/h(\mathbf{x})$. Now $h(\mathbf{x})$ does not involve θ, so we have

$$q(\theta; \mathbf{x}) \propto L(\theta; \mathbf{x})p(\theta)$$

where the constant of proportionality is fixed by the requirement that $q(\theta; \mathbf{x})$ must integrate to unity. This is often a convenient expression for determining the posterior distribution, especially if we can recognize the form of the distribution from the part depending on θ, since then the constant follows automatically. This is frequently possible in standard problems if the conjugate prior distribution is used.

Example 6.3

Let $X_1, X_2, \ldots, X_n$ be a random sample from $N(\theta, \sigma^2)$, σ^2 known, and let the prior information regarding θ be expressed as $\theta \sim N(\phi; \tau^2)$. The posterior distribution is obtained using

$$q(\theta; \mathbf{x}) \propto p(\theta)L(\theta; \mathbf{x}) = (2\pi\tau^2)^{-1/2} \exp\left\{-\frac{1}{2}\left(\frac{\theta-\phi}{\tau}\right)^2\right\}$$
$$\times \prod_{i=1}^{n} (2\pi\sigma^2)^{-1/2} \exp\left\{-\frac{1}{2}\left(\frac{x_i-\theta}{\sigma}\right)^2\right\}.$$

Note that in finding the posterior distribution, we can ignore the part which does not involve θ, and therefore forms part of the constant of proportionality. Thus

$$q(\theta; \mathbf{x}) \propto \exp\left\{-\frac{1}{2}\left[\theta^2\left(\frac{1}{\tau^2}+\frac{n}{\sigma^2}\right) - 2\theta\left(\frac{\phi}{\tau^2}+\frac{\Sigma x_i}{\sigma^2}\right)\right]\right\}$$

$$= \exp\left\{-\frac{(\sigma^2 + n\tau^2)}{2\sigma^2\tau^2}\left[\theta - \frac{\phi\sigma^2 + n\bar{x}\tau^2}{\sigma^2 + n\tau^2}\right]^2 + \text{constant}\right\}.$$

But this form shows that $q(\theta; \mathbf{x})$ is the p.d.f. of a normal distribution with mean $(\phi\sigma^2 + n\bar{x}\tau^2)/(\sigma^2 + n\tau^2)$ and variance $((\sigma^2 + n\tau^2)/\sigma^2\tau^2)^{-1}$.

In the last example, the reciprocal of the posterior variance was the sum of the reciprocals of the prior variance and the variance of the sample mean, respectively. Because of this reciprocal relationship, precision (= 1/variance) is sometimes quoted instead of variance. Also, the mean may be written

$$\left(\frac{\phi}{\tau^2} + \frac{\bar{\mathbf{x}}}{\sigma^2/n}\right)\bigg/\left(\frac{1}{\tau^2} + \frac{1}{\sigma^2/n}\right).$$

This illustrates a relationship that holds for several standard sampling distributions when a conjugate prior distribution is used: the posterior mean is a weighted average of the prior mean ϕ and sample mean $\bar{x}$, with weights proportional to the prior precision and the precision of the sample mean, respectively. This relationship holds, for example, for binomial, Poisson, negative binomial, gamma and exponential distributions, as well as the normal distribution. A general result is given by Ericson (1969).

Example 6.4

Consider a member of the k-parameter exponential family with p.d.f.

$$f(x; \boldsymbol{\theta}) = \exp\left\{\sum_{j=1}^{k} A_j(\boldsymbol{\theta})B_j(x) + C(x) + D(\boldsymbol{\theta})\right\}.$$

Suppose we take as a prior distribution, another member of the k-parameter exponential family with p.d.f.

$$p(\boldsymbol{\theta}) = \exp\left\{\sum_{j=1}^{k} A_j(\boldsymbol{\theta})\alpha_j + \alpha_{k+1}D(\boldsymbol{\theta}) + K(\alpha_1, \alpha_2, \ldots, \alpha_{k+1})\right\}$$

where $\alpha_1, \alpha_2, \ldots, \alpha_{k+1}$ are parameters of the prior distribution, and $K(., ., \ldots, .)$ is a function of these parameters alone.

The p.d.f. of the posterior distribution is then proportional to

$$p(\boldsymbol{\theta})L(\boldsymbol{\theta}; \mathbf{x}) \propto \exp\left\{\sum_{i=1}^{n}\sum_{j=1}^{k} A_j(\boldsymbol{\theta})B_j(x_i) + nD(\boldsymbol{\theta}) + \sum_{j=1}^{k} A_j(\boldsymbol{\theta})\alpha_j + \alpha_{k+1}D(\boldsymbol{\theta})\right\}$$

$$= \exp\left\{\sum_{j=1}^{k} A_j(\boldsymbol{\theta})\alpha'_j + \alpha'_{k+1}D(\boldsymbol{\theta})\right\}$$

where $\alpha'_j = \alpha_j + \Sigma_{i=1}^{n} B_j(x_i)$, for $j = 1, 2, \ldots, k$ and $\alpha'_{k+1} = \alpha_{k+1} + n$.

The posterior distribution has the same form as the prior, except for different values of the parameters. Hence we have found conjugate prior distributions for members of the k-parameter exponential family. We have already seen two special cases of this general result in Examples 6.1 and 6.3.

In Example 6.1, for the binomial distribution, $k = 1$, $A_1(\theta) = \ln(\theta/(1-\theta))$ and $D(\theta) = \ln(1-\theta)$. Hence

$$p(\theta) \propto \exp\left\{\alpha_1 \ln(\theta/(1-\theta)) + \alpha_2 \ln(1-\theta)\right\} = \theta^{\alpha_1}(1-\theta)^{\alpha_2-\alpha_1}$$

which is the form of a beta distribution.

In Example 6.3, where θ is the mean of a normal distribution with σ^2 known, $A_1(\theta) = \theta/\sigma^2$ and $D(\theta) = -\theta^2/2\sigma^2$, so

$$p(\theta) \propto \exp\left\{\frac{\alpha_1\theta}{\sigma^2} - \frac{\alpha_2\theta^2}{2\sigma^2}\right\}$$

$$\propto \exp\left\{-\frac{\alpha_2}{2\sigma^2}\left(\theta - \frac{\alpha_1}{\alpha_2}\right)^2\right\}.$$

Here the dependence of $p(\theta)$ on θ is precisely that of a normal distribution.

As a third special case, consider a Poisson distribution with mean θ. Here $A_1(\theta) = \ln(\theta)$ and $D(\theta) = -\theta$, so

$$p(\theta) \propto \exp\left\{\alpha_1 \ln(\theta) - \alpha_2\theta\right\} = \theta^{\alpha_1}\, e^{-\alpha_2\theta},$$

which is the form of a gamma distribution.

Uninformative prior distributions

Another way of constructing a prior distribution is to use an uninformative or 'flat' prior distribution, i.e. $p(\theta) = \text{constant}$ for all θ. Such priors can usually be obtained as special, or limiting, cases of conjugate priors. For example, setting $\alpha = \beta = 1$ in Example 6.1 gives a flat prior for the binomial parameter. Similarly, letting $\tau^2 \to \infty$ in Example 6.3 gives a flat prior for a normal mean.

Uninformative priors are the usual way of representing ignorance about θ, and they are frequently used in practice. It can be argued that they are more objective than a subjectively assessed prior distribution, since the latter may contain personal bias as well as background knowledge. Also, in some applications the amount of prior information available is far less than the information contained in the data. Then there seems little point in worrying about a precise specification of the prior distribution. If the prior distribution is known to be fairly constant over the range of $\boldsymbol{\theta}$ for which the likelihood is appreciable, and not too large for the range of $\boldsymbol{\theta}$ where the likelihood is small, then the posterior distribution will be approximately proportional to the likelihood. With an uninformative prior this will be the case.

When an uninformative prior distribution is used, we often find the inference based on decision theory becomes equivalent to classical procedures. For example, the mode of the posterior distribution is simply the MLE. An uninformative prior also may have the attraction of keeping any subsequent mathematics relatively simple.

There are, however, a number of problems associated with the use of uninformative priors. First, if the range of values of θ is infinite, an uninformative prior p.d.f. cannot integrate to 1 – it is said to be *improper*.

A second problem is whether the prior distribution should be flat for θ or for some function of θ, such as θ^2 or $\ln \theta$. In fact, it is fairly conventional to construct a prior which is flat for a function $\phi(\theta)$ of θ whose Fisher's information, I_ϕ, is constant. This idea is due to Jeffreys (1961), and leads to a prior distribution for θ which is proportional to $I_\theta^{1/2}$.

Example 6.1 (*continued*)

For the binomial distribution $I_\theta = n/[\theta(1-\theta)]$. Hence, the Jeffreys prior is proportional to $[\theta(1-\theta)]^{-1/2}$, which is a beta distribution with parameters $\alpha = \beta = \frac{1}{2}$.

Example 6.3 (*continued*)

For a normal mean, $I_\theta = n/\sigma^2$. This does not depend on θ, so the Jeffreys prior is flat for θ, rather than for any other function of θ.

Example 6.5

Let $X_1, X_2, \ldots, X_n$ be a random sample from $N(\mu, \theta^2)$, μ known. Then

$$l = -\frac{n}{2}\ln(2\pi\theta^2) - \frac{1}{2\theta^2}\sum_{i=1}^{n}(x_i - \mu)^2$$

and

$$I_\theta = -\mathrm{E}\left[\frac{\partial^2 l}{\partial\theta^2}\right] = -\mathrm{E}\left[\frac{n}{\theta^2} - \frac{3}{\theta^4}\Sigma(x_i - \mu)^2\right]$$

$$= -\frac{n}{\theta^2} + \frac{3n}{\theta^2}$$

so $I_\theta^{1/2} = (2n/\theta^2)^{1/2}$. Hence the Jeffreys prior is $p(\theta) \propto 1/\theta$, and the prior distribution for $\ln(\theta)$ is flat.

As the last two examples illustrate, $p(\theta) =$ constant is a common choice when the range of θ is $(-\infty, +\infty)$, whereas $p(\theta) \propto 1/\theta$ is a common choice when the range of θ is $(0, \infty)$. If $\theta \in [0, 1]$, three uninformative priors that are each often used are $p(\theta) \propto \theta^{-1/2}(1-\theta)^{-1/2}$, $p(\theta) \propto \theta^{-1}(1-\theta)^{-1}$ and $p(\theta) =$ constant.

A third problem with using uninformative priors is the purely practical one that we are likely to have *some* prior knowledge about θ, and if we do, we should use it, rather than claim ignorance. Such ignorance is somewhat illusory in any case; if we have a flat prior for one function of θ, then implicitly we have some knowledge (a non-flat prior) for other functions of θ.

6.5 Hypothesis testing

Hypothesis testing seems to be tailor-made for decision theory, since even in its classical setting it is a decision problem, i.e. choose between H_0 and H_1. To fit it into the decision theory framework we simply need to add a loss or utility function and possibly prior probabilities. In fact, choosing these is not particularly straightforward unless both null and alternative hypotheses are simple, so we concentrate on this special case.

Suppose that $X_1, X_2, \ldots, X_n$ are a random sample from a distribution with p.d.f. $f(x; \boldsymbol{\theta})$ and we wish to test $H_0: \boldsymbol{\theta} = \boldsymbol{\theta}_0$ vs. $H_1: \boldsymbol{\theta} = \boldsymbol{\theta}_1$. There are two possible decisions: accept or reject H_0, and two possible states of nature: either H_0 or H_1 holds. The obvious way to define the loss function is given by Table 6.1, where a, b are positive constants. For composite hypotheses we would probably want a loss function which depends on 'how far' from the true value of $\boldsymbol{\theta}$ is the wrongly chosen hypothesis.

Recall that the minimax procedure is the one that minimizes maximum possible risk, and the Bayes procedure minimizes expected (or Bayes) risk or, equivalently, expected loss where the expectation is determined with respect to the posterior distribution of θ.

In the context of hypothesis testing, let δ_c denote the decision to use a test with critical region C, and let α, β be the probabilities of Type I, Type II errors respectively for the test with critical region C. The risk function $R(\delta_c; \boldsymbol{\theta}_i)$ is the expected loss, using δ_c, when $\boldsymbol{\theta} = \boldsymbol{\theta}_i$, $i = 0, 1$, where expectation is with respect to the distribution of $\mathbf{X}$, conditional on $\boldsymbol{\theta}$.

It is clear that

$$R(\delta_c; \boldsymbol{\theta}_0) = \{\text{Loss when } X \in C \mid \boldsymbol{\theta}_0\} \Pr[\mathbf{X} \in C \mid \boldsymbol{\theta}_0]$$

$$+ \{\text{Loss when } \mathbf{X} \in \overline{C} \mid \boldsymbol{\theta}_0\} \Pr[\mathbf{X} \in \overline{C} \mid \boldsymbol{\theta}_0$$

$$= a\alpha + 0 = a\alpha.$$

Similarly $R(\delta_c; \boldsymbol{\theta}_1) = b\beta$. The expected risk is thus

$$p_0 a\alpha + p_1 b\beta.$$

The Bayes procedure (in this context, the Bayes test) chooses C to minimize this expected risk. The minimax test minimizes the maximum risk, i.e. minimizes $\text{Max}\{a\alpha, b\beta\}$.

Table 6.1 Loss function for simple vs. simple hypothesis testing

	H_0 holds	H_1 holds
Accept H_0	0	b
Reject H_0	a	0

The Neyman–Pearson lemma (Lemma 4.1) states that the best test of size α of H_0 *vs.* H_1 is a likelihood ratio (LR) test, i.e. it has a critical region of the form

$$\frac{L(\boldsymbol{\theta}_1;\ \mathbf{x})}{L(\boldsymbol{\theta}_0;\ \mathbf{x})} \geqslant A$$

for some non-negative constant A. We now give a lemma which states that Bayes tests are also LR tests, and hence best tests for some α.

Lemma 6.1

Bayes tests and LR tests are equivalent, i.e. every LR test is a Bayes test for some p_0, p_1 and every Bayes test is an LR test with

$$A = \frac{p_0 a}{p_1 b}.$$

Proof

The Bayes test minimizes

$$\begin{aligned} p_0 a\alpha + p_1 b\beta &= p_0 a \Pr[\mathbf{X} \in C \mid \boldsymbol{\theta}_0] + p_1 b \Pr[\mathbf{X} \in \bar{C} \mid \boldsymbol{\theta}_1 \\ &= p_o a \int_C L(\boldsymbol{\theta}_0;\ \mathbf{x})\, d\mathbf{x} + p_1 b \int_{\bar{C}} L(\boldsymbol{\theta}_1;\ \mathbf{x})\, d\mathbf{x}, \end{aligned}$$

assuming that X is continuous, and that the likelihood function satisfies appropriate regularity conditions,

$$\begin{aligned} &= p_0 a \int_C L(\boldsymbol{\theta}_0;\ \mathbf{x})\, d\mathbf{x} + p_1 b\left[1 - \int_C L(\boldsymbol{\theta}_1;\ \mathbf{x})\, d\mathbf{x}\right] \\ &= p_1 b + \int_C [p_0 a L(\boldsymbol{\theta}_0;\ \mathbf{x}) - p_1 b L(\boldsymbol{\theta}_1;\ \mathbf{x})]\, d\mathbf{x}. \end{aligned}$$

It is fairly clear that to minimize this quantity we must choose C to include all $\mathbf{x}$ for which the integrand is negative and to exclude all $\mathbf{x}$ for which the integrand is positive.

Hence $\mathbf{x} \in C$ if and only if $p_0 a L(\boldsymbol{\theta}_0;\ \mathbf{x}) - p_1 b L(\boldsymbol{\theta}_1;\ \mathbf{x}) \leqslant 0$ i.e. $\mathbf{x} \in C$ if and only if

$$\frac{L(\boldsymbol{\theta}_1;\ \mathbf{x})}{L(\boldsymbol{\theta}_0;\ \mathbf{x})} \geqslant \frac{p_0 a}{p_1 b}.$$

This proves our result, since for any A and any a, $b \geqslant 0$, we can choose p_0 and p_1 to satisfy $p_0/p_1 = Ab/a$. □

Turning now to minimax tests, there is a general result in decision theory stating that when there is a finite number of possible decisions and there is a unique minimax procedure then it is a Bayes procedure for some set of prior probabilities. This set of prior possibilities is the least favourable prior distribution. We shall not prove this result – see, for example, Ferguson (1967, p. 82).

Because the minimax test is Bayes for the least favourable prior distribution, the minimax test must be an LR test and hence a best test for some value of α. In Section 6.2 we noted that a Bayes procedure with risk constant over all states of nature is minimax. It follows that the minimax test is the LR test for which $a\alpha = b\beta$. We can alternatively justify this result by noting that α increases as β decreases, and vice versa, so that a decrease in one of $a\alpha$ or $b\beta$ will lead to an increase in the other, and hence increase maximum loss.

Example 6.6

We look again at Examples 4.1 and 4.2 which were used to illustrate the Neyman–Pearson lemma, and see what happens when they are tackled via decision theory. Suppose that $X_1, X_2, \ldots, X_n \sim N(\mu, \sigma^2)$ with σ^2 known, and that we wish to test $H_0\colon \mu = \mu_0$ vs. $H_1\colon \mu = \mu_1$, with $\mu_1 > \mu_0$. The critical region for the LR test

$$\frac{L(\mu_1;\, \mathbf{x})}{L(\mu_0;\, \mathbf{x})} \geqslant A \text{ becomes } \bar{x} \geqslant \frac{\sigma^2 \ln(A)}{n(\mu_1 - \mu_0)} + \tfrac{1}{2}(\mu_0 + \mu_1) = B, \text{ say.}$$

In the classical approach we were able to ignore the form of B (we simply used the distribution of $\overline{X}$ under H_0 to choose an appropriate value) but this cannot be done in the decision theory approach. Lemma 6.1 has shown that the Bayes test is the LR test with $A = p_0 a/p_1 b$, so we have to substitute this into the expression for B.

Consider now a numerical example where $\mu_0 = 0$, $\mu_1 = 1$, $\sigma^2 = 1$, $n = 4$, $a = 2$, $b = 1$, $p_0 = \frac{1}{4}$, $p_1 = \frac{3}{4}$. Then the Bayes test has critical region

$$\bar{x} \geqslant \tfrac{1}{4}\ln(\tfrac{1}{3} \times \tfrac{2}{1}) + \tfrac{1}{2} = \tfrac{1}{4}\ln(\tfrac{2}{3}) + \tfrac{1}{2} = 0.399.$$

For this test

$$\alpha = \Pr[\overline{X} \geqslant 0.399 \,|\, \overline{X} \sim N(0, \tfrac{1}{4})] = 1 - \Phi\left(\frac{0.399}{0.500}\right) = 0.212$$

where $\Phi(.)$ is the c.d.f. for $N(0, 1)$, and

$$\beta = \Pr[\overline{X} < 0.399 \,|\, \overline{X} \sim N(1, \tfrac{1}{4})] = \Phi\left(\frac{-0.601}{0.500}\right) = 0.115.$$

Using the standard approach of Chapter 4 and fixing α at $\alpha = 0.05$, say, would give $B = 1.645\sqrt{\frac{1}{4}} = 0.822$, so $\beta = \Pr[\overline{X} < 0.822 \,|\, \overline{X} \sim N(1, \frac{1}{4})] = 0.363$.

The Bayes test has therefore increased α considerably beyond any of the usual fixed levels but, at the same time, β has been much reduced. The fact that $\beta < \alpha$ follows from the inequality $p_1 b/p_0 a > 1$, which implies that it is more important to control β than α.

The minimax test, found by iteration, has critical region $\overline{X} \geqslant 0.613$, with $\alpha = 0.110$, $\beta = 0.220$; because $a/b = 2$ the minimax test is one for which $\beta/\alpha = 2$. Note that if the prior probabilities $p_0 = \frac{1}{4}$, $p_1 = \frac{3}{4}$ are correct, then the Bayes test incurs a Bayes risk of $p_0 a\alpha + p_1 b\beta = 0.192$. If we ignore the prior information and use the minimax procedure (which would be optimal for a different, least favourable, prior distribution) the Bayes risk rises to 0.220.

Example 6.7

Consider a binomial experiment with 10 trials, where we wish to test H_0: $p = 0.5$ vs. H_1: $p = 0.8$. Suppose again that $a = 2$, $b = 1$. Table 6.2 gives the risk function $a\alpha$, $b\beta$, for each of the twelve critical regions $X \geqslant B$, $B = 0, 1, 2, \ldots, 11$, where X is the number of successes in 10 trials – as we saw in Example 4.2 there is no point in looking at other critical regions.

Let δ_B denote the test with critical region $X \geqslant B$. The Bayes test is the test for which the Bayes risk $p_0 a\alpha + p_1 b\beta$ is smallest; δ_0 is optimal for very small p_0, δ_1 for slightly larger p_0, and so on, so that δ_{11} is optimal for values of p_0 close to one. To be more precise, δ_{11} is preferred to δ_{10} and is therefore optimal if the expected risk for δ_{11} is smaller than that for δ_{10}, i.e. $p_1 \leqslant 0.002p_0 + 0.893p_1$, or $0.002p_0 \geqslant 0.107p_1$ or $p_0 \geqslant 0.982$.

It can be similarly discovered when other δ_B are Bayes tests. For example, δ_7 is Bayes for $0.177 \leqslant p_0 \leqslant 0.462$ and δ_8 is Bayes when $0.462 \leqslant p_0 \leqslant 0.774$, and these two possibilities together cover a wide range of values for p_0. Bayes risks are easily determined. For example, if $p_0 = 0.2$ then δ_7 is the Bayes test and its expected risk is $p_0 a\alpha + p_1 b\beta = 0.2(0.344) + 0.8(0.121) = 0.1656$. Similarly if $p_0 = 0.6$ then δ_8 is the Bayes test with expected risk $0.6(0.110) + 0.4(0.322) = 0.1948$.

Examination of Table 6.2 reveals that δ_8 is the minimax test, with a maximum risk of 0.322. We can, however, reduce this maximum risk by using a randomized critical region, as described in Example 4.2. If we choose test δ_7 with probability 0.487 and δ_8 with probability 0.513, then $a\alpha = b\beta = 0.224$, so the maximum risk can be reduced substantially from that of δ_8. In Example 4.2 the use of randomized critical regions was criticized from a practical point of view. It seems strange that randomization should be of benefit if the choice of test has no influence on the state of nature. In turn, this casts doubt on the usefulness of minimax tests other than in competitive situations where the state of nature is influenced by an opponent. In contrast, a Bayes test cannot be improved by introducing randomization.

6.6 Interval estimation

It is fair to say that relatively little work has been done on the decision theory approach to interval estimation. A major stumbling block to the implementation of decision theory in this context is the choice of a loss function. We could, for example, take a (0–1) loss function of the form

Table 6.2 Risk function for various tests of H_0: $p = 0.5$ vs. H_1: $p = 0.8$

B	0	1	2	3	4	5	6	7	8	9	10	11
$a\alpha$	2.000	1.998	1.978	1.890	1.656	1.246	0.754	0.344	0.110	0.022	0.002	0.000
$b\beta$	0.000	~0.000	~0.000	~0.000	0.001	0.006	0.033	0.121	0.322	0.624	0.893	1.000

$$L_S(\boldsymbol{\theta}, S_X) = \begin{cases} 1, & \boldsymbol{\theta} \notin S_{\mathbf{X}} \\ 0, & \boldsymbol{\theta} \in S_{\mathbf{X}} \end{cases}.$$

The risk function for a rule which chooses $S_{\mathbf{X}}$ when $\mathbf{X} = \mathbf{x}$ is then $\Pr[\boldsymbol{\theta} \notin S_{\mathbf{X}}]$, and a Bayes procedure would minimize

$$\int_{\theta} p(\boldsymbol{\theta}) \Pr[\boldsymbol{\theta} \notin S_{\mathbf{X}}] \, d\boldsymbol{\theta}$$

whereas a minimax procedure would find the Bayes procedure for which $\Pr[\boldsymbol{\theta} \notin S_{\mathbf{X}}]$ is constant over $\boldsymbol{\theta}$. It is interesting to note that standard confidence intervals, as defined in Chapter 5, also have the property that $\Pr[\boldsymbol{\theta} \notin S_{\mathbf{X}}]$ is constant for all $\boldsymbol{\theta}$.

The (0–1) loss function above has a clear disadvantage – if we increase the physical size of $S_{\mathbf{X}}$, we get a 'better' $S_{\mathbf{X}}$ according to $L_S(\boldsymbol{\theta}, S_{\mathbf{X}})$. In the limit, if $S_{\mathbf{X}} = \Omega$ for all $\mathbf{x}$, then the risk function is uniformly zero, which is the best we can do. Clearly, the size of $S_{\mathbf{X}}$ has to be taken into account in any definition of a loss function. Meeden and Vardeman (1985) tackle this problem; they use the risk function defined above, but in addition they consider a function which measures the size of $S_{\mathbf{X}}$. These two functions are then combined in different ways to produce three different definitions of optimal sets. The properties of these three definitions are then investigated and compared with Bayesian credible intervals (see Chapter 7).

6.7 Bayesian sequential procedures

With a sequential procedure, one usually draws sample items one at a time, rather than taking a sample of predetermined size. After each sample there is the choice of stopping and making a decision (e.g. rejecting/accepting a hypothesis, estimating a parameter, etc.), or sampling may be continued. We first consider sequential procedures in general, and then describe the sequential probability ratio test (SPRT), which is the most common sequential procedure. It is possible to make sequential procedures more flexible, by taking $k_i \geqslant 1$ further observations on the ith sampling occasion. We shall restrict ourselves to the simplest case where all $k_i = 1$.

6.7.1 General

The cost of sampling must be taken into account when determining the loss or risk of a sequential procedure. If the cost of sampling n items one at a time is little more than the cost of a single sample of size n, then the flexibility of a sequential procedure often makes it preferable to taking a single, fixed-size sample. To illustrate, suppose each sampled item is classed as either a success or a failure, and we wish to test $H_0: \theta = 0.10$ vs. $H_1: \theta = 0.20$, where θ is the probability of success. If a single sample of fixed size is taken, it might become clear, on examining the data, that the sample was unnecessarily large or, alternatively, hindsight can reveal the sample was too small. For example, if a sample of size 100 contains 25 successes, then the evidence against H_0 was probably substantial well before the 100th item, and a smaller sample would have been adequate. On the other hand, if the sample of 100 items contains 15 successes, one would prefer to sample further items before choosing between H_0 and H_1. In either case, a sequential procedure would lead to a better sample size.

To introduce notation, suppose a stream of observations $X_1, X_2, \ldots$ may be taken one at a time and let c_i denote the cost of the ith observation. A sequential decision procedure involves a stopping rule for deciding whether to stop after observing data $\mathbf{x}_n = x_1, x_2, \ldots, x_n$, and a terminal decision rule for choosing the action to take after stopping. We denote the stopping rule by $S_n(\mathbf{x}_n)$ and assign it values as follows:

$$S_n(\mathbf{x}_n) = \begin{cases} 1, & \text{if, given } \mathbf{x}_n\text{, the rule decides to stop for the first} \\ & \text{time after the } n\text{th observation} \\ 0, & \text{otherwise.} \end{cases}$$

We let $d_n(\mathbf{x}_n)$ denote the action chosen if $\mathbf{x}_n$ represents the data when sampling stops. We assume the loss function is such that sampling costs may be simply added to other losses to give the total loss. Then, if $S_n(\mathbf{x}_n) = 1$, the total loss is

$$L_S(\theta; d_n(\mathbf{x}_n)) + \sum_{i=1}^{n} c_i.$$

Conditional on θ, the expectation of this loss with respect to n and $\mathbf{X}_n$ gives the risk,

$$R(\theta; S, d) = \sum_n \underset{\mathbf{X}_n|\theta}{\mathrm{E}} \left[S_n(\mathbf{X}_n) \left\{ L_S(\theta, d_n(\mathbf{X}_n)) + \sum_{i=1}^{n} c_i \right\} \right].$$

The Bayes risk is obtained by determining the expectation of the risk over the prior distribution of θ, and in accordance with Definition 6.4, the Bayes procedure must minimize this expected risk.

After deciding to stop, choosing the optimal action is straightforward. The Bayes risk is

$$\begin{aligned} B(S, d) &= \sum_n \underset{\theta}{\mathrm{E}} \underset{\mathbf{X}_n|\theta}{\mathrm{E}} \left[S_n(\mathbf{X}_n) \left\{ L_S(\theta, d_n(\mathbf{X}_n)) + \sum_{i=1}^{n} c_i \right\} \right] \\ &= \sum_n \underset{\mathbf{X}_n}{\mathrm{E}} \left[S_n(\mathbf{X}_n) \left\{ \underset{\theta|\mathbf{X}_n}{\mathrm{E}} [L_S(\theta, d_n(\mathbf{X}_n))] + \sum_{i=1}^{n} c_i \right\} \right]. \end{aligned}$$

As $S_n(\mathbf{X}_n)$ is non-negative, for each n we wish to minimize $\mathrm{E}_{\theta|\mathbf{X}_n}[L_S(\theta, d_n(\mathbf{X}_n))]$. However, if a sample of fixed size n had been taken, rather than a sequential sample, then $\mathrm{E}_{\theta|\mathbf{X}_n}[L_S(\theta, d_n(\mathbf{X}_n))]$ would again be the quantity that the Bayes rule minimized. Thus whether or not sampling was sequential does not affect what action is optimal after the decision to stop has been made.

The difficult part of a sequential procedure is deciding when to stop sampling. The principle is simple: after each observation one should compare the current Bayes risk from making an immediate decision with the expected Bayes risk if more observations are taken, stopping if the current Bayes risk is the smaller. However, applying this principle is generally very hard, exceptions sometimes occurring when

(1) for each n, the Bayes risk after taking n observations does not depend on the values of the observations (an example is given in Exercise 6.16); or
(2) the number of observations that can be taken has a finite upper bound, say m. In this latter case, backward induction may be used to find an optimal sequential procedure.

With backward induction, the Bayes risk is first determined for each possible set of values $X_1, X_2, \ldots, X_m$. The situation after $m-1$ observations have been taken is then considered. For each possible set of values of $X_1, X_2, \ldots, X_{m-1}$, one determines (1) whether X_m should be taken or whether sampling should be stopped and (2) the Bayes risk for the optimal procedure from that point. This task is simplified by knowing that sampling cannot continue past X_m. The previous stage with $m-2$ observations is next considered. Given the values of $X_1, \ldots, X_{m-2}$, the minimum Bayes risk after observing any possible value of X_{m-1} is now known. This is used to decide whether sampling should be stopped for a given set of values of $X_1, \ldots, X_{m-2}$, or whether X_{m-1} should be observed. Hence, given any $X_1, \ldots, X_{m-2}$, the Bayes risk for the optimal procedure from that point is found. Continuing backwards in this way, the Bayes risk for taking the first observation (X_1) is eventually found, together with the optimal procedure from that point. (It should be mentioned that the Bayes risk for taking the first observation can be greater than that for taking no observations, so sometimes the optimal strategy is to choose a terminal act without observing any data.) The following example illustrates the method. De Groot (1970) discusses backward induction in detail and gives further examples.

Example 6.8

Suppose up to three independent observations may be taken sequentially and each observation is either a success, with probability θ, or a failure, with probability $1-\theta$. One of two simple hypotheses is true, $H_0: \theta = \frac{1}{4}$ or $H_1: \theta = \frac{1}{2}$, and the prior distribution gives $\Pr[H_0] = \Pr[H_1] = \frac{1}{2}$. A decision as to which hypothesis holds must be made, the loss being 30 for an incorrect decision and 0 for a correct one. Each observation costs 1 and the task is to determine the optimal procedure.

If X denotes the number of successes from n observations, then

$$\Pr[H_0 \mid X, n] = \frac{\Pr[X \mid H_0, n]\ \Pr[H_0]}{\Pr[X \mid H_0, n]\ \Pr[H_0] + \Pr[X \mid H_1, n]\ \Pr[H_1]}$$

$$= \frac{\binom{n}{X}(\frac{1}{4})^X(\frac{3}{4})^{n-X}\frac{1}{2}}{\binom{n}{X}(\frac{1}{4})^X(\frac{3}{4})^{n-X}\frac{1}{2} + \binom{n}{X}(\frac{1}{2})^n\frac{1}{2}} = \frac{3^{n-X}}{3^{n-X} + 2^n}$$

and

$$\Pr[H_1 \mid X, n] = \frac{2^n}{3^{n-X} + 2^n}.$$

Also, the expected risk from the terminal decision is $30[\min(3^{n-X}, 2^n)]/(3^{n-X} + 2^n)$. Consequently, if all three observations are taken then the Bayes risk, including the sampling costs of 3, are as follows:

X	0	1	2	3
Bayes risk	9.857	17.118	11.182	6.333

Suppose now that exactly two observations have been taken and 1 success was observed. From the above formula, the expected risk from the terminal decision, if sampling is stopped, is 12.857 and the Bayes risk, with sampling costs, is 14.857. If a further observation were taken, the probability of a success is

$$\tfrac{1}{4}\Pr[H_0 \mid X = 1, n = 2] + \tfrac{1}{2}\Pr[H_1 \mid X = 1, n = 2] = \tfrac{1}{4}(\tfrac{3}{7}) + \tfrac{1}{2}(\tfrac{4}{7}) = 0.3929.$$

Thus, if sampling is continued, we reach 2 successes from 3 observations with probability 0.3929, and 1 success from 3 observations with probability $(1 - 0.3929)$. Hence the Bayes risk from continued sampling is

$$(0.3929)11.182 + (1 - 0.3929)17.118 = 14.786.$$

This is just less than the Bayes risk from making a decision immediately (14.857), so if 2 observations are taken and they give exactly one success, then a third observation should be taken and the expected risk is 14.786.

On the other hand, suppose two observations have been taken and two successes were observed. Similar calculations show that if sampling stops the Bayes risk, with sampling costs, is 8.0, while the Bayes risk if sampling is continued is

$$\{\tfrac{1}{4}\Pr[H_0 \mid X = 2, n = 2] + \tfrac{1}{2}\Pr[H_1 \mid X = 2, n = 2]\}6.333$$

$$+ \{1 - \tfrac{1}{4}\Pr[H_0 \mid X = 2, n = 2] - \tfrac{1}{2}\Pr[H_1 \mid X_2, n = 2]\}11.182$$

$$= (0.45)6.333 + (0.55)11.182 = 9.0.$$

Hence sampling should not be continued if two successes from two observations are observed, with a Bayes risk of 8.0. It can similarly be shown that if two observations are taken and no successes are observed, then sampling should be stopped with a Bayes risk of 11.23.

We have the following Bayes risks after X successes are obtained from two observations.

X	0	1	2
Bayes risk	11.23	14.786	8.0

The procedure can be continued to determine when to stop sampling after one observation has been taken and to determine whether the first observation should be taken. This is left as an exercise (Exercise 6.17).

When the number of potential observations is unbounded, backward induction cannot be used and various strategies have been proposed for approximating the optimal Bayes sequential procedure. An obvious strategy (called a *Bayes truncated procedure*) is to impose an upper bound on the number of observations that are taken, giving the case considered above. Another strategy (*the m-step look ahead procedure*) assumes at each stage that no more than a further m observations may be taken. Thus $n + m$ is treated as an upper bound for the number of observations when deciding whether to stop after observing values of $X_1, X_2, \ldots, X_n$. A third approach, called the *fixed sample size look ahead procedure*, compares the current Bayes risk from stopping with the expected risk from taking *exactly* m further observations. If, for every positive integer m, the current risk is always smaller then sampling is stopped. Otherwise the next observation is taken and the test for stopping is repeated. Berger (1985, Chapter 7) discusses these strategies and several others. The following example shows that the fixed-sample look-ahead procedure can be relatively easy to apply and can lead to a simple stopping rule.

Example 6.9

Suppose observations $X_1, X_2, \ldots$ can be taken sequentially from a Poisson distribution with mean θ and the task is to estimate θ under the quadratic loss function, loss$=(\theta - \hat{\theta})^2$. Assume that the prior distribution for θ is a gamma distribution

$$p(\theta) \propto \theta^{\alpha_0 - 1}\, e^{-\beta_0 \theta}, \qquad \theta > 0.$$

After taking n observations whose values are $\mathbf{x}_n = x_1, x_2, \ldots, x_n$, the posterior distribution is given by

$$\begin{aligned} q(\theta; \mathbf{x}_n) &\propto L(\theta; \mathbf{x}_n)p(\theta) \\ &\propto \left\{ \sum_{i=1}^{n} e^{-\theta}\theta^{x_i} \right\} \theta^{\alpha_0 - 1}\, e^{-\beta_0 \theta} \\ &= \theta^{\alpha_1 - 1}\, e^{-\beta_1 \theta} \end{aligned}$$

where $\alpha_1 = \alpha_0 + \Sigma x_i$ and $\beta_1 = \beta_0 + n$. Thus the posterior distribution is also a gamma distribution. If sampling is stopped at this point, the Bayes estimate of θ is the posterior mean, α_1/β_1, and the expected quadratic loss equals the posterior variance, α_1/β_1^2. Taking sampling costs into account, the Bayes risk from stopping, say $B_n(0)$, is

$$B_n(0) = \alpha_1/\beta_1^2 + \sum_{i=1}^{n} c_i.$$

Suppose we instead decide to take exactly m further observations. The Bayes risk from this strategy, say $B_n(m)$, is

$$B_n(m) = \mathrm{E}\left[\left(\alpha_1 + \sum_{i=n+1}^{n+m} X_i\right)/(\beta_1 + m)^2 \,|\, \mathbf{x}_n\right] + \sum_{i=1}^{n+m} c_i$$

$$= \left\{\alpha_1 + \sum_{i=n+1}^{n+m} \mathrm{E}[X_i \,|\, \mathbf{x}_n]\right\}/(\beta_1 + m)^2 + \sum_{i=1}^{n+m} c_i.$$

Now $\mathrm{E}[X_i \,|\, \theta] = \theta$ and $\mathrm{E}[\theta \,|\, \mathbf{x}_n] = \alpha_1/\beta_1$, so after observing the first n observations, $\mathrm{E}[X_i \,|\, \mathbf{x}_n] = \alpha_1/\beta_1$ for $i = n + 1, n + 2, \ldots$. Thus

$$B_n(m) = \{\alpha_1 + m\alpha_1/\beta_1\}/(\beta_1 + m)^2 + \sum_{i=1}^{n+m} c_i.$$

This gives

$$B_n(m) - B_n(0) = \sum_{i=n+1}^{n+m} c_i - \alpha_1 m/\{\beta_1^2(\beta_1 + m)\}.$$

The fixed-sample look-ahead procedure is to stop when $B_n(m) - B_n(0) > 0$ for every positive integer m. In particular, if each observation costs the same amount, say c, the rule is to stop after observing $\mathbf{x}_n$ if and only if $c > \alpha_1/\{\beta_1^2(\beta_1 + 1)\}$.

6.7.2 *Sequential probability ratio test*

The sequential probability ratio test (SPRT) can be regarded as both a frequentist (non-Bayesian) procedure and a Bayesian procedure, and we first describe it as a frequentist test. We again have a stream of observations $X_1, X_2, \ldots$ and now assume that these are independently and identically distributed with p.d.f. $f(x_i; \theta)$. One of two simple hypotheses is true, H_0: $\theta = \theta_0$ or H_1: $\theta = \theta_1$, and the task is to choose between them. The SPRT is based upon the likelihood ratio

$$LR(n) = \frac{L(\theta_1; \mathbf{x}_n)}{L(\theta_0; \mathbf{x}_n)} = \frac{\prod_{i=1}^{n} f(x_i; \theta_1)}{\prod_{i=1}^{n} f(x_i; \theta_0)}.$$

The test procedure specifies two numbers, say K_1 and K_2 that do not depend on n, with $0 < K_1 < K_2 < \infty$. After the nth observation, $LR(n)$ is calculated and one of the following actions taken.

(1) If $LR(n) \leqslant K_1$, then sampling is stopped and H_0 is accepted.
(2) If $LR(n) \geqslant K_2$, then sampling is stopped and H_1 is accepted.
(3) If $K_1 < LR(n) < K_2$, then another observation is taken.

This defines the test from a frequentist approach.

Now suppose that each observation costs the same, say c ($c > 0$), and the loss structure is that given in Table 6.1, i.e. the loss is a for incorrectly rejecting H_0 and b for incorrectly accepting H_0, with no loss for a correct decision. Then for appropriate choice of K_1 and K_2, the above SPRT is a Bayes procedure, provided that the option of taking no data and immediately choosing between H_0 and H_1 is allowed. Most of this section is used to prove this equivalence.

At any stage in the sequential sampling, the quantities relevant to whether sampling should be continued or stopped (and an immediate decision made) are:

(a) the cost (c) of each future observation;
(b) the losses (a and b) from incorrect decisions;
(c) the information about θ that future observations will provide;
(d) the current information about θ.

(Whether to stop does not depend on the number of observations already taken, since the cost of those observations is incurred regardless of whether or not one now stops sampling.) Obviously (a) and (b) do not depend on the number of observations already taken or their values. Nor does (c), since the information about θ will be given by the likelihood from the future observations, and these observations are independent of those already obtained. Only (d) depends upon the data obtained so far, say $\mathbf{x}$. However, θ can take one of only two values, θ_0 or θ_1, depending on whether H_0 or H_1 holds. If we put $p = \Pr[H_0 | \mathbf{x}]$ and $1 - p = \Pr[H_1 | \mathbf{x}]$, where $\Pr[H_0 | \mathbf{x}]$ and $\Pr[H_1 | \mathbf{x}]$ are the current probabilities that H_0 and H_1 are true, then information about θ is completely summarized by the value of p. Consequently, whether sampling should be continued depends only on the current value of p and on quantities that do not change as data are gathered.

We next consider the relationship between the value of p and whether sampling should be stopped immediately. Costs of observations already taken will be ignored in evaluating losses and risks, since they are incurred by all available procedures. Suppose first that sampling is stopped; then the expected losses are $b(1 - p)$ from choosing H_0 and ap from choosing H_1. Hence, if sampling stops, H_0 is chosen if $p > b/(a + b)$ and H_1 if $p < b/(a + b)$, and the Bayes risk for this decision rule is given by the solid line in Figure 6.3.

Suppose instead that sampling is continued and let δ be any decision that takes at least one more observation. (For any additional sample size $m \geqslant 1$, δ specifies the action to take for each possible combination of values of the next m observations.) We write $B(\delta; p)$ for the Bayes risk of δ, to show its dependence on $p = \Pr[H_0 | \mathbf{x}]$. Define

$$B^*(p) = \min_{\delta} (\delta; p)$$

so $B^*(p)$ is the minimum risk for procedures that take at least one more observation. Clearly $B^*(p) \geqslant c$, the cost of the next observation. Also, $B^*(0) = B^*(1) = c$ since the correct hypothesis will be chosen with certainty when p equals 0 or 1. We next show

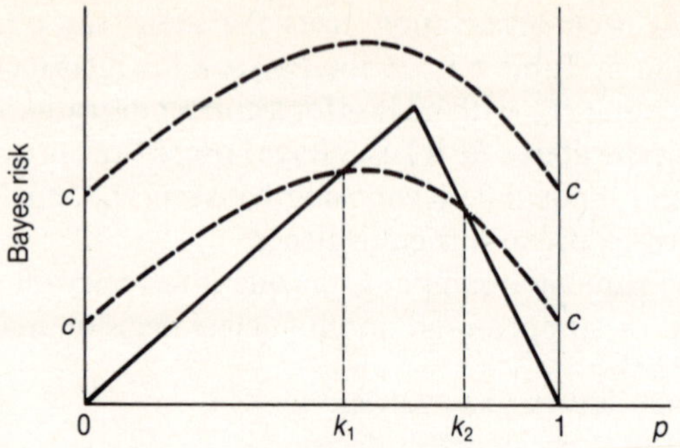

Fig. 6.3 Bayes risks for two different procedures.

that $B^*(p)$ is a concave function and hence has a shape similar to the dotted lines in Figure 6.3. Now

$$B(\delta;\, p) = pR(\delta;\, \theta_0) + (1 - p)R(\delta;\, \theta_1)$$

where $R(\delta;\, \theta_i)$ is δ's risk when $\theta = \theta_i$, $i = 0, 1$. Thus if λ, p_1 and p_2 are any three values from the interval [0, 1],

$$\begin{aligned} B(\delta;\, \lambda p_1 + (1-\lambda)p_2) &= \{\lambda p_1 + (1-\lambda)p_2\}R(\delta;\, \theta_0) \\ &\quad + \{1 - \lambda p_1 - (1-\lambda)p_2\}R(\delta;\, \theta_1) \\ &= \lambda\{p_1 R(\delta;\, \theta_0) + (1-p_1)R(\delta;\, \theta_1)\} + (1-\lambda)\{p_2 R(\delta;\, \theta_0) + (1-p_2)R(\delta;\, \theta_1)\} \\ &= \lambda B(\delta;\, p_1) + (1-\lambda)B(\delta;\, p_2). \end{aligned}$$

Hence

$$\begin{aligned} B^*(\lambda p_1 + (1-\lambda)p_2) &= \min_{\delta}[\lambda B(\delta;\, p_1) + (1-\lambda)B(\delta;\, p_2)] \\ &\geqslant \lambda \min_{\delta}[B(\delta;\, p_1)] + (1-\lambda)\min_{\delta}[B(\delta;\, p_2)] \\ &= \lambda B^*(p_1) + (1-\lambda)B^*(p_2), \end{aligned}$$

which establishes that $B^*(p)$ is a concave function of p.

We can now state the relationship between p and whether sampling should stop. As mentioned before, the solid line in Figure 6.3 is the risk if sampling is stopped and a decision made. The curve for $B^*(p)$ will either (1) always be above this line or (2) will intersect it at two points. (This is because $B^*(p)$ is concave with $B^*(0) = B^*(1) = c$.) If (1) happens then sampling should stop for any value of p, as the aim is to minimize expected risk. If (2) happens then sampling should continue only if $k_2 < p < k_1$, where k_1 and k_2 are the points of intersection, as indicated in Figure 6.3. Thus the Bayes procedure may be expressed as

(1) stop sampling and accept H_0 if $p \geqslant k_1$;
(2) stop sampling and accept H_1 if $p \leqslant k_2$;
(3) continue sampling if $k_2 < p < k_1$;

where we put $k_1 = k_2 = b/(a + b)$ if there are no points of intersection.

To establish the equivalence of this procedure to the SPRT (given at the start of this section), note that

$$p = \Pr[H_0 \mid \mathbf{x}] = \frac{f(\mathbf{x} \mid H_0)\ \Pr[H_0]}{f(\mathbf{x} \mid H_0)\ \Pr[H_0] + f(\mathbf{x} \mid H_1)\ \Pr[H_1]}$$

where $\Pr[H_0]$ and $\Pr[H_1]$ are the probabilities as to which hypothesis is true before observing any data. Hence,

$$p = \frac{\Pr[H_0]}{\Pr[H_0] + LR(n)\ \Pr[H_1]}$$

since $LR(n)$, the likelihood ratio, equals $\Pi f(x_i;\, \theta_1)/\Pi f(x_i;\, \theta_0)$. The Bayes procedure continues sampling if

$$\begin{aligned} k_2 < p < k_1 &\equiv \frac{1}{k_1} < \frac{1}{p} < \frac{1}{k_2} \\ &\equiv \frac{1}{k_1} < 1 + LR(n)\frac{\Pr[H_1]}{\Pr[H_0]} < \frac{1}{k_2} \\ &\equiv \left(\frac{1 - k_1}{k_1}\right)\frac{\Pr[H_0]}{\Pr[H_1]} < LR(n) < \left(\frac{1 - k_2}{k_2}\right)\frac{\Pr[H_0]}{\Pr[H_1]}. \end{aligned}$$

Putting $K_1 = (1 - k_1)\Pr[H_0]/\{k_1 \Pr[H_1]\}$ and $K_2 = (1 - k_2)\Pr[H_0]/\{k_1 \Pr[H_1]\}$ gives the SPRT rule for sampling to continue. For this choice of K_1 and K_2, the Bayes procedure and the SPRT also agree as to which of H_0 and H_1 should be accepted when sampling is stopped. Hence the SPRT is the Bayes procedure for appropriate choice of K_1 and K_2.

The SPRT largely stems from the work of Wald (1947) and a good description of it is given from a Bayesian perspective by Berger (1985). We now summarize some of its properties; it is helpful to introduce the variable

$$Z = \ln\left[\frac{f(X \mid H_1)}{f(X \mid H_0)}\right].$$

Now $f(x \mid H_0)$ and $f(x \mid H_1)$ may be equal for some x values, and then $Z = 0$. For H_0 and H_1 to be distinguishable, we require $\Pr[Z = 0] < 1$. The first property relates to the number of observations taken before the test terminates, say N, and states that sampling stops with probability 1 (clearly desirable) and that all moments of N are finite.

Property 6.1

If $\Pr[Z = 0] < 1$, then (a) $\Pr[N < \infty] = 1$ and (b) $E[N^s] < \infty$ for all $s = 1, 2, \ldots$.

Approximate values of $E[N \mid H_i]$ are given by the following property and are useful for estimating the cost of sampling, and for comparing the SPRT with tests of fixed sample size.

Property 6.2

Let $\mu_0 = E[Z \mid H_0]$ and $\mu_1 = E[Z \mid H_1]$. Then, approximately,

$$E[N \mid H_0] \approx \frac{(K_2 - 1)\ln(K_1) + (1 - K_1)\ln(K_2)}{(K_2 - K_1)\mu_0}$$

and

$$E[N \mid H_1] \approx \frac{K_1(K_2 - 1)\ln(K_1) + K_2(1 - K_1)\ln(K_2)}{(K_2 - K_1)\mu_1}.$$

The above properties hold when the SPRT is considered as either a Bayesian procedure or as a frequentist test (proofs of the properties may be found, for example, in De Groot (1970, pp. 311–14)). Differences between Bayesian and non-Bayesian methods arise, however, in choosing appropriate values for K_1 and K_2.

The frequentist criteria for choosing K_1 and K_2 is that error rates should have specific sizes. Suppose the desired sizes are $\alpha = \Pr[\text{reject } H_0 \mid H_0]$ and $\beta = \Pr[\text{reject } H_1 \mid H_1]$. Then the following property gives the usual choices of K_1 and K_2.

Property 6.3

Given α and β, put $K_1 = \beta/(1 - \alpha)$ and $K_2 = (1 - \beta)/\alpha$. Then

$$\Pr[\text{reject } H_0 \mid H_0] \approx \alpha \text{ and } \Pr[\text{reject } H_1 \mid H_1] \approx \beta.$$

Proof

The following is a heuristic proof. (For a rigorous proof, see Wald 1947, pp. 44–48.) For a specified K_1 and K_2, let $\alpha' = \Pr[\text{reject } H_0 \mid H_0]$ and $\beta' = \Pr[\text{reject } H_1 \mid H_1]$. From Property 6.1, the test terminates with probability 1 so $1 - \beta' = \Pr[\text{reject } H_0 \mid H_1]$. Hence, if C is the rejection region of H_0,

$$\int_{\mathbf{x} \in C} f(\mathbf{x} \mid H_0)\,d\mathbf{x} = \alpha' \quad \text{and} \quad \int_{\mathbf{x} \in C} f(\mathbf{x} \mid H_1)\,d\mathbf{x} = 1 - \beta'.$$

Also, for any $\mathbf{x} \in C$,

$$\frac{f(\mathbf{x} \mid H_1)}{f(\mathbf{x} \mid H_0)} \geqslant K_2$$

since this defines the rejection region of H_0. Consequently, $(1 - \beta')/\alpha' \geqslant K_2$. Similar arguments imply that $\beta'/(1 - \alpha') \leqslant K_1$.

Usually the SPRT test statistic gradually approaches a threshold, K_1 or K_2, and does not overshoot the threshold by very much when the test stops. Hence, when it stops, $f(\mathbf{x} \mid H_1)/f(\mathbf{x} \mid H_0)$ approximately equals K_2 if H_0 is rejected, or K_1 if H_1 is rejected. Thus $(1 - \beta')/\alpha' \approx K_2$ and $\beta'/(1 - \alpha') \approx K_1$, and Property 6.3 follows. □

The Bayesian criterion for choosing K_1 and K_2 is that the Bayes risk should be minimized. If H_0 holds, the expected cost of sampling is $c\ \mathrm{E}(N \mid H_0)$ and the expected loss from the possibility of rejecting H_0 is αa, so the total expected loss is $\alpha a + c\ \mathrm{E}(N \mid H_0)$. Similarly, if H_1 holds the expected loss is $\beta b + c\ \mathrm{E}(N \mid H_1)$. Thus the Bayes risk is

$$\Pr[H_0]\{\alpha a + c\ \mathrm{E}[N \mid H_0]\} + \Pr[H_1]\{\beta b + c\ \mathrm{E}[N \mid H_1]\}.$$

From Property 6.3, $\alpha \approx (1 - K_1)/(K_2 - K_1)$ and $\beta \approx K_1(K_2 - 1)/(K_2 - K_1)$. Using these approximations with Property 6.2. leads to

Property 6.4

The Bayes risk of the SPRT is approximately

$$\Pr[H_0]\left[\frac{(1 - K_1)a}{K_2 - K_1} + c\left\{\frac{(K_2 - 1)\ \ln K_1 + (1 - K_1)\ \ln K_2}{(K_2 - K_1)\mu_0}\right\}\right]$$

$$+ \Pr[H_1]\left[\frac{K_1(K_2 - 1)b}{K_2 - K_1} + c\left\{\frac{K_1(K_2 - 1)\ \ln K_1 + K_2(1 - K_1)\ \ln K_2}{(K_2 - K_1)\mu_1}\right\}\right].$$

Values of K_1 and K_2 that minimize the above approximation to the Bayes risk can be determined relatively easily using numerical methods.

When the cost of each observation (c) is small, a large number of observations are usually taken. The consequence is that K_1 will be small and K_2 large. It can then be shown (Exercise 6.20) that

$$K_1 \approx \frac{-\Pr[H_0]c}{\Pr[H_1]b\mu_0} \quad \text{and} \quad K_2 \approx \frac{\Pr[H_0]a\mu_1}{\Pr[H_1]c}.$$

The SPRT tends to overshoot the thresholds, K_1 and K_2, when it stops. That is, the test stops with $LR(n) < K_1$ or $LR(n) > K_2$, and not $LR(n) = K_1$ or $LR(n) = K_2$. It is because of the 'overshoot' that results given in Properties 6.2–6.4 are approximations, rather than exact equalities. The approximations will usually be quite accurate as the overshoot is generally small.

Exercises

6.1 Suppose $\theta_1, \theta_2, \ldots, \theta_k$ are the only possible values of θ and that $p(\theta_i) > 0$ for $i = 1, 2, \ldots, k$ (k finite), where $p(\theta)$ is the prior distribution. Show that the Bayes estimator is admissible.

6.2 Suppose θ^* is the Bayes estimator of θ both if $p_1(\theta)$ is the prior distribution and if $p_2(\theta)$ is the prior distribution. Show that θ^* is then also the Bayes estimator for the prior distribution $p(\theta) = \alpha p_1(\theta) + (1 - \alpha)p_2(\theta)$, $0 \leqslant \alpha \leqslant 1$.

6.3 A baker has to decide how many loaves to bake tomorrow. If he bakes too many, he loses k_1 pence for each loaf he has left. If he bakes too few, he loses k_2 pence in profit on each loaf that he could have sold. Suppose his opinion about θ, the number of loaves he could sell tomorrow (if he does not run out) is represented by the probability density function $f(\theta)$ whose distribution function is $F(\theta)$. Show that $\hat{\theta}$, the number of loaves he should bake tomorrow, satisfies $F(\hat{\theta}) = k_2/(k_1 + k_2)$ if $k_2/(k_1 + k_2)$ is an integer.

6.4 Evaluate the following integrals by comparing them to standard distributions

(a) $\int_0^\infty \frac{\Gamma(\alpha)}{\Gamma(\alpha + \beta)} x^\alpha \, e^{-\beta x + 3} \, dx$

(b) $\int_0^1 15x^{10}(1 - x)^5 \, dx$.

6.5 Suppose X is an observation from the distribution

$$f(x; \theta) = (x - 1)\theta^2(1 - \theta)^{x-2}, \qquad x = 2, 3, \ldots; \quad 0 < \theta < 1.$$

The prior distribution for θ is a Beta distribution:

$$p(\theta) = \frac{\Gamma(7)}{\Gamma(4)\Gamma(3)} \theta^3(1 - \theta)^2, \qquad 0 \leqslant \theta \leqslant 1$$

and the loss from estimating θ by $\hat{\theta}$ is $(\theta - \hat{\theta})^2$.

(a) Find the posterior distribution, $q(\theta; x)$.

(b) Show that the Bayes estimator of θ is $6/(7 + x)$.

(c) Find the Bayes risk associated with this estimator.

6.6 Let $X_1, X_2, \ldots, X_n$ be a random sample from $N(\theta, \sigma^2)$, σ^2 known, and let the prior distribution be $\theta \sim N(\phi, \tau^2)$. Suppose θ is to be estimated under a quadratic loss function.

(a) State the Bayes estimator of θ and show that this estimator has constant risk as $\tau \to \infty$.

(b) Deduce the minimax estimator of θ and state its risk.

6.7 Suppose $\theta > 0$ and the prior distribution for θ is to be modelled by a gamma distribution. Give the distribution if the prior mean is 20 and the prior standard deviation is 10.

6.8 Suppose θ is the probability of 'success' in a trial and the prior distribution for θ is to be modelled by a beta distribution. Before observing any trials, the prior mean of θ is 0.4 while, after observing ten independent trials in which there are seven successes, the posterior mean of θ is 0.5. Determine the prior distribution (before observing the ten trials).

6.9 The time to failure of a particular type of component follows an exponential distribution with mean $1/\theta$. In a trial, the times to failure (in hours) of five components were 4, 1, 3, 3 and 6, and a sixth component was still working after 10 h when the trial was terminated. Prior to the trial, opinion about the value of θ corresponded to the gamma distribution:

$$p(\theta) = \tfrac{1}{2}\theta^2 \, e^{-\theta}, \qquad \theta > 0.$$

Show that the posterior distribution of θ is

$$q(\theta \mid \text{data}) = \frac{28^8}{7!} \theta^7 \, e^{-28\theta}, \qquad \theta > 0.$$

6.10 Suppose $X_1, X_2, \ldots, X_n$ are random independent observations from a Poisson distribution with unknown mean θ.

(a) Give the likelihood for θ.

(b) Form the conjugate prior distribution and identify it.

(c) Determine the posterior distribution.

(d) State the mean and variance of the prior distribution and the prior expectation of the variance of $\overline{X} = \Sigma X_i/n$. Show that the posterior mean is a weighted average of the prior mean and the sample mean, with weights proportional to the reciprocals of the prior variance and $\mathrm{E}[\mathrm{Var}(\overline{X})]$.

(e) Suggest an uninformative prior distribution for θ.

6.11 A large population must be screened for the presence of a certain antibody in the blood. An infallible test on a blood sample from a single person costs c_1 and gives a positive result if the antibody is present and a negative result otherwise. If a batch of blood samples from n people are mixed together and then tested, the test costs $c_1 + c_2 n$ and gives a negative result only if none of the n samples contained the antibody. If a positive result is obtained all n samples then have to be tested individually. The proportion θ of people who have the antibody has prior distribution

$$p(\theta) = \beta(1-\theta)^{\beta-1} \quad \text{with } \beta \text{ large.}$$

(a) Show that individual testing gives a lower expected cost than testing in batches of size n if

$$c_2 > c_1\left(\frac{\beta}{n+\beta} - \frac{1}{n}\right).$$

(b) Show that if batch testing is adopted, the optimal value of n is approximately $\sqrt{\beta}$.

(c) As screening goes along, suggest how increasing knowledge about θ can be used to modify the value of n used in successive screening tests.

6.12 Suppose we have a single observation, x_1, which comes from a distribution with p.d.f. $f(x)$, and we want to test

$$H_0: f(x) = 3(1-x)^2, \qquad 0 \leqslant x \leqslant 1$$

against

$$H_1: f(x) = 2x, \qquad 0 \leqslant x \leqslant 1.$$

Show that the best critical region for the test of H_0 vs. H_1 is given by $x_1 \geqslant B$ for some constant B. Show also that this test is unbiased for any value of B. Given that the losses incurred when Type I and Type II errors occur are equal, find the values of B which give

(a) the minimax procedure;

(b) the Bayes procedure, when the prior probabilities are $\frac{1}{4}, \frac{3}{4}$ for H_0 and H_1, respectively.

Find the values of the Type I and Type II errors in (a) and (b), and find also the prior probabilities of H_0 and H_1 for which the Bayes procedure is equivalent to the minimax procedure.

6.13 A random sample $X_1, X_2, \ldots, X_n$ $(n \geqslant 1)$ is to be taken from a uniform distribution whose range is $(0, \theta)$. One of the hypotheses, $H_0: \theta = 1$ or $H_1: \theta = 3$, holds and the prior probabilities of these hypotheses are $\frac{1}{5}$ and $\frac{4}{5}$, respectively. On the basis of a sample of size n, H_0 must be accepted or rejected. There is zero loss for a correct decision, a loss of 1 if H_0 is incorrectly rejected and a loss of 10 if H_0 is incorrectly accepted.

(a) Show that for $n \leqslant 3$, the Bayes test always rejects H_0 while, for $n \geqslant 4$, H_0 is accepted if no observation in the range (1, 3) is observed.
(b) For each n, determine the expected risk for the Bayes test.

6.14 Two outwardly identical measuring instruments M_1 and M_2 have normally distributed errors with variance σ_1^2 and σ_2^2 respectively, where $\sigma_1^2 < \sigma_2^2$. A technician brings you one of the instruments, but can shed no light on which instrument he has brought. You therefore make n observations $x_1, x_2, \ldots, x_n$ of a fixed quantity θ and calculate $U = \Sigma(x_i - \bar{x})^2$. Suppose the loss from incorrectly deciding the instrument is M_1 equals the loss from incorrectly deciding it is M_2. Show that you will decide it is M_2 if and only if

$$U > 2(n-1)(\ln(\sigma_2) - \ln(\sigma_1))/(\sigma_1^{-2} - \sigma_2^{-2}).$$

An alternative approach to the problem is to argue on intuitive grounds that the decision rule should take the form: decide it is M_2 if $U > k$, and choose k to minimize the probability of making a mistake. Show this leads to the same rule as above. (Note $U/\sigma_i^2 \sim \chi^2_{(n-1)}$ if M_i is the instrument being used.)

6.15 Consider a binomial experiment with two trials, where H_0: $p = 0.2$ is to be tested against H_1: $p = 0.7$. Suppose the loss is 0 for a correct decision and 5 for an incorrect decision (wrongly rejecting H_0 or wrongly accepting H_0). Let p_0 be the prior probability that H_0 is true. State the possible critical regions and, for each region, determine the range of values for p_0 for which that region is the critical region. If $p_0 = 0.4$, what is the expected risk for the Bayes test?

6.16 A stream of observations $X_1, X_2, \ldots$ may be taken one at a time at a cost of c for each observation. The X_i are i.i.d. with $X_i \sim N(\theta, \sigma^2)$, σ^2 known, and the prior distribution for θ is $\theta \sim N(\phi, \tau^2)$. The value of θ must be estimated under the quadratic loss function, $L_S(\theta, \hat{\theta}) = (\hat{\theta} - \theta)^2$, where $\hat{\theta}$ is the estimate of θ.

Suppose n observations, $x_1, x_2, \ldots, x_n$, are taken.
(a) Give the posterior distribution of θ and the Bayes risk (including sampling costs) from stopping at this stage.
(b) Hence show that sampling should stop when

$$n \approx \frac{\sigma}{\sqrt{c}} - \frac{\sigma^2}{\tau}.$$

(Notice that we know how many observations to take before sampling starts, because the Bayes risk does not depend on their values.)

6.17 Continue Example 6.8 and determine
(a) when sampling should be continued after one observation has been taken;
(b) whether the first observation should be taken.

6.18 Consider the problem presented in Exercise 6.13. If $x_i \in (0, 1)$ for $i = 1, 2, \ldots, n$ and $\mathbf{x} = (x_1, x_2, \ldots, x_n)^{\mathrm{T}}$, show that

$$\Pr[H_1 \mid \mathbf{x}] = 4(\tfrac{1}{3})^n/\{1 + 4(\tfrac{1}{3})^n\}.$$

Suppose a fixed-size sample is not taken but, instead, observations are taken sequentially at a cost of c for each observation, where c is small and positive. Show that the fixed-sample look-ahead procedure stops when either
(a) an observation in the range (1, 3) is obtained (in which case H_1 is accepted), or

(b) n^* observations in the range (0, 1) are obtained, where n^* is the smallest integer for which

$$c > 80/(12 + 3^{n^*+1})$$

(in which case H_0 is accepted).

6.19 Suppose $X_1, X_2, \ldots$ is a sequential sample of i.i.d. observations with

$$f(x; \theta) = \frac{1}{x \ln(\theta)} \left(\frac{\theta - 1}{\theta}\right)^x, \qquad x = 1, 2, \ldots$$

and we wish to test H_0: $\theta = 2$ against H_1: $\theta = 4$.

(a) From Property 6.3, find an SPRT for which the probabilities of incorrectly rejecting H_0 and incorrectly rejecting H_1 are both 0.1.

(b) For this SPRT, determine the expected number of observations that will be taken if H_0 is true and if H_1 is true.

6.20 Suppose $1/K_1$ and K_2 are both large and of similar orders of magnitude. Show that the Bayes risk given in Property 6.4 is then minimized when $K_1 \approx -\Pr[H_0]c/\{\Pr[H_1]b\mu_0\}$ and $K_2 \approx \Pr[H_0]a\mu_1/\{\Pr[H_1]c\}$.

6.21 Suppose $X_1, X_2, \ldots$ is a sequential sample from a Poisson distribution with mean θ and H_0: $\theta = 2$ is to be tested against H_1: $\theta = 3$. The prior probabilities are $\frac{3}{4}$ that $\theta = 2$ and $\frac{1}{4}$ that $\theta = 3$. The loss in incorrectly accepting H_0 is 50, from incorrectly accepting H_1 is 25, and there is no loss from a correct decision. The cost of each observation is $c = 0.001$. Using the approximations in Exercise 6.20, determine

(a) K_1 and K_2 for the Bayes SPRT,

(b) the Bayes risk for this SPRT.

Chapter Seven

Bayesian inference

7.1 Introduction

The previous chapter introduced some elements of Bayesian inference. These will be discussed further now, together with additional topics. In some cases the Bayesian approach and the frequentist (or classical) approach have clear similarities. For example, in the next section we give the Bayesian definition of sufficiency and show its equivalence to the frequentist definition. We also find that Bayesian interval estimates are often numerically equal to frequentist confidence intervals when an uninformative prior distribution is used. With other topics, such as hypothesis testing, we find marked differences between the two approaches.

The form of the prior distribution plays an important role in much of this chapter. Improper prior distributions can cause problems in hypothesis testing. Hierarchical prior distributions (introduced in Section 7.7) can be used to model structural relationships between parameters, illustrating the flexibility that results from treating unknown parameters as random variables. Empirical Bayes methods are also discussed, in which past data are used to construct a prior distribution.

7.2 Sufficiency

The Bayesian definition of sufficiency is based on the distribution of θ but, as we will show, it is equivalent to the frequentist definition given earlier. The notation used is the same as in Chapter 6.

Definition 7.1

A statistic $T(X_1, X_2, \ldots, X_n)$ is sufficient for $\boldsymbol{\theta}$ if and only if the posterior distribution of $\boldsymbol{\theta}$ given $X_1, X_2, \ldots, X_n$ is the same as the posterior distribution of $\boldsymbol{\theta}$ given T.

Theorem 7.1 Definitions 2.5 and 7.1 are equivalent.

Proof

Suppose that T satisfies Definition 2.5. Then, leaving out terms which do not depend on $\boldsymbol{\theta}$, we have

$$f(\mathbf{x};\,\boldsymbol{\theta}) = g(\mathbf{x}\,|\,t,\,\boldsymbol{\theta})h(t\,|\,\boldsymbol{\theta}) \propto h(t\,|\,\boldsymbol{\theta})$$

because g does not depend on $\boldsymbol{\theta}$.

Hence

$$q(\boldsymbol{\theta};\,\mathbf{x}) \propto f(\mathbf{x};\,\boldsymbol{\theta})p(\boldsymbol{\theta}) \propto h(t\,|\,\boldsymbol{\theta})p(\boldsymbol{\theta}) \propto q(\boldsymbol{\theta};\,t)$$

where $q(\boldsymbol{\theta};\mathbf{x})$ and $q(\boldsymbol{\theta};t)$ are the posterior distributions for $\boldsymbol{\theta}$ given $\mathbf{x}$ and t, respectively. Because $q(\boldsymbol{\theta};\,\mathbf{x})$ and $q(\boldsymbol{\theta};\,t)$ are both p.d.f.s for $\boldsymbol{\theta}$, and must therefore both integrate to 1, proportionality of the two functions implies identity. Thus Definition 7.1 is satisfied.

Now assume that T is sufficient according to Definition 7.1. We have that

$$f(\mathbf{x};\,\boldsymbol{\theta}) = \frac{q(\boldsymbol{\theta};\,\mathbf{x})h(\mathbf{x})}{p(\boldsymbol{\theta})} = \frac{q(\boldsymbol{\theta};\,t)h(\mathbf{x})}{p(\boldsymbol{\theta})}$$

$$= K_1[t\,|\,\boldsymbol{\theta}]K_2[\mathbf{x}]$$

where

$$K_1[t\,|\,\boldsymbol{\theta}] = \frac{q(\boldsymbol{\theta};\,t)}{p(\boldsymbol{\theta})} \text{ and } K_2[\mathbf{x}] = h(\mathbf{x}).$$

It follows from the Factorization Theorem 2.1 that T satisfies Definition 2.5. □

A Bayes procedure (Definition 6.4) depends on the posterior distribution for $\boldsymbol{\theta}$. Definition 7.1 implies that all the information regarding $\boldsymbol{\theta}$ which is contained in the posterior distribution depends on the data $\mathbf{x}$ only through the sufficient statistic T. In fact, it is clear from the earlier definition (Definition 2.1) of minimal sufficiency, and Theorem 7.1, that $q(\boldsymbol{\theta};\mathbf{x})$ depends on $\mathbf{x}$ only through the minimal sufficient statistic.

We saw in earlier chapters that sufficiency plays a central role in frequency-based inference. Definition 6.4, together with Theorem 7.1, shows that the same concept is important for Bayes procedures. Indeed, its importance is much wider than this; it has a crucial role to play throughout Bayesian inference.

One method of forming conjugate prior distributions is to express the part of the likelihood that involves $\boldsymbol{\theta}$ in terms of sufficient statistics. Then replace quantities dependent on the sample by prior parameters and multiply by a constant so that the resulting quantity integrates to 1. The following examples illustrate this.

Example 7.1

Let $X_1, X_2, \ldots, X_n$ be a random sample from a Poisson distribution with mean θ. Then

$$L(\theta;\,\mathbf{x}) = \theta^{\Sigma x_i}\,\mathrm{e}^{-n\theta}/\Pi(x_i!) \propto \theta^{\Sigma x_i}\,\mathrm{e}^{-n\theta}. \tag{7.1}$$

As $\mathbf{x}$ influences the posterior distribution of θ only through the likelihood, Σx_i is a sufficient statistic and the only other quantity dependent on the sample is n. Replacing Σx_i and n by parameters, say α_1 and α_2, gives

$$p(\theta) \propto \theta^{\alpha_1}\,\mathrm{e}^{-\alpha_2\theta}.$$

This is the form of a Gamma distribution (as was found in Section 6.4.2) so a conjugate prior distribution is

$$p(\theta) = \alpha_2^{(\alpha_1+1)}\theta^{\alpha_1}\,\mathrm{e}^{-\alpha_2\theta}/\Gamma(\alpha_1+1), \qquad \alpha_1 > -1, \quad \alpha_2 > 0.$$

Example 7.2

As an example where the sampling distribution is not a member of the exponential family, consider a random sample of size n from the uniform distribution, $U[0, \theta]$.

$$L(\theta; \mathbf{x}) = \begin{cases} \theta^{-n}, & 0 \leqslant x_{(1)} \leqslant x_{(2)} \leqslant \cdots \leqslant x_{(n)} \leqslant \theta \\ 0, & \text{elsewhere.} \end{cases}$$

Thus $x_{(n)}$ is a sufficient statistic, so we replace n and $x_{(n)}$ by parameters, α and β say. This gives

$$p(\theta) \propto \theta^{-\alpha}, \qquad \theta \geqslant \beta$$

which has the form of a Pareto distribution, $p(\theta) = (\alpha - 1)\beta^{\alpha-1}\theta^{-\alpha}$ for $\theta \geqslant \beta$ ($\beta > 0$, $\alpha > 1$). The posterior distribution is

$$\begin{aligned} q(\theta; \mathbf{x}) &\propto p(\theta)L(\theta; \mathbf{x}) \\ &= \theta^{-(\alpha+n)} \text{ for } \theta \geqslant \max[\beta, x_{(n)}]. \end{aligned}$$

This is again a Pareto distribution, demonstrating that $p(\theta)$ is a conjugate prior distribution.

The advantage of conjugate prior distributions is their mathematical convenience, but we would also like them to be as flexible as possible, so that they can model a wide variety of prior opinions. Thus, in the last two examples α_1, α_2 and α are not constrained to be integers, even though they replace statistics that are necessarily integer valued.

In similar vein, there are occasions where a sample statistic occurs in more than one place in the likelihood. Then in the prior distribution, the statistic is replaced by different parameters in the different places if mathematical convenience is not reduced. This is illustrated in the following example.

Example 7.3

$X_1, X_2, \ldots, X_n \sim N(\mu, \theta^{-1})$ with both μ and θ, the precision (= variance^{-1}), unknown.

$$\begin{aligned} L(\mu, \theta; \mathbf{x}) &= (2\pi)^{-n/2}\theta^{n/2} \exp\left[-\frac{\theta}{2}\Sigma(x_i - \mu)^2\right] \\ &\propto \theta^{n/2} \exp\left[-\frac{\theta}{2}\{r + n(\bar{x} - \mu)^2\}\right] \end{aligned} \tag{7.2}$$

where $r = \Sigma(x_i - \bar{x})^2$. The conjugate prior distribution is usually taken as

$$p(\mu, \theta) \propto \theta^{\nu_1/2} \exp\left[-\frac{\theta}{2}\{\alpha_1 + \nu_2(\alpha_2 - \mu)^2\}\right]$$

where, in equation (7.2), n has been replaced by v_1 in one place and by v_2 in another. This prior distribution is reasonably tractable, for it may be written as $p(\mu, \theta) = p_1(\mu \mid \theta)p_2(\theta)$, where

$$p_1(\mu \mid \theta) \propto \theta^{1/2} \exp\left[-\frac{\theta v_2}{2}(\mu - \alpha_2)^2\right]$$

and

$$p_2(\theta) \propto \theta^{(v_1 - 1)/2} \exp[-\theta\alpha_1/2].$$

Thus $p_1(\mu \mid \theta)$ is a normal distribution [mean $= \alpha_2$, variance $= (\theta v_2)^{-1}$] and $p_2(\theta)$ is a gamma distribution [parameters $(v_1 + 1)/2$ and $\alpha_1/2$]. For this reason $p(\mu, \theta)$ is often called a normal-gamma prior.

7.3 Credible intervals

We next consider region and interval estimates for a parameter. In Bayesian statistics, the region estimate is commonly termed a *credible region* or a *Bayesian confidence region.* It is defined as follows.

Definition 7.2

Suppose $\boldsymbol{\theta} \in \Omega$ and $S_{\mathbf{x}}$ is a subset of Ω. Then $S_{\mathbf{x}}$ is a $1 - \alpha$ credible region for $\boldsymbol{\theta}$ if

$$\int_{S_{\mathbf{x}}} q(\boldsymbol{\theta} \mid \mathbf{x}) \mathrm{d}\boldsymbol{\theta} = 1 - \alpha. \tag{7.3}$$

This definition states that $1 - \alpha$ is the probability that the *fixed* region $S_{\mathbf{x}}$ contains the *random* variable $\boldsymbol{\theta}$. This contrasts with the concept of a $1 - \alpha$ confidence region, $S_{\mathbf{x}}^*$ say, which states that $1 - \alpha$ is the probability that the random region $S_{\mathbf{x}}^*$ contains the fixed value of $\boldsymbol{\theta}$.

For any given α, generally there are many regions that have a credible level of $1 - \alpha$, so some criterion is needed to choose between them. We want to divide the possible values of $\boldsymbol{\theta}$ into a 'plausible' region and a 'less plausible' region, so it is natural to require

$$q(\boldsymbol{\theta}_1 \mid \mathbf{x}) \geqslant q(\boldsymbol{\theta}_2 \mid \mathbf{x}) \qquad \text{for all} \quad \boldsymbol{\theta}_1 \in S_{\mathbf{x}}, \quad \boldsymbol{\theta}_2 \notin S_{\mathbf{x}}.$$

Then any value of $\boldsymbol{\theta}$ included in $S_{\mathbf{x}}$ is at least as probable as any excluded value. A $1 - \alpha$ credible region that satisfies this requirement is called a $1 - \alpha$ *highest posterior density* (*HPD*) *credible region.*

An HPD interval might not be unique for some values of α if there are regions over which $q(\boldsymbol{\theta} \mid \mathbf{x})$ is constant. In Figure 7.1, the intervals (a, a'), (b, b'), (c, c') and infinitely many others are all 90% HPD regions, while (d, d') is the unique 99% HPD region.

An HPD credible region has the following desirable quality.

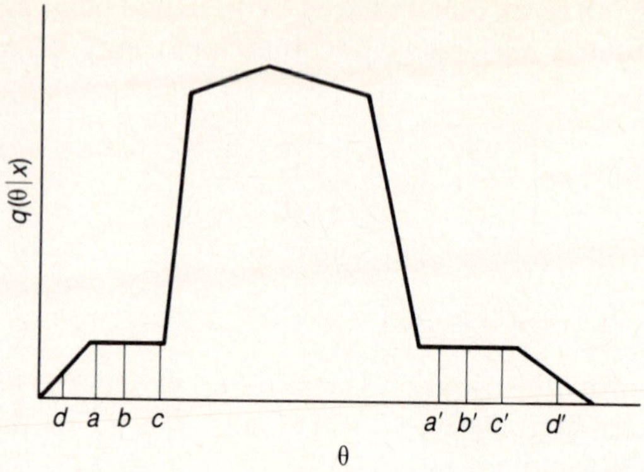

Fig. 7.1 HPD credible intervals.

Theorem 7.2

The hypervolume of a $1-\alpha$ HPD credible region is as small as that of any $1-\alpha$ credible region.

Proof

The proof is similar to the proof of the Neyman–Pearson lemma (Lemma 4.1). Let S_x be a $1-\alpha$ HPD credible interval and let S_x^* be any $1-\alpha$ credible interval. Then

$$1-\alpha = \int_{S_x} q(\boldsymbol{\theta}; \mathbf{x})d\boldsymbol{\theta} = \int_{S_x^*} q(\boldsymbol{\theta}; \mathbf{x})d\boldsymbol{\theta}.$$

Deleting the region common to S_x and S_x^* yields

$$\int_{\bar{S}_x^* \cap S_x} q(\boldsymbol{\theta}; \mathbf{x})d\theta = \int_{\bar{S}_x \cap S_x^*} q(\boldsymbol{\theta}; \mathbf{x})d\theta.$$

As S_x is an HPD credible region, $q(\boldsymbol{\theta}_1; \mathbf{x}) \geqslant q(\boldsymbol{\theta}_2; \mathbf{x})$ for all $\boldsymbol{\theta}_1 \in \bar{S}_x^* \cap S_x$, $\boldsymbol{\theta}_2 \in \bar{S}_x \cap S_x^*$. Thus

$$\text{hypervolume}(\bar{S}_x^* \cap S_x) \leqslant \text{hypervolume}(\bar{S}_x \cap S_x^*)$$

so, adding the hypervolume of $S_x \cap S_x^*$ to both sides,

$$\text{hypervolume}(S_x) \leqslant \text{hypervolume}(S_x^*).$$ □

When θ is a scalar, a credible region is often an interval on the real line and is then called a *credible interval.* For an HPD interval, the end points often satisfy $q(a; \mathbf{x}) = q(b; \mathbf{x})$, where (a, b) is the interval. This is the case, in particular, when $q(\theta; \mathbf{x})$ is unimodal and continuous over the real line.

An example where $q(\theta; \mathbf{x})$ is not unimodal is illustrated in Figure 7.2. Here $q(\theta; \mathbf{x})$ is truncated at a [$q(\theta; \mathbf{x}) = 0$ for $\theta < a$], $q(b) = q(c) = q(d)$ and

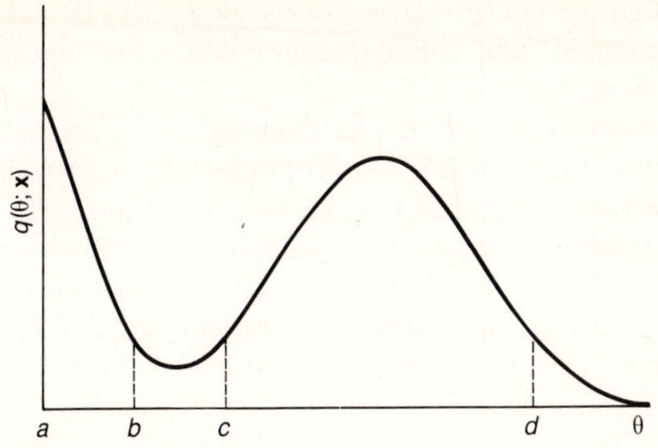

Fig. 7.2 Posterior probability distribution of θ.

$$\int_a^b q(\theta; \mathbf{x})d\theta + \int_c^d q(\theta; \mathbf{x})\, d\theta = 0.90.$$

From the diagram, the 90% HPD credible region for θ is the union of the intervals (a, b) and (c, d).

Example 7.4

Suppose $X_1, X_2, \ldots, X_n$ is a random sample from $N(\theta, \sigma^2)$, σ^2 known, and the prior distribution for θ is $\theta \sim N(\phi, \tau^2)$. From Example 6.3, the posterior distribution is normal with mean

$$\psi = \frac{\phi\sigma^2 + n\bar{x}\tau^2}{\sigma^2 + n\tau^2} \tag{7.4}$$

and variance

$$\gamma^2 = \left(\frac{\sigma^2 + n\tau^2}{\sigma^2\tau^2}\right)^{-1}. \tag{7.5}$$

Clearly

$$\int_{\psi - z_{\alpha/2}\gamma}^{\psi + z_{\alpha/2}\gamma} q(\theta; \mathbf{x})\, d\theta = 1 - \alpha,$$

where $q(\theta; \mathbf{x})$ is the posterior p.d.f. To verify that $\psi \pm z_{\alpha/2}\gamma$ is the $1 - \alpha$ HPD credible interval one may note either that $q(\theta; \mathbf{x})$ is a strictly monotonic decreasing function of $|\theta - \psi|$, or that $q(\psi - z_{\alpha/2}\gamma; \mathbf{x}) = q(\psi + z_{\alpha/2}\gamma; \mathbf{x})$ as $q(\theta; \mathbf{x})$ is symmetric about its mean.

As τ increases, the prior distribution for θ becomes more diffuse so that, *a priori*, a larger range of values for θ are considered plausible. As an uninformative prior distribution we might let $\tau \to \infty$. Then $\psi = \bar{x}$ and $\gamma^2 = \sigma^2/n$, so the $1 - \alpha$ HPD credible interval becomes $\bar{x} \pm z_{\alpha/2}\sqrt{\sigma^2/n}$. This interval is identical to a $1 - \alpha$ confidence interval, even though the interpretations of the two intervals differ. This agreement between credible intervals and confidence intervals is not uncommon when prior distributions are uninformative.

HPD credible intervals have some disadvantages. If $g(.)$ is a strictly monotonic increasing function and (a, b) is a $1 - \alpha$ HPD credible interval, then $(g(a), g(b))$ is a $1 - \alpha$ credible interval for $g(\theta)$, but in general it is not an HPD interval. Thus, for example, HPD intervals for a variance and standard deviation are not directly related (see also Exercise 7.7). Also, calculating HPD intervals can require numerical iteration. Mainly because they are simpler to determine, *equal-tailed* credible intervals are often quoted instead; (a, b) is such an interval if $\Pr[\theta < a] = \Pr[b < \theta] = \alpha/2$.

Example 7.5

Let $X_1, X_2, \ldots, X_n$ be a random sample from $N(\mu, \theta)$, μ known, and let $p(\theta) \propto 1/\theta$ be the prior distribution. Then the likelihood is

$$L(\theta; \mathbf{x}) = (2\pi)^{-n/2}\theta^{-n/2} \exp[-\Sigma(x_i - \mu)^2/(2\theta)]$$

and the posterior distribution is

$$\begin{aligned} q(\theta; \mathbf{x}) &\propto p(\theta)L(\theta; \mathbf{x}) \\ &\propto \theta^{-(n/2)-1} \exp[-\Sigma(x_i - \mu)^2/(2\theta)]. \end{aligned}$$

i.e. $\quad \Sigma(x_i - \mu)^2/\theta \sim \chi^2_n.$

Suppose $n = 15$, $\Sigma(x_i - \mu)^2 = 12.6$ and we want an interval estimate of the variance, θ. For the 95% equal-tailed credible interval we use tabulated 0.025 and 0.975 critical values of χ^2_{15}. These are 6.262 and 27.49, respectively, so

$$0.025 = \Pr[6.262 > 12.6/\theta] = \Pr[12.6/\theta > 27.49]$$

and a 95% equal-tailed credible interval for θ is (12.6/27.49, 12.6/6.262). i.e. (0.458, 2.012).

For the 95% HPD interval we want an interval (a, b) for which

$$a^{-17/2} \exp[-12.6/(2a)] = b^{-17/2} \exp[-12.6/(2b)],$$

so that $q(a; \mathbf{x}) = q(b; \mathbf{x})$ and for which

$$\int_{12.6/b}^{12.6/a} f(y)\, dy = 0.95$$

where $f(.)$ is the p.d.f. of a χ^2_{15} variate. A numerical search gives (0.377, 1.769) as the 95% HPD credible interval, which is 10% shorter than the equal-tailed interval.

7.4 Hypothesis testing

We suppose one of two hypotheses H_0 and H_1 is true. $\Pr[H_i]$ will denote the prior probability that H_i is the true hypothesis and, after sample data $\mathbf{x}$ have been observed,

$\Pr[H_i | \mathbf{x}]$ will denote the posterior probability that H_i is true. The Bayesian aim in hypotheses testing is to determine the posterior odds,

$$Q^* = \frac{\Pr[H_0 | \mathbf{x}]}{\Pr[H_1 | \mathbf{x}]}. \tag{7.6}$$

Hence the conclusion from a Bayesian analysis might be a statement of the form 'H_0 is Q^* times more likely to be true than H_1.' Alternatively, as $\Pr[H_0 | \mathbf{x}] + \Pr[H_1 | \mathbf{x}] = 1$, we might conclude that '$Q^*/(1 + Q^*)$ and $1/(1 + Q^*)$ are the respective probabilities that H_0 is true and that H_1 is true.' Notice that the hypotheses have equal status in the Bayesian approach. This contrasts with the frequentist approach, where the null and alternative hypotheses have differing roles, with the null hypothesis being retained unless there is evidence against it.

Now

$$\Pr[H_0 | \mathbf{x}] = \frac{\Pr[H_0] f(\mathbf{x} | H_0)}{h(\mathbf{x})}$$

so

$$Q^* = \frac{\Pr[H_0]}{\Pr[H_1]} \cdot \frac{f(\mathbf{x} | H_0)}{f(\mathbf{x} | H_1)} \tag{7.7}$$

$$= Q . B$$

where $Q = \Pr[H_0]/\Pr[H_1]$ are the prior odds and $B = f(\mathbf{x} | H_0)/f(\mathbf{x} | H_1)$ is called the *Bayes factor*. The prior odds represent our feelings, before collecting data, as to which hypothesis is true. Often they are impartially set equal to 1, so that each hypothesis is considered equally likely to hold, *a priori*. Interest usually centres on B, since this determines how the data have changed our belief as to which hypothesis is true.

In general H_0 specifies a sampling model, $f_0(\mathbf{x}; \boldsymbol{\theta})$, say, together with a prior distribution for $\boldsymbol{\theta}$, while H_1 specifies a model $f_1(\mathbf{x}; \boldsymbol{\phi})$ and a prior distribution for $\boldsymbol{\phi}$. Hence the above formulation can be used for a wide variety of problems, for example, whether to model survival time data by a Weibull or lognormal distribution, or which of two sets of regressor variables should be used in a linear model.

Here, however, we restrict attention to the situation where $f(\mathbf{x}; \theta)$ is the sampling model for both hypotheses, but the hypotheses differ in the values they specify for θ. The cases to be considered are (1) both H_0 and H_1 are simple hypotheses, (2) both H_0 and H_1 are composite hypotheses, and (3) H_0 is a simple hypothesis and H_1 is composite.

Case (1): Simple hypotheses

If H_0 and H_1 are both simple, they have the form

$$H_0: \theta = \theta_0; \qquad H_1: \theta = \theta_1$$

and the Bayes factor is just

$$B = \frac{f(\mathbf{x};\ \theta_0)}{f(\mathbf{x};\ \theta_1)}. \tag{7.8}$$

As an example, suppose $X_1, X_2, \ldots, X_n$ are observations from an exponential distribution with parameter θ. Then $f(\mathbf{x};\ \theta) = \Pi\theta \exp(-\theta x_i) = \theta^n \exp(-\theta\Sigma x_i)$. So $B = \theta_0^n \exp(-\theta_0 \Sigma x_i)/\theta_1^n \exp(-\theta_1 \Sigma x_i)$ and the posterior odds are

$$Q^* = \frac{\Pr[H_0]}{\Pr[H_1]} \cdot \left(\frac{\theta_0}{\theta_1}\right)^n \exp\{(\theta_1 - \theta_0)\Sigma x_i\}.$$

Case (2): Composite hypotheses

In this case the hypotheses are

$$H_0\colon \theta \in \omega; \qquad H_1\colon \theta \in \Omega - \omega.$$

For each hypothesis a prior distribution for θ must be specified. Denote these by $p_0(\theta \mid H_0)$ and $p_1(\theta \mid H_1)$. Then $f(\mathbf{x} \mid H_0) = \int_\omega f(\mathbf{x};\ \theta)p_0(\theta \mid H_0)\ \mathrm{d}\theta$ and similarly for $f(\mathbf{x} \mid H_1)$. Thus the Bayes factor is

$$B = \int_\omega f(\mathbf{x};\ \theta)p_0(\theta \mid H_0)\ \mathrm{d}\theta \Bigg/ \int_{\Omega - \omega} f(\mathbf{x};\ \theta)p_1(\theta \mid H_1)\ \mathrm{d}\theta. \tag{7.9}$$

Case (3): A simple and a composite hypothesis

We have

$$H_0\colon \theta = \theta_0; \qquad H_1\colon \theta \neq \theta_0.$$

Combining the numerator and denominator from equations (7.8) and (7.9), the Bayes factor is

$$B = f(\mathbf{x};\ \theta_0) \Bigg/ \int_{\theta \neq \theta_0} f(\mathbf{x};\ \theta)p(\theta \mid H_1)\ \mathrm{d}\theta. \tag{7.10}$$

Example 7.1 (*continued*)

We have a random sample $X_1, X_2, \ldots, X_n$ from a Poisson distribution with mean θ. Suppose we wish to test the hypothesis $H_0\colon \theta = \theta_0$ against $H_1\colon \theta \neq \theta_0$ and that, under H_1, the prior distribution for θ is the conjugate distribution

$$p(\theta \mid H_1) = \alpha_2^{(\alpha_1 + 1)}\theta^{\alpha_1}\ \mathrm{e}^{-\alpha_2\theta}/\Gamma(\alpha_1 + 1). \tag{7.11}$$

To obtain the Bayes factor we require

$$f(\mathbf{x};\ \theta_0) = \theta_0^{\Sigma x_i}\ \mathrm{e}^{-n\theta_0}/\Pi(x_i!)$$

and $\int_{\theta \neq \theta_0} f(\mathbf{x};\ \theta)p(\theta \mid H_1)\ \mathrm{d}\theta$. From equation (7.1), the latter equals

$$\int_0^\infty [\theta^{\Sigma x_i}\ \mathrm{e}^{-n\theta}/\Pi(x_i!)].[\alpha_2^{(\alpha_1 + 1)}\theta^{\alpha_1}\ \mathrm{e}^{-\alpha_2\theta}/\Gamma(\alpha_1 + 1)]\ \mathrm{d}\theta$$

$$= \frac{\alpha_2^{(\alpha_1+1)}\Gamma(\alpha_1 + \Sigma x_i + 1)}{(n+\alpha_2)^{\alpha_1+\Sigma x_i+1}\Pi(x_i!)\Gamma(\alpha_1+1)} \int_0^\infty \frac{(n+\alpha_2)^{\alpha_1+\Sigma x_i+1}\theta^{\alpha_1+\Sigma x_i}\,\mathrm{e}^{-(n+\alpha_2)}\mathrm{d}\theta}{\Gamma(\alpha_1+\Sigma x_i+1)}$$

where the fact that $\theta \neq \theta_0$ has been ignored, as this does not affect the value of the integral. The terms in the integral have been chosen so as to form a gamma density, so the integral equals 1. (Forming density functions is a common device for performing integration in Bayesian statistics – an advantage from using a natural conjugate prior.) Thus

$$\int_{\theta \neq \theta_0} f(\mathbf{x};\theta)p(\theta \mid H_1)\,\mathrm{d}\theta = \frac{\alpha_2^{(\alpha_1+1)}\Gamma(\alpha_1+\Sigma x_i+1)}{(n+\alpha_2)^{\alpha_1+\Sigma x_i+1}\Pi(x_i!)\Gamma(\alpha_1+1)}$$

and, from (7.10),

$$B = \theta_0^{\Sigma x_i}\,\mathrm{e}^{-n\theta_0}(n+\alpha_2)^{\alpha_1+\Sigma x_i+1}\Gamma(\alpha_1+1)/\{\alpha_2^{(\alpha_1+1)}\Gamma(\alpha_1+\Sigma x_i+1)\}. \tag{7.12}$$

As a specific illustration suppose (1) the random sample consists of six observations whose values are 3, 1, 6, 2, 5, 2, (2) H_0 specifies $\theta = 2.0$, (3) H_1 specifies the parameter values in the conjugate distribution are $\alpha_1 = 2.6$ and $\alpha_2 = 0.6$ and (4) before observing the data we believe H_0 is half as likely to be true as H_1, so that $\Pr[H_0]/\Pr[H_1] = 0.5$. Then substitution into the above formula gives $B = 0.77$, so the posterior odds are 0.385 and we have firmer belief that H_1 is the true hypothesis.

It should be noted that the posterior odds are influenced by the choice of prior distribution. In the last example the prior mean and variance of θ are 6 and 10. If instead, H_1 had specified $\alpha_1 = 35$ and $\alpha_2 = 6$ (so the prior mean is again 6 but the variance is 1) then the Bayes factor would have equalled 3.16. This would suggest H_0 is the true hypothesis. Hence to some extent at least, Bayesian hypothesis testing compares *prior distributions*, rather than simpler statements such as $\theta = \theta_0$ and $\theta \neq \theta_0$. To examine the sensitivity of conclusions to the prior distribution, it is prudent to determine posterior odds for a selection of prior distributions.

When comparing a simple and a composite hypothesis, the following theorem is often applicable and gives a useful alternative expression for the Bayes factor.

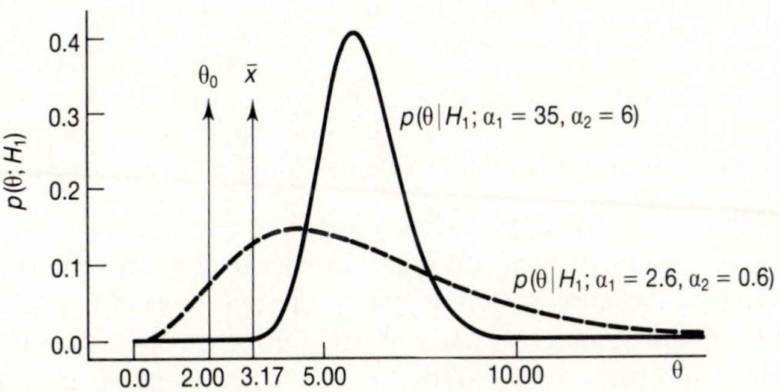

Fig. 7.3 Prior distributions considered in Example 7.2.

Theorem 7.3

Assume Case (3). If $f(\mathbf{x};\theta)$ is continuous at $\theta=\theta_0$, then

$$B=\lim_{\theta\to\theta_0}\frac{q(\theta\,|\,\mathbf{x},H_1)}{p(\theta\,|\,H_1)} \tag{7.13}$$

where $p(\theta\,|\,H_1)$ and $q(\theta\,|\,\mathbf{x},H_1)$ are the conditional prior and posterior distributions of θ.

Proof

H_1 holds if $\theta\neq\theta_0$ so, using the continuity condition,

$$f(\mathbf{x}\,|\,H_0)=\lim_{\theta\to\theta_0}f(\mathbf{x};\theta)=\lim_{\theta\to\theta_0}f(\mathbf{x}\,|\,\theta,H_1).$$

But $f(\mathbf{x}\,|\,\theta,H_1)=f(\mathbf{x}\,|\,H_1)q(\theta\,|\,x,H_1)/p(\theta\,|\,H_1)$, so

$$\frac{f(\mathbf{x}\,|\,H_0)}{f(\mathbf{x}\,|\,H_1)}=\lim_{\theta\to\theta_0}\frac{q(\theta\,|\,\mathbf{x},H_1)}{p(\theta\,|\,H_1)}.$$

□

Example 7.1 (*continued*)

We have a random sample $X_1, X_2, \ldots, X_n$ from a Poisson distribution with mean θ and hypotheses $H_0:\theta=\theta_0$, $H_1:\theta\neq\theta_0$. Under H_1, we assume the prior distribution for θ is the conjugate distribution given by equation (7.11),

$$p(\theta\,|\,H_1)=\alpha_2^{(\alpha_1+1)}\theta^{\alpha_1}\,\mathrm{e}^{-\alpha_2\theta}/\Gamma(\alpha_1+1).$$

The likelihood is $L(\theta;\mathbf{x})\propto\theta^{\Sigma x_i}\,\mathrm{e}^{-n\theta}$, so the posterior distribution is

$$q(\theta\,|\,\mathbf{x},H_1)\propto\theta^{\alpha_1+\Sigma x_i}\,\mathrm{e}^{-(n+\alpha_2)\theta}.$$

This is a gamma distribution so

$$q(\theta\,|\,\mathbf{x},H_1)=(n+\alpha_2)^{(\alpha_1+\Sigma x_i+1)}\theta^{(\alpha_1+\Sigma x_i)}\,\mathrm{e}^{-(n+\alpha_2)\theta}/\Gamma(\alpha_1+\Sigma x_i+1).$$

Hence, from (7.13),

$$B=\lim_{\theta\to\theta_0}\frac{(n+\alpha_2)^{(\alpha_1+\Sigma x_i+1)}\theta^{(\alpha_1+\Sigma x_i)}\,\mathrm{e}^{-(n+\alpha_2)\theta}}{\alpha_2^{(\alpha_1+1)}\theta^{\alpha_1}\,\mathrm{e}^{-\alpha_2\theta}}\cdot\frac{\Gamma(\alpha_1+1)}{\Gamma(\alpha_1+\Sigma x_i+1)}$$

$$=\theta_0^{\Sigma x_i}\,\mathrm{e}^{-n(\theta_0)}(n+\alpha_2)^{(\alpha_1+\Sigma x_i+1)}\Gamma(\alpha_1+1)/\{\alpha_2^{(\alpha_1+1)}\Gamma(\alpha_1+\Sigma x_i+1)\}.$$

This agrees with equation (7.12) and its calculation is relatively easy. □

Theorem 7.3 shows that the Bayes factor is the ratio of posterior and prior densities of θ at θ_0, where these densities are conditional on H_1. This contrasts with the classical approach to hypothesis testing, where the *tail areas* of densities are used.

The result also highlights the difficulty of using improper or infinitely diffuse prior distributions in hypothesis tests. With an improper prior distribution such as $p(\theta)\propto\theta^{-1}$ for $\theta>0$, or $p(\theta)\propto c$ for $-\infty<\theta<\infty$, the constant of proportionality cannot be chosen so that the distribution integrates to one; the integral of an improper distribution is infinity. Instead, the constant of proportionality is given an arbitrary positive value, so $\lim_{\theta \to \theta_0} p(\theta \mid H_1)$ and the Bayes factor are ill-defined when $p(\theta \mid H_1)$ is improper. As an example of the problem with an infinitely diffuse prior distribution, suppose the prior distribution under H_1 specifies that $\theta \sim N(\phi, \tau^2)$ and $\tau \to \infty$ (cf. Example 7.4). Then $p(\theta \mid H_1) = 0$, while $q(\theta \mid \mathbf{x}, H_1) > 0$ if there are adequate sample data. Hence the Bayes factor and the posterior odds in favour of H_0 will both tend to infinity.

It should be noted that determining $\Pr[H_0 \mid \mathbf{x}]$ and $\Pr[H_1 \mid \mathbf{x}]$ can be an intermediate step in the analysis. If the aim is to choose between H_0 and H_1 and the loss structure for an incorrect choice is specified, then the expected loss can be minimized. For example, with the loss structure given earlier in Table 6.1, H_0 should be chosen if $b\Pr[H_1 \mid \mathbf{x}] < a\Pr[H_0 \mid \mathbf{x}]$. Another possibility is that the aim is to estimate θ, in which case there is no need to choose between the hypotheses. Instead, the estimate might be based on the posterior distribution of θ, which is the mixture distribution

$$q(\theta \mid \mathbf{x}) = q(\theta \mid \mathbf{x}, H_0)\Pr[H_0 \mid \mathbf{x}] + q(\theta \mid \mathbf{x}, H_1)\Pr[H_1 \mid \mathbf{x}]. \tag{7.14}$$

This formula extends naturally if there are more than two hypotheses.

7.5 Nuisance parameters

7.5.1 Marginal distributions

Quite commonly in multiparameter problems, at any one time only a subset of the parameters are of interest. The remaining parameters are termed *nuisance* parameters. When possible, one integrates with respect to these so as to form the marginal distribution of the parameters of interest.

Specifically, suppose we have parameters $\boldsymbol{\theta}$ and $\boldsymbol{\lambda}$ (not necessarily scalars) and our interest is in $\boldsymbol{\theta}$. Then $\boldsymbol{\lambda}$ is a nuisance parameter so from the joint posterior distribution, $q(\boldsymbol{\theta}, \boldsymbol{\lambda} \mid \mathbf{x})$, we would form the marginal distribution

$$q_1(\boldsymbol{\theta} \mid \mathbf{x}) = \int_\Lambda q(\boldsymbol{\theta}, \boldsymbol{\lambda} \mid \mathbf{x})\, \mathrm{d}\boldsymbol{\lambda}.$$

Then $q_1(\boldsymbol{\theta} \mid \mathbf{x})$ would be used to test hypotheses about $\boldsymbol{\theta}$ or form point or region estimates, etc.

Example 7.3 (*continued*)

$X_1, X_2, \ldots, X_n \sim N(\mu, \theta^{-1})$, both μ and θ unknown. Since $q(\mu, \theta \mid \mathbf{x}) \propto p(\mu, \theta)L(\mu, \theta; \mathbf{x})$, the joint posterior distribution is

$$q(\mu, \theta \mid \mathbf{x}) \propto \theta^{\nu_1/2} \exp\left[-\frac{\theta}{2}\{\alpha_1 + \nu_2(\alpha_2 - \mu)^2\}\right].\theta^{n/2} \exp\left[-\frac{\theta}{2}\{r + n(\bar{x} - \mu)^2\}\right]$$

$$\propto \theta^{(\nu^* - 1)/2} \exp[-\{\beta + \gamma(\mu^* - \mu)^2\}\theta]$$

where $v^* = v_1 + n + 1$; $\beta = (\alpha_1 + r)/2$; $\gamma = (v_2 + n)/2$ and $\mu^* = (\alpha_2 v_2 + n\bar{x})/(v_2 + n)$. Suppose we wish to make inferences about μ. Then we obtain its marginal distribution by 'integrating out' the nuisance parameter, θ:

$$q_1(\mu \mid \mathbf{x}) \propto \int_0^\infty \theta^{(v^*-1)/2}\, e^{-\{\beta + \gamma(\mu^* - \mu)^2\}\theta}\, d\theta$$

$$\propto [\beta + \gamma(\mu^* - \mu)^2]^{-(v^*+1)/2} \int_0^\infty \frac{\{\beta + \gamma(\mu^* - \mu)^2\}^{(v^*+1)/2}\theta^{(v^*-1)/2} e^{-\{\beta + \gamma(\mu^* - \mu)^2\theta\}}\, d\theta}{\Gamma(\{v^* + 1\}/2)}$$

$$\propto \left[1 + \frac{\gamma}{\beta}(\mu^* - \mu)^2\right]^{-(v^*+1)/2}$$

since the integral equals 1 as the integrand is a gamma density. Thus $(\mu^* - \mu)/(\beta/\gamma v^*)^{1/2}$ has a standard t-distribution on v^* degrees of freedom. This could be used, for example, to construct a credible interval for μ: $\mu^* \pm t_{v^*}(\beta/\gamma v^*)^{1/2}$.

Bayesian methods typically require integrals to be evaluated, even if only to determine the constant of proportionality in the equation, posterior $\propto$ prior $\times$ likelihood. The presence of nuisance parameters can greatly increase the number and complexity of integrations that must be performed, so it is worth mentioning here that analytic approximations and numerical methods are often needed to evaluate the relevant integrals. Recent advances in computationally intensive methods, such as Gibbs sampling, have made it much easier to perform numerical integration in many cases. Gibbs sampling is described in Chapter 9.

7.5.2 *Predictive distributions*

Suppose $X_1, X_2, \ldots, X_n$ are observations from $f(x; \boldsymbol{\theta})$ and the predictive distribution of a further independent observation, X_{n+1}, is required. Then the posterior distribution, $q(\boldsymbol{\theta} \mid \mathbf{x}_n)$, is first formed, where $\mathbf{x}_n = (x_1, x_2, \ldots, x_n)'$. This yields the posterior joint distribution of x_{n+1} and $\boldsymbol{\theta}$: $f(x_{n+1}; \boldsymbol{\theta})q(\boldsymbol{\theta} \mid \mathbf{x}_n)$ since, given $\boldsymbol{\theta}$, x_{n+1} and $\mathbf{x}_n$ are independent. We treat $\boldsymbol{\theta}$ as a nuisance parameter and obtain the required posterior predictive distribution from

$$g(x_{n+1} \mid \mathbf{x}_n) = \int_\Omega f(x_{n+1}; \boldsymbol{\theta})q(\boldsymbol{\theta} \mid \mathbf{x})d\theta. \qquad (7.15)$$

If a point prediction of x_{n+1} were required, we would use the mean, median, mode or some other function of this distribution, depending on the loss function for inaccuracy (cf. Section 6.3).

Example 7.6

Suppose observations are from the exponential distribution $f(x_i; \theta) = \theta\, e^{-\theta x}$, $i = 1, \ldots, n$; $x_i > 0$; $\theta > 0$, and the prior distribution is the gamma distribution, $p(\theta) = \alpha_2^{\alpha_1+1}\theta^{\alpha_1}\, e^{-\alpha_2\theta}/\Gamma(\alpha_1 + 1)$. Then the posterior distribution of θ is

$$q(\theta\,|\,\mathbf{x}_n) \propto \theta^n\, \mathrm{e}^{-\theta\Sigma x_i}\alpha_2^{(\alpha_1+1)}\theta^{\alpha_1}\, \mathrm{e}^{-\alpha_2\theta}/\Gamma(\alpha_1+1)$$
$$\propto \theta^{n+\alpha_1}\, \mathrm{e}^{-\theta(\alpha_2+\Sigma x_i)}.$$

Hence,

$$q(\theta\,|\,\mathbf{x}_n) = (\alpha_2+\Sigma x_i)^{(n+\alpha_1+1)}\theta^{n+\alpha_1}\, \mathrm{e}^{-(\alpha_2+\Sigma x_i)\theta}/\Gamma(\alpha_1+n+1)$$

since $q(\theta\,|\,\mathbf{x}_n)$ is clearly a gamma distribution. From (7.15), the posterior predictive distribution of a future observation, X_{n+1}, is

$$g(x_{n+1}\,|\,\mathbf{x}_n) = \int_{\theta=0}^{\infty} \theta\, \mathrm{e}^{-\theta x_{n+1}}.q(\theta\,|\,\mathbf{x}_n)\, \mathrm{d}\theta$$
$$= \int_0^{\infty} (\alpha_2+\Sigma x_i)^{n+\alpha_1+1}\theta^{n+\alpha_1+1}\, \mathrm{e}^{-(\alpha_2+\Sigma x_i+x_{n+1})\theta}/\Gamma(\alpha_1+n+1)\, \mathrm{d}\theta.$$

Now

$$\int_0^{\infty} (\alpha_2+\Sigma x_i+x_{n+1})^{\alpha_1+n+2}\theta^{n+\alpha_1+1}\, \mathrm{e}^{-(\alpha_2+\Sigma x_i+x_{n+1})\theta}/\Gamma(\alpha_1+n+2)\, \mathrm{d}\theta = 1.$$

Thus the predictive distribution is

$$g(x_{n+1}\,|\,\mathbf{x}_n) = \frac{(\alpha_1+n+1)(\alpha_2+\Sigma x_i)^{(n+\alpha_1+1)}}{(\alpha_2+\Sigma x_i+x_{n+1})^{n+\alpha_1+2}}, \qquad x_{n+1} > 0.$$

7.5.3 *Comparing two populations*

Suppose θ_1 and θ_2 are corresponding parameters of two populations and we wish to make inferences about their relative values. A direct approach is

(1) Form a function of θ_1 and θ_2, say η. When θ_1 and θ_2 are population means, usually we would put $\eta = \theta_1 - \theta_2$, while if θ_1 and θ_2 are population variances (or precisions) then $\eta = \theta_1/\theta_2$ is more likely to give tractable mathematics.
(2) From the joint posterior distribution of θ_1 and θ_2, determine the joint distribution of η and θ_2 through a change of variables.
(3) Obtain the marginal distribution of η by integrating out the nuisance parameter θ_2. Inferences are then based on this marginal distribution.

As an example, suppose we take independent samples from two normal distributions with known means, μ_1 and μ_2, and unknown precisions, θ_1 and θ_2. Let n_1 and n_2 be the sample sizes and, for $i = 1, 2$, put $r_i = \Sigma_j(x_{ij}-\mu_i)^2$, where $\{x_{i1}, x_{i2}, \ldots, x_{in_i}\} = \mathbf{x}_i$ are the observations from population i. Then

$$L(\theta_i;\, \mathbf{x}_i) = (2\pi)^{-n_i/2}\theta_i^{n_i/2} \exp\{-r_i\theta_i/2\}.$$

We shall suppose prior knowledge about θ_1 is independent of knowledge about θ_2 and that each is represented by a conjugate prior distribution,

$$p_i(\theta_i) \propto \theta_i^{\nu_i/2} \exp\{-\alpha_i\theta_i/2\}, \qquad i = 1, 2. \tag{7.16}$$

We want to find the posterior marginal distribution of $\eta = \theta_1/\theta_2$.
The joint posterior distribution for θ_1, θ_2 is given by

$$q(\theta_1, \theta_2; \mathbf{x}_1, \mathbf{x}_2) \propto \prod_i L(\theta_i; \mathbf{x}_i)p_i(\theta_i)$$

$$\propto \prod_i \theta^{(n_i+v_i)/2} \exp\{-(r_i + \alpha_i)\theta_i/2\}.$$

For the change of variables $(\theta_1, \theta_2) \to (\eta, \theta_2)$ with $\eta = \theta_1/\theta_2$, the Jacobian equals θ_2. Hence

$$q(\eta, \theta_2; \mathbf{x}_1, \mathbf{x}_2) \propto \theta_2(\eta\theta_2)^{(n_1+v_1)/2}\theta_2^{(n_2+v_2)/2} \exp\{-(r_1+\alpha_1)\eta\theta_2/2 - (r_2+\alpha_2)\theta_2/2\}$$
$$= \eta^{(n_1+v_1)/2}\theta_2^{(n_1+n_2+v_1+v_2+2)/2} \exp - \tfrac{1}{2}\theta_2\{(r_1+\alpha_1)\eta + (r_2+\alpha_2)\}.$$

To integrate this expression with respect to θ_2, we note its relationship to a gamma distribution and obtain

$$q(\eta; \mathbf{x}_1, \mathbf{x}_2) \propto \eta^{(n_1+v_1)/2}[\tfrac{1}{2}\{(r_1+\alpha_1)\eta + (r_2+\alpha_2)\}]^{-\{n_1+v_1+n_2+v_2+4\}/2}.$$

This yields the result that

$$\eta.\frac{(r_1+\alpha_1)/(n_1+v_1+2)}{(r_2+\alpha_2)/(n_2+v_2+2)} \sim F_{(n_1+v_1+2),\,(n_2+v_2+2)}. \qquad (7.17)$$

In some common problems the above procedure can be shortened substantially by using well-known distributional results. For instance, in the last example it is obvious that the posterior distributions of θ_1 and θ_2 are gamma distributions. Using standard distribution theory to first relate gamma distributions to χ^2 distributions, and then to relate a ratio of independent χ^2 variables to an F-distribution, gives the above result without changing variables or the subsequent integration (see Exercise 7.15).

For this example, the common choice of non-informative prior is obtained by putting $\alpha_i = 0$ and $v_i = -2$ in equation (7.16). Then (7.17) becomes

$$\eta.\frac{r_1/n_1}{r_2/n_2} \sim F_{n_1, n_2}.$$

It should be mentioned that this is identical to the distributional result obtained with the frequentist approach, except that η is then viewed as an unknown fixed constant and $(r_1/n_1)/(r_2/n_2)$ is the variable.

7.6 Non-informative stopping

Assume we have a sequence of independent Bernoulli trials, in each of which the probability of success (S) is θ and the probability of failure (F) is $(1-\theta)$. Suppose the observed sequence of outcomes in ten trials is

F F S F F F F S F S

and that we wish to estimate θ. With the frequentist approach to statistics, the estimate will depend on the reason exactly ten outcomes were observed. Perhaps, before sampling began, it was decided to perform $n = 10$ trials. Then the number of successes, r, follows a binomial $(10, \theta)$ distribution and an unbiased estimate of θ is $r/n = 3/10$. On the other hand, if it had been decided to perform trials until the third success was observed, then the number of trials would be the random variable and would follow a negative binomial distribution. The unbiased estimate of θ would then be 2/9. (The estimate is $(r-1)/(n-1)$ if n trials are required to obtain r successes.) The point of this example is that, with the frequentist approach, the unbiased estimator depends upon the stopping rule used to decide the number of trials. Hypothesis tests and confidence intervals for θ would also depend on the stopping rule.

To examine the role of stopping rules in Bayesian inference, suppose $X_1, X_2, \ldots$ are a sequence of potential observations, where the X_i are i.i.d. with probability density function $f(x; \boldsymbol{\theta})$. Given the first i observations, $x_1, x_2, \ldots, x_i$, let δ_{i+1} be a function that indicates whether sampling is stopped or continued. Specifically,

$$\delta_{i+1} = \begin{cases} 1 \text{ indicates 'Stop sampling'.} \\ 0 \text{ indicates 'Take observation } X_{i+1}\text{'.} \end{cases}$$

The function δ_{i+1} may be random or deterministic and may depend on the parameter $\boldsymbol{\theta}$, the data $x_1, x_2, \ldots, x_i$, and other factors unrelated to $\boldsymbol{\theta}$, such as whether it is time for lunch. If sampling stops after data $x_1, x_2, \ldots, x_n$ have been observed (and not before), the likelihood for $\boldsymbol{\theta}$ is

$$L(\boldsymbol{\theta}; \mathbf{x}_n, \boldsymbol{\delta}) = \left[\sum_{i=1}^{n} \Pr[\delta_i = 0 \mid \boldsymbol{\theta}, x_1, \ldots, x_{i-1}] f(x_i; \boldsymbol{\theta}) \right] \times \Pr[\delta_{n+1} = 1 \mid \boldsymbol{\theta}, x_1, \ldots, x_n] \tag{7.18}$$

where $\boldsymbol{\delta}$ denotes the observed values of $\delta_1, \ldots, \delta_{n+1}$ [i.e. $\boldsymbol{\delta} = (0, 0, \ldots, 0, 1)$] and $\mathbf{x}_n = (x_1, x_2, \ldots, x_n)$.

In Bayesian inference a stopping rule is said to be *non-informative* if the posterior distribution for $\boldsymbol{\theta}$, given $\mathbf{x}_n$, does not depend on the rule used to stop sampling. Suppose now that

(a) $\Pr[\delta_i = 0 \mid \boldsymbol{\theta}, x_1, \ldots, x_{i-1}] = \Pr[\delta_i = 0 \mid x_1, \ldots, x_{i-1}]$ for $i = 1, 2, \ldots$ and any $x_1, \ldots, x_{i-1}$.

Thus $\boldsymbol{\theta}$ does not influence the decision to stop, other than through the observed data. Suppose also that

(b) the prior distribution for $\boldsymbol{\theta}$ does not depend on the stopping rule. From (a),

$$L(\boldsymbol{\theta}; \mathbf{x}_n, \boldsymbol{\delta}) \propto \prod_{i-1}^{n} f(x_i; \boldsymbol{\theta})$$

as the other terms in equation (7.18) do not depend on $\boldsymbol{\theta}$. Consequently,

$$L(\boldsymbol{\theta}; \mathbf{x}_n, \delta) = L(\boldsymbol{\theta}; \mathbf{x}_n),$$

so the likelihood does not depend on the stopping rule. From (b), the same is true of the prior distribution. Hence, when (a) and (b) hold, the stopping rule will not affect the posterior distribution for $\boldsymbol{\theta}$, other than through $\mathbf{x}_n$, and the stopping rule is non-informative.

Condition (a) holds in the great majority of situations. For instance, it holds if the sample size is fixed in advance, if the reason sampling was stopped is unrelated to θ, or if the data $x_1, \ldots, x_i$ completely determine whether X_{i+1} is taken. The last case applies in the example given earlier, where a sequence of Bernoulli trials were performed until the third success was observed. It also applies in the Bayesian sequential procedures considered in Section 6.7. Situations where condition (b) fails to hold are rare. Before any data are gathered, knowledge about the stopping rule is unlikely to affect knowledge about $\boldsymbol{\theta}$, so it should seldom affect the prior distribution for $\boldsymbol{\theta}$. As (a) and (b) hold in most situations, stopping rules are usually non-informative in Bayesian statistics and the reason that sampling was terminated is generally irrelevant in a Bayesian analysis.

Interim analysis is one important case where it is contentious as to whether or not stopping rules should be ignored. Interim analyses are commonly performed, for example, during the course of a medical trial: if there is strong evidence that one treatment is better than another then the trial is terminated early, for knowingly allocating patients to a poorer treatment would be unethical. With the frequentist approach, the final analysis of the data at the end of the trial must take into account the interim analyses that were performed along the way. In marked contrast, the Bayesian approach ignores interim analyses completely. Useful discussion papers on this topic are Jennison and Turnbull (1989), where the frequentist approach is adopted, and Berger and Berry (1988), where the Bayesian approach is strongly advocated.

7.7 Hierarchical models

Bayesian inference treats unknown parameters as variables, so parameters can have probability distributions. Hierarchical models exploit the flexibility this gives. Suppose data $\mathbf{x}$ have density

$$f(\mathbf{x}; \boldsymbol{\theta}) \qquad \text{[stage 1]}$$

where $\boldsymbol{\theta}$ is a vector of unknown parameters. Denote the prior distribution for $\boldsymbol{\theta}$ by

$$p(\boldsymbol{\theta}; \psi) \qquad \text{[stage 2]}.$$

In earlier sections we have assumed that ψ is a fixed parameter: its value has been chosen to reflect prior knowledge. If, however, we treat ψ as an unknown parameter, then it will itself have a prior distribution,

$$g(\psi) \qquad \text{[stage 3]}.$$

The resulting model is termed a (three-stage) hierarchical model. If the prior distribution of $\boldsymbol{\psi}$ involves unknown parameters, then they would be given a prior distribution, forming a four-stage hierarchical model. This could obviously be extended but, in practice, three-stage models are the most common. A seminal paper on hierarchical models is by Lindley and Smith (1972).

Hierarchical models are particularly useful when $\boldsymbol{\theta}$ can be subdivided into vectors (or scalars), some of which are *exchangeable*. Suppose $\boldsymbol{\theta} = (\boldsymbol{\theta}_1, \boldsymbol{\theta}_2, \ldots, \boldsymbol{\theta}_k)$ and $\boldsymbol{\theta}_1$ are the parameters associated with one sub-population, $\boldsymbol{\theta}_2$ with a second sub-population and so forth. The $\boldsymbol{\theta}_i$ are assumed to play the same role in each sub-population (e.g. $\boldsymbol{\theta}_i$ might be the mean and/or variance of the ith sub-population). If prior knowledge does not distinguish between the sub-populations, then

$$p(\boldsymbol{\theta}_1, \boldsymbol{\theta}_2, \ldots, \boldsymbol{\theta}_k; \boldsymbol{\psi})$$

might not depend on which sub-population is associated with $\boldsymbol{\theta}_1$, which with $\boldsymbol{\theta}_2$, etc. When this is true, $\boldsymbol{\theta}_1, \boldsymbol{\theta}_2, \ldots, \boldsymbol{\theta}_k$ are said to be exchangeable.

More generally, vectors $\boldsymbol{\theta}_1, \boldsymbol{\theta}_2, \ldots, \boldsymbol{\theta}_k$ are exchangeable if any permutation of their indices does not change their joint distribution, so that

$$p(\boldsymbol{\theta}_1, \boldsymbol{\theta}_2, \ldots, \boldsymbol{\theta}_k; \boldsymbol{\psi}) = p(\boldsymbol{\theta}_{(1)}, \boldsymbol{\theta}_{(2)}, \ldots, \boldsymbol{\theta}_{(k)}; \boldsymbol{\psi})$$

where $(\boldsymbol{\theta}_{(1)}, \boldsymbol{\theta}_{(2)}, \ldots, \boldsymbol{\theta}_{(k)})$ is any permutation of $(\boldsymbol{\theta}_1, \boldsymbol{\theta}_2, \ldots, \boldsymbol{\theta}_k)$. Often when we have this property,

$$p(\boldsymbol{\theta}_1, \boldsymbol{\theta}_2, \ldots, \boldsymbol{\theta}_k; \boldsymbol{\psi}) = \prod_{i=1}^{k} p^*(\boldsymbol{\theta}_i; \boldsymbol{\psi}).$$

i.e. Given $\boldsymbol{\psi}$, the $\boldsymbol{\theta}_i$ are independent and identically distributed.

Example 7.7

Suppose $\theta_1, \theta_2, \ldots, \theta_k$ are the average reading abilities of seven-year-old children in each of k different schools. Samples of seven-year-olds are to be given reading tests to estimate the θ_i. Let $X_{i1}, X_{i2}, \ldots, X_{in_i}$ be the reading abilities of a sample of n_i children from school i. Suppose

$$X_{ij} \sim N(\theta_i, \sigma^2)$$

where σ^2 is the same for all schools. Before obtaining the data we might have no prior reason to believe that any specified school is better than another and hence treat the θ_i as exchangeable. If σ^2 is known and the θ_i are assumed to be normally distributed, the second stage to the model is

$$p(\theta_1, \theta_2, \ldots, \theta_k; \boldsymbol{\psi}) = \prod_{i=1}^{k} (2\pi\tau^2)^{-1/2} \exp\left\{-\frac{1}{2\tau^2}(\theta_i - \phi)^2\right\}$$

where $\boldsymbol{\psi} = (\phi, \tau)$. Here ϕ is the mean 'average reading ability' in the schools and τ^2 is the variance of their averages. The last stage of the model is formed by giving a prior distribution to ϕ and τ^2. An uninformative prior distribution would be $g(\phi, \tau^2) \propto \tau^{-1}$.

Notice that information about some of the θ_i provides information about the remainder. This is a consequence of the model structure. For example, if we obtained sample data from the first $k-1$ schools, then we could estimate $\theta_1, \theta_2, \ldots, \theta_{k-1}$. Hence we could estimate ϕ and τ^2, the parameters of the distribution of θ_k.

After obtaining sample data the prescription for obtaining the posterior distribution is the same as usual:

$$\text{posterior} \propto \text{prior} \times \text{likelihood}.$$

The unknown parameters are $\boldsymbol{\theta}$ and $\boldsymbol{\psi}$ and their prior distribution is $p(\boldsymbol{\theta}; \boldsymbol{\psi})g(\boldsymbol{\psi})$. The sampling distribution of the data, $\mathbf{x}$, depends only on $\boldsymbol{\theta}$ and not on $\boldsymbol{\psi}$,

$$L(\boldsymbol{\theta}, \boldsymbol{\psi}; \mathbf{x}) = f(\mathbf{x}; \boldsymbol{\theta}).$$

Hence the joint posterior distribution of the parameters is

$$q(\boldsymbol{\theta}, \boldsymbol{\psi}; \mathbf{x}) \propto f(\mathbf{x}; \boldsymbol{\theta})p(\boldsymbol{\theta}; \boldsymbol{\psi})g(\boldsymbol{\psi}). \tag{7.19}$$

Integration with respect to $\boldsymbol{\psi}$, which is usually a nuisance parameter, yields the posterior density of $\boldsymbol{\theta}$. Further integration can yield marginal distributions of individual components of $\boldsymbol{\theta}$.

For many hierarchical models, performing the relevant integrations cannot be done analytically and computationally intensive methods are needed, such as Gibbs sampling (see Section 9.6). Approximate analytical results may be obtained by taking the modes of posterior distributions as estimators when the necessary integrals cannot be obtained in closed form.

Example 7.7 (*continued*)

The data are $X_{ij} \sim N(\theta_i, \sigma^2)$ with σ^2 known, $i = 1, 2, \ldots, k$; $j = 1, 2, \ldots, n_i$. The model states $\theta_i \sim N(\phi, \tau^2)$, independently for each i, $\boldsymbol{\psi} = (\phi, \tau^2)$ and $g(\boldsymbol{\psi}) \propto \tau^{-1}$. Hence the joint posterior distribution is

$$q(\boldsymbol{\theta}, \phi, \tau^2; \mathbf{x}) \propto \left[\prod_{i=1}^{k} \exp\left\{-\frac{1}{2\sigma^2}\sum_{j=1}^{n_i}(x_{ij} - \theta_i)^2\right\}\right]$$
$$\times \left[\prod_{i=1}^{k} \tau^{-1} \exp\left\{-\frac{1}{2\tau^2}(\theta_i - \phi)^2\right\}\right] . \tau^{-1}$$

This must be integrated w.r.t. ϕ and τ^2 to obtain the posterior distribution of $\boldsymbol{\theta}$. Putting $\bar{\theta} = (1/k)\,\Sigma\theta_i$, we have $\Sigma(\theta_i - \phi)^2 = k(\phi - \bar{\theta})^2 + \Sigma(\theta_i - \bar{\theta})^2$. Hence,

$$\tau^{-1}\int_{-\infty}^{\infty}\left[\prod_{i=1}^{k} \tau^{-1} \exp\left\{-\frac{1}{2\tau^2}(\theta_i - \phi)^2\right\}\right]\mathrm{d}\phi$$
$$= \tau^{-(k+1)}\int_{-\infty}^{\infty} \exp\left\{-\frac{k}{2\tau^2}(\phi - \bar{\theta})^2 - \frac{1}{2\tau^2}\Sigma(\theta_i - \bar{\theta})^2\right\}\mathrm{d}\phi$$
$$\propto \tau^{-k} \exp\left\{-\frac{1}{2\tau^2}\Sigma(\theta_i - \bar{\theta})^2\right\}.$$

Substituting $h = \tau^{-2}$ and comparison with a gamma distribution yields

$$\int_0^\infty \tau^{-k} \exp\left\{-\frac{1}{2\tau^2}\Sigma(\theta_i - \bar{\theta})^2\right\} d\tau \propto [\Sigma(\theta_i - \bar{\theta})^2]^{-(k-1)/2}.$$

Thus the posterior distribution of $\boldsymbol{\theta}$ is

$$q(\boldsymbol{\theta}; \mathbf{x}) \propto \left[\prod_{i=1}^{k} \exp\left\{-\frac{1}{2\sigma^2}\sum_{j=1}^{n_i}(x_{ij} - \theta_i)^2\right\}\right].[\Sigma(\theta_i - \bar{\theta})^2]^{-(k-1)/2}$$

$$\propto \left[\exp\left\{-\frac{1}{2\sigma^2}\sum_{i=1}^{k} n_i(\theta_i - \bar{x}_i)^2\right\}\right].[\Sigma(\theta_i - \bar{\theta})^2]^{-(k-1)/2} \qquad (7.20)$$

where $\bar{x}_i = \Sigma_j x_{ij}/n_i$.

Let $\hat{\theta}_j$ denote the modal estimate of θ_j and put $\hat{\theta}^* = \Sigma\hat{\theta}_j/k$. Differentiating $q(\boldsymbol{\theta}; \mathbf{x})$ w.r.t. θ_j and equating the result to 0 shows that

$$\hat{\theta}_j = (\omega_1\bar{x}_j + \omega_2\hat{\theta}^*)/(\omega_1 + \omega_2) \qquad (7.21)$$

where $\omega_1 = (\sigma^2/n_j)^{-1}$ and $\omega_2 = \{\Sigma_i(\hat{\theta}_i - \hat{\theta}^*)^2/(k-1)\}^{-1}$ (see Exercise 7.18). If the θ_i were unrelated, $\bar{x}_j$ would be the estimate of θ_j. We see the model modifies this estimate by pulling it towards the mean of the estimated θ_i's.

7.8 Empirical Bayes

The Bayesian approach to statistics treats unknown parameters as random variables. Prior distributions model information about parameters and they may be based on expert opinion, past data, or some combination of the two, or they might take some special form, such as an uninformative prior distribution. In contrast, the classical approach to statistics has no need of prior distributions as it treats unknown parameters as fixed constants. Empirical Bayes (EB) is an approach to statistics that lies somewhere between the two. Unknown parameters are treated as fixed constants (as in the classical approach) except when they can be given a prior distribution that has a frequency interpretation. Example 7.7 can be used to clarify this. The parameters $\theta_1, \theta_2, \ldots, \theta_k$ were the average reading abilities in different schools and relate to a frequency distribution: θ is the average reading ability in a school and took the (unobserved) values $\theta_1, \theta_2, \ldots, \theta_k$ in a random sample of k schools. Hence θ has a prior distribution (sometimes called a mixing distribution). If we suppose $\theta \sim N(\phi, \tau^2)$, then ϕ and τ^2 admit no frequentist interpretation and so EB treats them as unknown fixed constants.

To consider a standard problem addressed by EB, suppose each element of a sequence is a bivariate random variable (x_i, θ_i), $i = 1, 2, \ldots, k$. We suppose these variables are independent and identically distributed, $f(x_i; \theta_i)$ being the probability density of each x_i given θ_i, and $p(\theta_i)$ denoting the marginal distribution of each θ_i. We assume $f(x_i; \theta_i)$ is known while $p(\theta_i)$ is unknown. The θs are unobservable and the task is to estimate *one* of them, say the last one (θ_k), or make inferences about it.

The only observable quantities are $x_1, x_2, \ldots, x_k$ and they each have the marginal distribution

$$f(x) = \int f(x;\,\theta)p(\theta)\,\mathrm{d}\theta.$$

Using this equation, $p(\theta)$ is usually identifiable (i.e. determined uniquely) from $f(x)$ and $f(x;\,\theta)$, provided some assumptions are made about the forms of $p(\theta)$ and $f(x;\,\theta)$. The main strategy in EB is to estimate $p(\theta)$ from the data and use this estimate, $\hat{p}(\theta)$ say, as a prior distribution. Using $\hat{p}(\theta)$ and the single datum x_k, standard Bayesian methods are then applied to make estimates or inferences about θ_k. These are based on the posterior distribution

$$q(\theta_k;\,x_k) \propto f(x_k;\,\theta_k)\hat{p}(\theta_k). \tag{7.22}$$

Note the role of $x_1, x_2, \ldots, x_{k-1}$. They only affect inferences about θ_k through the estimated prior distribution, $\hat{p}(\theta)$.

We first describe some common estimates of $p(\theta)$ and then discuss estimation of θ_k and inference.

7.8.1 Estimating $p(\theta)$

One commonly used estimate of $p(\theta)$ is obtained by assuming $p(\theta)$ has a particular parametric form and using the method of moments or maximum likelihood to estimate its parameters. Let $p(\theta;\,\psi)$ denote the probability density of θ, where ψ is the vector of unknown parameters. For example, $p(\theta;\,\psi)$ might be the conjugate prior distribution for $f(x;\,\theta)$, which would often be mathematically convenient. If ψ consists of one or two scalar parameters, their moment estimators can sometimes be determined easily from the relationships

$$\mathrm{E}[X] = \mathrm{E}_\psi[\mathrm{E}_\theta[X \mid \theta]] \tag{7.23}$$

and

$$\mathrm{Var}[X] = \mathrm{E}_\psi[\mathrm{Var}_\theta(X \mid \theta)] + \mathrm{Var}_\psi[\mathrm{E}_\theta(X \mid \theta)]. \tag{7.24}$$

Example 7.8

Suppose $X \sim \text{Poisson}(\theta)$ and $\theta \sim \text{gamma}(\alpha,\,\beta)$. Then, with $\psi = (\alpha,\,\beta)$,

$$\mathrm{E}[X] = \mathrm{E}_\psi[E_\theta(X \mid \theta)] = E_\psi[\theta] = \alpha/\beta$$

and

$$\begin{aligned}\mathrm{Var}[X] &= \mathrm{E}_\psi[\mathrm{Var}_\theta(X \mid \theta)] + \mathrm{Var}_\psi(\mathrm{E}(X \mid \theta)] \\ &= \mathrm{E}_\psi[\theta] + \mathrm{Var}_\psi[\theta] = \alpha/\beta + \alpha/\beta^2.\end{aligned}$$

Hence if $\bar{x}$ and s^2 are the sample mean and variance of $x_1, x_2, \ldots, x_k$, then, using the method of moments, estimates of α and β are $\hat{\alpha} = \bar{x}^2/(s^2 - \bar{x})$ and $\hat{\beta} = \bar{x}/(s^2 - \bar{x})$.

Maximum likelihood estimation is preferable to the method of moments, but performing the required calculations is more difficult. The joint density of x and θ is $f(x; \theta)p(\theta; \psi)$, so ψ is determined by maximizing the likelihood,

$$L(\psi; \mathbf{x}) = \prod_{i=1}^{k} \left\{ \int f(x_i; \theta)p(\theta; \psi)\, \mathrm{d}\theta \right\}. \tag{7.25}$$

Presuming that $p(\theta)$ has a particular parametric form is a strong assumption. An appealing alternative is to let $p(\theta)$ be a probability mass density,

$$\Pr[\theta = \phi_j] = p_j$$

for $j = 1, 2, \ldots, m$, with $\Sigma p_j = 1$. Usually either $\phi_1, \phi_2, \ldots, \phi_m$ or $p_1, p_2, \ldots, p_m$ are fixed at set values. In either case, the resulting cumulative distribution function converges (weakly) as $m \to \infty$ to the true c.d.f. of θ, provided the set values are appropriately chosen.

When the p_j are fixed, the most common choice for their values is $p_j = 1/m$ for $j = 1, 2, \ldots, m$. Then the likelihood of $\boldsymbol{\phi} = (\phi_1, \phi_2, \ldots, \phi_m)$ is

$$L(\boldsymbol{\phi}; \mathbf{x}) = \prod_{i=1}^{k} \left\{ \sum_{j=1}^{m} f(x_i; \phi_j)/m \right\}.$$

Maritz and Lwin (1989, p. 55) describe the use of the EM algorithm (Dempster *et al.*, 1977) to determine the values $\phi_1 \leqslant \phi_2 \leqslant \cdots \leqslant \phi_m$ that maximize this likelihood. They also show how a similar approach can be used to estimate $p_1, \ldots, p_m$ when $\phi_1, \ldots, \phi_m$ are fixed at set values.

7.8.2 *Estimation and inference*

As mentioned earlier, the general EB approach to making estimates or inferences about θ_k is to first estimate $p(\theta)$ and then use standard Bayesian methods with $\hat{p}(\theta)$ as the prior distribution. However, in special cases there can be alternatives. In particular, it is not always necessary to estimate $p(\theta)$ in order to obtain a point estimator of θ_k.

To illustrate, suppose we have a squared error loss function and $f(x; \theta)$ is a Poisson distribution. Then the posterior distribution of θ, given an observation x, is

$$q(\theta; x) = \frac{\{\mathrm{e}^{-\theta}\theta^x/x!\}p(\theta)}{\int \{\mathrm{e}^{-\theta}\theta^x/x!\}p(\theta)\, \mathrm{d}\theta}$$

where the denominator is determined by the requirement that $q(\theta; x)$ integrates to 1. With squared error loss, the best estimate of θ_k is its expected value,

$$\mathrm{E}[\theta_k \mid x_k] = \frac{\int \theta\{\mathrm{e}^{-\theta}\theta^{x_k}/x_k!\}p(\theta)\, \mathrm{d}\theta}{\int \{\mathrm{e}^{-\theta}\theta^{x_k}/x_k!\}p(\theta)\, \mathrm{d}\theta}.$$

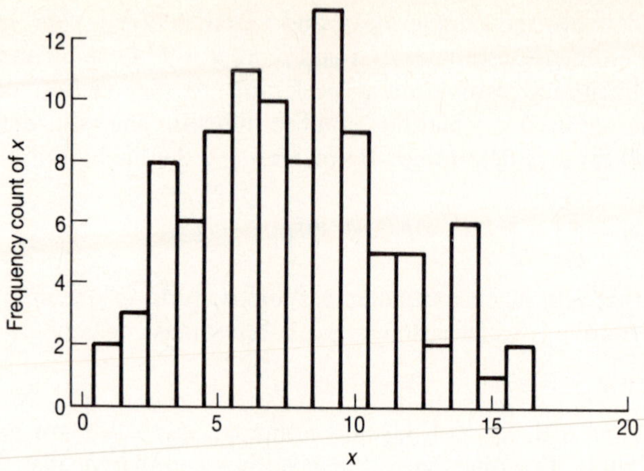

Fig. 7.4 Frequency distribution of simulated data from Poisson distributions.

Now $f(x) = \int f(x;\, \theta)p(\theta)\, d\theta = \int \{e^{-\theta}\theta^x/x!\}p(\theta)\, d\theta$.

Similarly, $f(x+1) = \int \{e^{-\theta}\theta^{x+1}/(x+1)!\}p(\theta)\, d\theta$. Hence

$$E[\theta_k \mid x_k] = (x_k + 1)f(x_k + 1)/f(x_k). \tag{7.26}$$

From the observed data, the ratio $f(x_k + 1)/f(x_k)$ can be estimated directly by

$$\frac{\text{number of } x_1, x_2, \ldots, x_k \text{ equal to } x_k + 1}{\text{number of } x_1, x_2, \ldots, x_k \text{ equal to } x_k}$$

i.e. as a ratio of frequency counts. The probability density $p(\theta)$ does not enter into this estimate.

Similar results may be obtained with other distributions, such as when $f(x;\, \theta)$ is the geometric or negative binomial distribution (Exercise 7.21). However, although such estimators have good asymptotic properties, they can be poor even for moderately large sample sizes and, for small samples, the estimator will not be a smooth function of x_k. To illustrate this, suppose $p(\theta)$ is a gamma distribution with mean 8 and variance 4 (i.e. $\Gamma(16,2)$). One hundred observations were generated by simulation from this distribution, giving values $\theta_1, \theta_2, \ldots, \theta_{100}$. For each θ_i an observation x_i from Poisson (θ_i) was also generated. The frequency distribution of the resultant *x*s is given in Figure 7.4. The distribution is far from smooth so the estimate of the next θ (θ_{101}) is not a smooth function of x_{101}. For example, the estimate of θ_{101} is 6.4, 14.6 or 6.9 when x_{101} equals 7, 8 or 9 respectively. To avoid this type of erratic behaviour, point estimators are usually obtained by first determining a smooth estimate of $p(\theta)$.

Turning to hypothesis testing, $\theta_1, \theta_2, \ldots, \theta_k$ again have a common distribution, $p(\theta)$, and we suppose the hypothesis relates to the particular parameter θ_k. In the case of two simple hypotheses, θ can take one of two values, θ_0^* and θ_1^* say, and some of $\theta_1, \ldots, \theta_k$ take the value θ_0^* while the remainder equal θ_1^*. As the θs are i.i.d. we may put

$$\Pr[\theta_i = \theta_0^*] = p \qquad \text{and} \qquad \Pr[\theta_i = \theta_1^*] = 1 - p$$

for $i = 1, 2, \ldots, k$. The question that is addressed is whether θ_k equals θ_0^* or θ_1^*, so the hypotheses are H_0: $\theta_k = \theta_0^*$ and H_1: $\theta_k = \theta_1^*$.

To estimate p from the data $x_1, x_2, \ldots, x_k$, one can maximize the likelihood,

$$L(p; \mathbf{x}) = \prod_{i=1}^{k} \{pf(x_i; \theta_0^*) + (1 - p)f(x_i; \theta_1^*)\}.$$

Taking the logarithm of this quantity and differentiating, $\hat{p}$ is a solution of

$$\sum_{i=1}^{k} \frac{f(x_i; \theta_0^*) - f(x_i; \theta_1^*)}{pf(x_i; \theta_0^*) + (1 - p)f(x_i; \theta_1^*)} = 0. \tag{7.27}$$

In the case where $\mathrm{E}[X \mid \theta] = \theta$, a simpler estimate of p is given by the method of moments. Then $\hat{p}$ solves

$$\bar{x} = p\theta_0^* + (1 - p)\theta_1^*$$

where $\bar{x}$ is the sample mean. Both this method and maximum likelihood can lead to an estimate of p that is negative, in which case we set $\hat{p} = 0$, and the estimate can be greater than 1, when we set $\hat{p} = 1$. With the moment estimator, this happens if $\bar{x}$ is not in the range bounded by θ_0^* and θ_1^*.

Suppose the loss function for testing the simple hypotheses is the one given in Table 6.1, i.e. a and b are the losses from incorrectly rejecting H_0 and H_1, respectively, and there is no loss for a correct decision. Then, from Lemma 6.1, the EB test is to reject H_0 if

$$\frac{L(\theta_1^*; x_k)}{L(\theta_0^*; x_k)} \geqslant \frac{\hat{p}a}{(1 - \hat{p})b}. \tag{7.28}$$

Example 7.9

For $i = 1, 2, \ldots, k$, suppose $X_i \sim N(\theta_i, \sigma^2)$ with σ^2 known and the hypotheses are H_0: $\theta_k = \mu_0$ vs H_1: $\theta_k = \mu_1$, with $\mu_1 > \mu_0$. From Example 6.6, H_0 is rejected if

$$x_k > \frac{1}{2}(\mu_0 + \mu_1) + \frac{\sigma^2}{(\mu_1 - \mu_0)} \ln\left(\frac{\hat{p}a}{(1 - \hat{p})b}\right),$$

provided $\hat{p} \in (0, 1)$. If $\hat{p} = 0$, then H_0 is rejected regardless of the value of x_k, while if $\hat{p} = 1$, then H_0 is always accepted.

A disadvantage of many EB methods is that the estimate $\hat{p}(\theta)$ is treated as if it were actually $p(\theta)$; no allowance is made for the uncertainty of the estimate. In the last example, for instance, if $\bar{x} > \mu_1$ then $\hat{p} = 0$ and H_0 is rejected for any value of x_k. This is unreasonable if $\bar{x}$ is based on a small set of data.

Forming confidence intervals is one task where allowance is sometimes made for the uncertainty in $\hat{p}(\theta)$. Let $I(\hat{p}(\theta))$ be the $\alpha\%$ Bayesian credible interval for θ_k when $\hat{p}(\theta)$ is taken as the prior distribution for θ. The *coverage* of the interval,

$C[I(\hat{p}(\theta))] = \Pr[\theta \in I(\hat{p}(\theta))]$, is the posterior probability, given $p(\theta)$ and the data, that the interval actually contains θ. One useful measure of the level of confidence to associate with $I(\hat{p}(\theta))$ is its expected coverage,

$$E[C[I(\hat{p}(\theta))]]$$

where the expectation is with respect to variation of $\hat{p}(\theta)$. Determining the expected coverage is usually difficult, but approximations using Taylor expansions are possible when $p(\theta)$ has a known parametric form. Some examples are given in Maritz and Lwin (1989, Chapter 6) and a practical application is given by Martz and Zimmer (1992).

In discussing Empirical Bayes we have assumed throughout that a single observation, x_i, is obtained for each θ_i. Clearly the methods can be extended straightforwardly to cases where there are several observations for each θ_i.

In situations where EB can be used, an obvious alternative is to use hierarchical Bayesian models, as the θ_i are exchangeable. When the two methods use the same prior distribution, $p(\theta)$, they typically give similar estimators, especially if the hierarchical method uses modal estimates of parameters.

Exercises

7.1 Suppose $X_1, X_2, \ldots, X_n$ form a random sample from a distribution parameterized by $\boldsymbol{\theta}$.
(a) Suppose also that, in the likelihood, the Xs can be replaced by functions $T_1(X_1, \ldots, X_n)$ and $T_2(X_1, \ldots, X_n)$. Show that T_1 and T_2 are sufficient for $\boldsymbol{\theta}$.
(b) If observations are from the gamma distribution, $X_i \sim \Gamma(\alpha, \beta)$, determine sufficient statistics for $\boldsymbol{\theta} = (\alpha, \beta)$ and give a conjugate prior distribution for θ up to a constant of proportionality.

7.2 Show that

$$f(x; \theta) = \frac{\theta^2}{\theta + 1}(x + 1)\,e^{-x\theta}, \qquad x > 0$$

is, for any $\theta > 0$, a p.d.f. A random sample of size n is taken from this p.d.f. giving values $x_1, x_2, \ldots, x_n$.
(a) What is the likelihood function (ignore constants of proportionality)?
(b) Give function(s) of $x_1, \ldots, x_n$ which are sufficient for θ.
(c) Give the conjugate prior distribution for θ.
(d) Obtain the posterior distribution for θ using this prior and compare its form with that of the prior distribution.

7.3 Suppose $X_1, X_2, \ldots, X_n$ form a random sample from the uniform distribution $U(\theta - \frac{1}{2}, \theta + \frac{1}{2})$, where $\theta \in (-\infty, \infty)$ is unknown.
(a) Show that the smallest and largest of $X_1, \ldots, X_n$ are jointly sufficient for θ.
(b) Given that $p(\theta) = \text{constant}$, $\theta \in (-\infty, \infty)$, is the prior distribution of θ, find its posterior distribution.

7.4 Suppose (a, b) is the $1 - \alpha$ HPD credible interval for θ and $\phi = c + k\theta$ $(c \geqslant 0, k > 0)$. Show that $(c + ka, c + kb)$ is the $1 - \alpha$ HPD credible interval for ϕ.

7.5 A random sample of size 25 from $N(\theta, 1)$ has mean 0.30.
(a) Given that $p(\theta) \propto$ constant, $\theta \in (-\infty, \infty)$, determine a 95% HPD credible interval for θ.
(b) Given that θ is known to be positive, so that

$$\begin{array}{ll} p(\theta) \propto \text{constant}, & \theta > 0 \\ p(\theta) = 0, & \theta \leqslant 0, \end{array}$$

determine a 95% HPD credible interval for θ.

7.6 Prior information on the mean θ of a Poisson distribution is such that the prior mean and prior variance of θ are both unity.
(a) Determine a conjugate prior density having these properties.
(b) Find the corresponding posterior density of θ given a random sample of n observations from the Poisson distribution.
(c) Derive px ressions, in terms of n, the sample observations, and probability points of a χ^2 distribution, for the limits of the $1 - \alpha$ equal-tailed credible interval for θ.

7.7 Suppose the posterior distribution of θ is

$$q(\theta) = \begin{cases} \theta, & 0 \leqslant \theta \leqslant 1 \\ 2 - \theta, & 1 < \theta \leqslant 2 \end{cases}$$

(a) Show that $(\sqrt{0.05}, 2 - \sqrt{0.05})$ is a 95% HPD credible interval for θ.
(b) Let $\phi = \theta^2$.
  (i) Show that $(0.05, \{2 - \sqrt{0.05}\}^2)$ is a 95% credible interval for ϕ and verify that it is not a 95% HPD interval.
  (ii) Find a 95% HPD interval for ϕ.
  (iii) Explain why there are an infinite number of 40% HPD credible intervals for ϕ.

7.8 Suppose that X is the number of successes in a binomial experiment with n trials and probability of success θ. Either $H_0\colon \theta = \frac{1}{2}$ or $H_1\colon \theta = \frac{3}{4}$ is true. Show that the posterior probability that H_0 is true is greater than the prior probability for H_0 if and only if

$$x \ln(3) < n \ln(2).$$

7.9 Let $X_1, X_2, \ldots, X_n$ be a random sample from an exponential distribution, $f(x; \theta) = \theta\, e^{-\theta x}$, $x > 0$, $\theta > 0$. Suppose we wish to test $H_0\colon \theta = 1$ against $H_1\colon \theta \neq 1$, where $\Pr[H_0] = p$ and $\Pr[H_1] = 1 - p$. If the prior distribution for θ, given H_1, is

$$p_1(\theta \mid H_1) = \beta^\alpha \theta^{\alpha-1}\, e^{-\beta\theta}/\Gamma(\alpha), \qquad \theta \neq 1,$$

determine the posterior odds $Q^* = \Pr[H_0 \mid \mathbf{x}]/\Pr[H_1 \mid \mathbf{x}]$ using
(a) equaton (7.10),
(b) equation (7.13).

7.10 Let $X_1, X_2, \ldots, X_n$ be a random sample from $N(\theta, 1)$. There are two hypotheses, $H_0\colon \theta = 1$ and H_1, where $\Pr[H_0] = p$ and $\Pr[H_1] = 1 - p$.
(a) Given that H_1 specifies $\theta = -1$, show that

$$\Pr[H_0 \mid \mathbf{x}] = \frac{p\, e^{\Sigma x_i}}{p\, e^{\Sigma x_i} + (1-p)\, e^{-\Sigma x_i}}.$$

(b) Given that H_1 specifies $\theta \neq 1$ and gives θ the prior distribution

$$p(\theta \mid H_1) = \frac{1}{\sqrt{2\pi}} \exp(-\theta^2/2), \qquad \theta \neq 1,$$

determine $\Pr[H_0 \mid \mathbf{x}]$ when $\Sigma x_i = n$.

7.11 In order to measure the intensity, θ, of a source of radiation in a noisy environment a measurement X_1 is taken without the source present and a second, independent measurement X_2 is taken with it present. It is known that X_1 is $N(\mu, 1)$ and X_2 is $N(\mu + \theta, 1)$, where μ is the mean noise level. The prior distribution for μ is $N(\mu_0, 1)$ while the prior distribution for θ is constant. (Thus μ is known with some accuracy while little is known about θ.)

(a) Write down (apart from a constant of proportionality) the joint posterior distribution of μ and θ.

(b) Hence obtain the posterior marginal distribution of θ.

(c) The usual estimate of θ is $x_2 - x_1$. Explain why $\frac{1}{2}(2x_2 - x_1 - \mu_0)$ might be better.

7.12 The number of phone calls a man receives in a week follows a Poisson distribution with mean θ. At one point in time, the man's opinion about the value of θ corresponds to the gamma distribution:

$$p(\theta) = \tfrac{1}{54}\theta^2 \, e^{-\theta/3}, \qquad \theta > 0.$$

In the next 4 weeks the man received 3, 7, 6 and 10 phone calls, respectively. Determine the distribution that should now represent his opinion about θ and find the predictive distribution for the number of calls he will receive in week 5.

7.13 Independent observations $X_1, X_2, \ldots, X_{10}$ from a power distribution with p.d.f.

$$f(x; \theta) = \theta x^{\theta - 1}, \qquad 0 < x < 1, \qquad \theta > 0$$

take the values 0.91, 0.41, 0.99, 0.05, 0.90, 0.76, 0.59, 0.98, 0.78, 0.51. If the prior distribution for θ is such that $\theta/0.075$ has a χ^2_{20} distribution (i.e. $p(\theta) \propto \theta^9 \, e^{-6.67\theta}$), find the posterior distribution of θ.

Find also the predictive distribution for a further independent observation, X_{11}. Given that m is the best prediction of X_{11} under a linear loss function, show that m satisfies

$$\left(\frac{12.51}{12.51 - \ln(m)}\right)^{20} = 0.5.$$

7.14 The unknown means of two Poisson distributions are θ_1 and θ_2, and these have independent and identical prior distributions

$$p(\theta) \propto \theta^{\alpha_1} \, e^{-\alpha_2 \theta}, \qquad \alpha_1 > -1, \quad \alpha_2 > 0.$$

A single observation from each Poisson distribution is taken, giving values x_1 and x_2, respectively. If $\eta = \theta_1/(\theta_1 + \theta_2)$, show that the posterior distribution of η is $\text{Beta}(x_1 + \alpha_1 + 1, x_2 + \alpha_1 + 1)$.

7.15 Suppose independent samples of sizes n_1 and n_2 are taken from two normal distributions with known means, μ_1 and μ_2, and unknown precisions, θ_1 and θ_2. For $i = 1, 2$, let $r_i = \Sigma_j (x_{ij} - \mu_i)^2$ where $\{x_{i1}, x_{i2}, \ldots, x_{in_i}\} = \mathbf{x}_i$ are the observations from population i. Show that

$$(r_i + \alpha_i)\theta_i \sim \chi^2_{(n_i+v_i+2)} \quad \text{for} \quad i = 1, 2.$$

Hence derive equation (7.17).

7.16 Suppose $X_1, X_2, \ldots$ is a sequential sample from a B(1, θ) distribution and we have an uninformative prior distribution for θ, $p(\theta) = \theta^{-1}(1-\theta)^{-1}$. Suppose also that θ is to be estimated under a quadratic loss function.
(a) After observing $x_1, x_2, \ldots, x_n$, show that $\bar{x}_n = (1/n)\,\Sigma x_i$ is the Bayes estimate of θ.
(b) Suppose the following stopping rule is used. Take X_1 and stop if $X_1 = 1$. If $X_1 = 0$, take X_2 and then stop, regardless of the value of X_2. Show that, using frequentist methods, $\text{E}[\bar{X}] = \theta + \frac{1}{2}\theta(1-\theta)$. (Thus, from a frequentist perspective, the Bayes estimator is biased under this stopping rule. A Bayesian might argue that it is not important for an estimator to satisfy the frequentist definition of unbiasedness.)

7.17 Suppose k different people repeatedly perform a test that results in success or failure. Let X_i be the number of successes the ith person obtains in n_i trials. Assume $X_i \sim B(n_i, p_i)$, where the p_i vary between people but assume that, *a priori*, the p_i are exchangeable. Construct a hierarchical model in which the p_i are independent values from a beta distribution whose parameters have an uninformative prior distribution. Form an equation equivalent to (7.19) and explain how estimates of the p_i could be obtained.

7.18 Derive equation (7.21) from equation (7.20).

7.19 Suppose (x_i, θ_i), $i = 1, 2, \ldots, k$ is a sequence of i.i.d. bivariate random variables for which only the x_i are observable. Suppose also that each x_i has a binomial distribution, $B(n, \theta_i)$, and $\theta_i \sim \text{Beta}(\alpha, \beta)$. Show that, using the method of moments, the empirical Bayes estimates of α and β are

$$\alpha = \frac{n\bar{x}^2 - \bar{x}^3 - \bar{x}s^2}{ns^2 - n\bar{x} + \bar{x}^2}, \qquad \beta = \frac{(n-\bar{x})(n\bar{x} - \bar{x}^2 - s^2)}{ns^2 - n\bar{x} + \bar{x}^2}$$

where $\bar{x}$ and s^2 are the sample mean and variance of $x_1, \ldots, x_k$.

7.20 Suppose $X_i \mid \theta_i \sim N(\theta_i, 1)$ where $X_1, \ldots, X_k$ are independent and $\theta_i \sim N(\mu, \sigma^2)$, where $\theta_1, \ldots, \theta_k$ are independent. Determine empirical Bayes estimates of μ and σ^2
(a) using the method of moments;
(b) using maximum likelihood.

7.21 Suppose $f(x_i; \theta_i)$ is the geometric distribution $f(x_i; \theta_i) = \theta_i(1-\theta_i)^{x_i-1}$, $i = 1, 2, \ldots, k$, and θ_k must be estimated under a squared-error loss function. Show that $\{f(x_k) - f(x_k+1)\}/f(x_k)$ is a non-parametric EB estimator of θ_k. Given that the θ_i are independent and identically distributed, explain how this expression can be used to estimate θ_k from $x_1, x_2, \ldots, x_k$.

7.22 Suppose an observation comes from the uniform distribution $U(0, \theta_0)$ with probability p and from $U(0, \theta_1)$ with probability $1-p$, where $\theta_0 < \theta_1$. Out of k independent observations, n were in the range $(0, \theta_0)$ while the remaining $k - n$ were in the range (θ_0, θ_1). Show that the maximum likelihood estimate of p is

$$\hat{p} = \begin{cases} \dfrac{n\theta_1 - k\theta_0}{k(\theta_1 - \theta_0)}, & \text{if } n > k\theta_0/\theta_1 \\ 0, & \text{if } n \leqslant k\theta_0/\theta_1 \end{cases}.$$

Given that $n > k\theta_0/\theta_1$, predict the value of the next observation when error in prediction is penalized by a linear loss function.

7.23 For $i = 1, 2, \ldots$, suppose θ_i can take one of two values, 0.2 or 0.5 (some θ_i equal 0.2, while others equal 0.5). Let $p = \Pr[\theta_i = 0.2]$ and $1 - p = \Pr[\theta_i = 0.5]$. Suppose we have 20 independent observations $x_1, x_2, \ldots, x_{20}$, where $f(x_i; \theta_i) = \theta_i \, e^{-\theta_i x_i}$ for $x_i > 0$, and that $\Sigma x_i = 60$.

(a) Use the method of moments to estimate p.

(b) Calculate the probability that the next observation, X_{21}, will exceed 4.

(c) Suppose one must choose between H_0: $\theta_{21} = 0.2$ and H_1: $\theta_{21} = 0.5$, and the loss for incorrectly rejecting H_0 is 10 and for incorrectly rejecting H_1 is 5. For what values of x_{21} should H_0 be accepted and for what values should it be rejected?

Chapter Eight

Non-parametric and robust inference

8.1 Introduction

In much of inference, for both the frequentist and Bayesian approaches, there is an assumption that we know the form of the p.d.f. $f(x; \theta)$, apart from the values of one or more parameters θ. In practice, this is often an unrealistic assumption. For example, a distribution may be approximately normal, but it is rarely exactly so. In more extreme cases, we have little confidence that we can correctly specify a parametric form for a p.d.f. This occurs particularly when only a small sample of observations is available, so that the shape of the distribution is not easily estimated.

It is therefore desirable to construct methods of inference which do not depend on distributional assumptions for their validity, or which are relatively insensitive to any distributional assumptions which are made. These two requirements lead us to the topics of non-parametric inference and robust inference, which are discussed in this chapter. There is also a brief description of semiparametric methods which occupy an intermediate position between non-parametric inference and approaches described in earlier chapters. (Approaches that make full distributional assumptions are termed parametric methods.) Some types of computationally intensive methods are non-parametric in nature. Among these, permutation tests are introduced in this chapter, but discussion of other computationally intensive approaches is deferred until Chapter 9.

8.2 Non-parametric inference

Suppose, as usual, that we have a random sample $(x_1, x_2, \ldots, x_n)$ from a probability distribution with p.d.f. $f(x; \theta)$, but that we do not know $f(x; \theta)$. In previous chapters the form of $f(x; \theta)$ was assumed known, except for the value of the parameter(s), θ, and our objective was to make inferences about θ. If we do not know the form of $f(x; \theta)$ then it is not immediately obvious what we mean by 'inference about θ'. However, there are certain parameters of a distribution such as its median, or a percentile, which do not need explicit knowledge of $f(x; \theta)$ in order for the parameter to be well-defined. In this chapter we deal with inference regarding such parameters.

If $f(x; \theta)$ is unknown, then one possible approach is to devise procedures which make no assumption about the form of $f(x; \theta)$, so-called *distribution-free* procedures. It is often impossible to get away with no assumptions at all, but procedures are available whose assumptions are minimal, for example assuming only continuity of $f(x; \theta)$. The term *non-parametric inference* strictly refers to inferences which do not make statements about parameter values, such as the goodness-of-fit tests discussed in Section 8.2.3. However, the adjective 'non-parametric' is often taken to be synonymous with 'distribution-free', and in much of what follows we adopt this convention.

The topic of non-parametric inference is vast. There are many good text books on the subject. Some, such as Sprent (1989), are aimed at non-statisticians and are very practically orientated. More theoretical background is supplied by texts such as Gibbons (1985) and Pratt and Gibbons (1981).

For virtually every parametric procedure there is at least one non-parametric alternative, and often many. In Sections 8.2.1 and 8.2.2 we concentrate, for hypothesis testing and estimation respectively, on looking in detail at what is perhaps the simplest context for non-parametric inference, namely inference for the location of a single distribution. Although simple, this example illustrates many aspects of non-parametric inference. A second example is also discussed, concerning the analysis of data from so-called randomized block experiments. This example demonstrates that many of the ideas introduced in the simple case carry over to more complex situations.

Both of the examples are concerned with distribution-free inference about parameters, and so are not non-parametric in the strictest sense. Section 8.2.3 discusses goodness-of-fit tests which are genuinely non-parametric in this strict sense.

8.2.1 Hypothesis testing

In parametric inference we devoted more space to point estimation than to hypothesis testing or interval estimation, reflecting the emphasis in the literature. The reverse is true for non-parametric inference – much more has been written on hypothesis testing than on point estimation or interval estimation, so we begin the chapter by discussing various approaches to hypothesis testing within a non-parametric framework. We illustrate some of the possible approaches to non-parametric hypothesis testing using the simple example which is introduced next.

Example 8.1

Suppose that $X_1, X_2, \ldots, X_n$ are a random sample of observations from a distribution with p.d.f. $f(x; \theta)$, where θ is a parameter describing the location of the distribution. Typically in non-parametric inference θ is the median, but it need not be. It is required to test H_0: $\theta = \theta_0$ against some alternative. For definiteness we shall take the alternative to be H_1: $\theta \neq \theta_0$, but H_1 could equally well be one-sided or even simple.

If we assume that the distribution is normal, and that θ is the mean of the distribution, the usual test statistic is the familiar t-statistic $T = (\bar{X} - \theta_0)/(S/\sqrt{n})$.

This statistic is optimal in a number of ways (see Example 4.6), provided that the assumption of normality is true. In particular, it is then asymptotically efficient within a broad class of tests. But such desirable properties will fail to hold if the assumption of normality is invalid. If we cannot substitute any firm alternative distributional assumption, then a distribution-free or non-parametric approach may be appropriate. We now discuss a number of such approaches.

Permutation tests

Permutation tests are a form of computationally intensive procedure and will be discussed further in Chapter 9. However, the ideas which underlie them are commonly used in non-parametric inference, so we introduce them briefly here. The first step is to choose a plausible test statistic for whatever pair of hypotheses we are testing. We then find a family of permutations of the data such that the probability of each permutation can be found under the null hypothesis, H_0. Usually the family is chosen so that all permutations within it are equally likely under H_0. The value of the test statistic is then calculated for each possible permutation. Finally, a p-value is evaluated by comparing the observed value of the test statistic with the probability distribution of the statistic over all possible permutations. In practice, we often only need to look at a subset of the permutations, namely those corresponding to the 'tail(s)' of the distribution of the test statistic. This is illustrated below when we consider the signed-rank test in the context of Example 8.1.

Example 8.1 (*continued*)

To simplify matters slightly, take $\theta_0 = 0$ in our earlier example, so that we are testing H_0: $\theta = 0$ vs. H_1: $\theta \neq 0$, based on $X_1, X_2, \ldots, X_n$ from a distribution with location parameter θ. The first step is to choose a test statistic, and the same statistic as before, which reduces to $T = \bar{X}/(S/\sqrt{n})$, seems a sensible choice.

Suppose that we have observed $x_1, x_2, \ldots, x_n$, and that the distribution is symmetric about θ. Under H_0, it is then equally likely that x_i or $-x_i$ could have been observed, $i = 1, 2, \ldots, n$. We therefore have 2^n equally likely 'permutations' of the data set, obtained by attaching either a positive or negative sign to each of the n observations. For each of these permutations, the value of T can be calculated, thus building up a permutation distribution for T. The position of the observed value of T within this distribution determines the p-value for the test. For example, if $n = 10$, and the observed value of T is the fifth largest among the 2^{10} calculated values, then the p-value is $2 \times 5/2^{10} = 0.0098$ for a two-sided test. This reasoning is illustrated numerically when Example 8.1 is continued further below.

Although with increasing computer power it is becoming ever easier to perform computationally intensive tests such as this, it can still be impossible for large n. In such cases, we can take a random sample of the possible permutations. The null distribution of a test statistic, and hence a p-value, can then be estimated from the values of the statistic for this random sample. Such a test is called a *randomization test*.

It should be noted that there are usually various possible permutation or randomization tests in any particular example. Different test statistics, for example $\bar{X}$ rather than T in Example 8.1, could be chosen. Also, there may be a choice of 'permutations', such as, in Example 8.1, a restriction to those with the same number of positive and negative values as the original sample, though the latter choice has little to recommend it in this case.

Finally, note that the test in Example 8.1 is strictly neither non-parametric in its strict sense (it involves the parameter θ) nor completely distribution-free (we assumed symmetry). Such abuse of terminology is commonplace in non-parametric inference.

Tests based on ranks

Many non-parametric tests adapt a parametric procedure by ranking the observations in some appropriate way, and then replacing the observations by their ranks in the original test statistic.

Example 8.1 (*continued*)

Return to the general pair of hypotheses H_0: $\theta = \theta_0$, H_1: $\theta \neq \theta_0$. The well-known signed-rank test ranks the values of $|X_i - \theta_0|$, $i = 1, 2, \ldots, n$, and takes as its test statistic the sum of the ranks of the data for which $(X_i - \theta_0)$ is negative. This is equivalent to the earlier T statistic with observations replaced by ranks (Exercise 8.3), because of the fixed values for the mean and variance of the ranked observations, which are simply the integers $1, 2, \ldots, n$, assuming there are no ties among observations.

The ideas underlying permutation tests are used in finding the null distribution of this, and many other, non-parametric test statistics. Under H_0, each rank has equal probability of being assigned a positive or negative sign and the value of the test statistic is evaluated for each of these 2^n possibilities. As earlier, an assumption of symmetry is needed. For large n, normal approximations to the null distributions of many nonparametric test statistics have traditionally been used, but greater computer power means that it is increasingly feasible to use the 'exact' distributions – see, for example, Agresti (1992). As a numerical example, consider the following measurements of the resistance, in ohms, of 10 resistors sampled from a batch of supposed 100 ohm resistors.

99.7 99.9 100.4 100.6 100.7 101.1 101.3 101.6 102.1 102.3

If the null hypothesis is H_0: $\theta = 100$, the corresponding values of $X_i - \theta_0\,(= X_i - 100)$ are

-0.3 -0.1 0.4 0.6 0.7 1.1 1.3 1.6 2.1 2.3

The ranks of $|X_i - 100|$ are, respectively,

2 1 3 4 5 6 7 8 9 10,

and the test statistic, W, the sum of the ranks for which $X_i - 100$ is negative, takes the value $2 + 1 = 3$. W can be the sum of any of the 2^n subsets of the integers 1 to

10, and each of these 2^{10} possibilities is equally likely under H_0. Hence the p-value corresponding to the observed value, $W = 3$, is calculated by finding the number of ways in which W can be at least as extreme as 3, and dividing by 2^{10}. Now W can take the value 0 in one possible way, the value 1 in one possible way, the value 2 in one possible way, and the value 3 in two possible ways. Hence $\Pr[W \leqslant 3 \mid H_0] = 5/2^{10} = 0.0049$. This is the p-value for a one-sided alternative $H_1: \theta > 100$, but should be doubled for a two-sided alternative $H_1: \theta \neq 100$.

Example 8.2

Consider a randomized block design, with no replication, so that we have measurements on the random variables Y_{ij}, $i = 1, 2, \ldots, k$; $j = 1, 2, \ldots, b$, where k is the number of treatments and b is the number of blocks. It is assumed that $\mathrm{E}[Y_{ij}] = \mu + \alpha_i + \beta_j$ where μ is the overall mean, α_i is the ith treatment effect and β_j is the jth block effect. This is a special case of the general linear model – see Example 2.12. To uniquely determine the values of the treatment and block effects, the constraints $\Sigma_{i=1}^{k} \alpha_i = \Sigma_{j=1}^{b} \beta_j = 0$ are conventionally imposed. The null hypothesis is $H_0: \alpha_1 = \alpha_2 = \cdots = \alpha_k$.

The detailed analysis of this model (two-way analysis of variance) will be familiar to many readers – see, for example, Hogg and Tanis (1993, Section 8.2). In outline, the total sum of squares of the Y_{ij} about their overall mean is decomposed into contributions, S_T, S_B, S_R, due to treatments, due to blocks, and due to residual variation, respectively. Associated with these contributions are degrees of freedom $(k-1)$, $(b-1)$ and $(k-1)(b-1)$ respectively, and $M_T = S_T/(k-1)$, $M_R = S_R/[(b-1)(k-1)]$ are defined as the so-called treatment and residual mean squares. The standard approach assumes that the Y_{ij} are normally distributed. The test statistic for testing H_0 is $F = M_T/M_R$ and, with the assumption of normality, $F \sim F_{(k-1),(k-1)(b-1)}$ under H_0.

It would be possible to construct a permutation test based on the F-statistic, or alternatively construct a so-called rank-transform version of the F-statistic by simply replacing each Y_{ij} with its overall rank, wherever Y_{ij} occurs in the calculations. However, neither of these approaches gives the best-known non-parametric procedure for randomized blocks, Friedman's test. In Friedman's test, each observation is replaced by its rank *within its block*, and the test statistic is then equivalent to the earlier F-statistic with these ranks, although it is not usually expressed in that form. As in Example 8.1, the null distribution of any of these non-parametric test statistics is constructed using the ideas underlying permutation tests. For example, in Friedman's test there are $k!$ equally-likely permutations of ranks within each block under H_0, so the null distribution is built up by evaluating the test statistic for each of the $(k!)^b$ equally likely permutations.

As a numerical example, consider the measurements of yields given in Table 8.1, in kilograms, from 12 plots of potatoes, consisting of three blocks each containing one plot for each of four varieties. The usual F-statistic for variety effects has a value of 5.60 on 3 and 6 degrees of freedom, corresponding to a p-value of 0.036.

For Friedman's test, the measurements are ranked within each block to give Table 8.2. The Friedman test statistic is equivalent to an F-test using these ranks, and has a p-value, obtained by looking at permutations rather than via an approximate F or χ^2 distribution, of 0.05. There are $(k!)^b = (4!)^3 = 13\,824$ possible permutations, but it is often possible to avoid consideration of all of these. For instance, in the present example, the values in Blocks 1 and 3 are each obtained by a single switch in adjacent ranks from the ordering in Block 2. Permutations which involve switches which are clearly more complicated will lead to less extreme values of the test statistic and need not be considered.

Table 8.1 Yields of potatoes

		Variety			
		A	B	C	D
	1	68	67	71	77
Block	2	82	83	86	89
	3	56	59	64	60

If we replace the original measurements by their overall ranks we obtain Table 8.3.

Table 8.2 Ranking within blocks

		Variety			
		A	B	C	D
	1	2	1	3	4
Block	2	1	2	3	4
	3	1	2	4	3

The F-statistic in this case has a value 9.26, and using a F-distribution with 3 and 6 degrees of freedom to approximate the null distribution of this statistic gives a p-value of 0.011. The discrepancy between this value and those found above casts some doubt on the safety of the last approach, or at least on the validity of the F-distribution as an approximation to the distribution of the test statistic. However, for this last statistic, as well as for the F-statistic based on the original measurements, it would also be possible to perform a permutation test, based on the $(k!)^b$ permutations within blocks of the measurements or their ranks.

Table 8.3 Overall ranking

		Variety			
		A	B	C	D
	1	6	5	7	8
Block	2	9	10	11	12
	3	1	2	4	3

Comparison of tests

We have given two general approaches to the construction of non-parametric tests. There are many other ways of producing such tests and in any particular setting there will usually be a number of possible non-parametric procedures. A natural question which then arises is how to compare competing test procedures. The most usual way of doing so is by looking at the asymptotic relative efficiency (ARE) of tests, a concept which was introduced in Definition 4.8. A result which is frequently useful in calculating AREs is as follows.

Lemma 8.1

Suppose that T_{1n}, T_{2n} are two possible test statistics for the pair of hypotheses H_0: $\theta = \theta_0$ vs. $H_1: \theta > \theta_0$, based on n observations. Let $E[T_{in}] = \mu_{in}(\theta)$,$\text{var}[T_{in}] = \sigma_{in}^2(\theta)$, $i = 1, 2$. Then, under certain regularity conditions, the ARE of T_2 compared to T_1 is

$$\text{ARE}(T_2, T_1) = \lim_{n\to\infty} \left[\frac{(\mathrm{d}\mu_{2n}(\theta)/\mathrm{d}\theta)^2}{\sigma_{2n}^2(\theta)}\bigg|_{\theta=\theta_0} \bigg/ \frac{(\mathrm{d}\mu_{1n}(\theta)/\mathrm{d}\theta)^2}{\sigma_{1n}^2(\theta)}\bigg|_{\theta=\theta_0} \right].$$

Details of the regularity conditions, and a proof of the lemma can be found in Gibbons (1985, Section 14.2). When expressed in this form, the ARE is sometimes known as the *Pitman efficiency*.

Example 8.1 (*continued*)

Here it is convenient to simplify to the pair of hypotheses H_0: $\theta = 0$ vs. $H_1: \theta > 0$, where θ is the mean (or median) of the (symmetric) distribution of $X_1, X_2, \ldots, X_n$. Because we are dealing with asymptotic properties, the t-statistic $T_{1n} = \bar{X}/(S/\sqrt{n})$ is equivalent to $\bar{X}/(\sigma/\sqrt{n})$, where $\sigma^2 = \text{var}(X_i)$, $i = 1, 2, \ldots, n$. Then

$$\mathrm{E}[T_{1n}] = \frac{\theta\sqrt{n}}{\sigma}, \qquad \text{Var}(T_{1n}) = 1 \qquad \text{and} \qquad \frac{\mathrm{dE}[T_{1n}]}{\mathrm{d}\theta} = \frac{\sqrt{n}}{\sigma}.$$

To illustrate the use of Lemma 8.1, we consider as an alternative to the t-test the *sign test*. The test statistic, T_{2n}, in this case is simply the number of negative observations. Then T_{2n} has a binomial distribution with n trials and probability of success p equal to the probability of an observation being negative. Hence $\mathrm{E}[T_{2n}] = np$, $\text{Var}[T_{2n}] = np(1-p)$, and $\mathrm{dE}[T_{2n}]/\mathrm{d}\theta = n\,\mathrm{d}p/\mathrm{d}\theta$. But $p = F(0; \theta)$, where $F(x; \theta)$ is the c.d.f. of an observation. If θ is a location parameter, we can write $F(x;\theta) = F(x-\theta)$, and $\mathrm{d}p/\mathrm{d}\theta = -f(x-\theta) = -f(0; \theta)$. Now using Lemma 8.1, we have

$$\begin{aligned}\text{ARE}(T_2, T_1) &= \lim_{n\to\infty} \left[\frac{(\mathrm{d}\mu_{2n}(\theta)/\mathrm{d}(\theta)^2}{\sigma_{2n}^2(\theta)}\bigg|_{\theta=0} \bigg/ \frac{(\mathrm{d}\mu_{1n}(\theta)/\mathrm{d}\theta)^2}{\sigma_{1n}^2(\theta)}\bigg|_{\theta=0} \right] \\ &= \lim_{n\to\infty} \left[\frac{n^2 f^2(0;\,0)}{n/4} \bigg/ \frac{n/\sigma^2}{1} \right] \\ &= [4\sigma^2 f^2(0;\,0)].\end{aligned}$$

When a particular test is most powerful either globally or within a restricted class of tests it is clearly the most efficient test within that class, so its ARE relative to any other test in that class will be $\geqslant 1$. The t-test holds that position in the present example for normally distributed data, and when the normal assumption holds $\text{ARE}[T_2, T_1] = 4\sigma^2\,(1/(2\pi\sigma^2)) = 2/\pi$. Given that the sign test throws away a lot of information compared to the t-test, this result is perhaps surprisingly favourable to the sign test. For some long-tailed distributions, $\text{ARE}[T_2, T_1]$ can actually exceed 1.

Implementing Lemma 8.1 for the signed-rank test is somewhat lengthier (see Randles and Wolfe, 1979, Section 5.4), but denoting the signed-rank test statistic by T_3, it can be shown that $\text{ARE}\,[T_3, T_1] = 12\sigma^2[\int_{-\infty}^{\infty} f^2(x;\theta)\,\mathrm{d}x]^2$. For normally distributed data this becomes

$$12\sigma^2\left[\frac{1}{2\sigma\sqrt{\pi}}\int_{-\infty}^{\infty}\frac{1}{\sqrt{2\pi(\sigma^2/2)}}\exp\left\{\frac{-x^2}{\sigma^2}\right\}\mathrm{d}x\right]^2 = \frac{3}{\pi}.$$

In fact, it can be shown that $\text{ARE}(T_3, T_1) \geqslant 0.864$ for any continuous $f(x;\theta)$ satisfying the conditions under which Lemma 8.1 holds. Also, the ARE can be very much larger than 1 for long-tailed distributions. Thus it is possible here, and in many other circumstances, to find non-parametric alternatives to standard parametric techniques which are much better than the parametric procedure when its assumptions are invalid and only a little worse when the assumptions hold.

Example 8.2 (*continued*)

In the randomized block example, calculation of AREs is a non-trivial matter and has been the subject of a number of papers, for example Hora and Iman (1988), Thompson and Ammann (1989). The results have, so far, been mixed; neither Friedman's test, the rank transform test, or various other tests suggested for this situation, have uniformly good AREs.

8.2.2 *Point and Interval Estimation*

As noted earlier, there has been a considerable emphasis on hypothesis testing within non-parametric inference. When confidence intervals are found, they are usually constructed by inverting a test of hypothesis, as described in Section 5.2.3. Point estimates are even less frequently cited than confidence intervals. When they are, they have usually been derived from the same reasoning as a non-parametric confidence interval.

Example 8.1 (*continued*)

Consider the sign test for the null hypothesis $H_0\colon \theta = \theta_0$, where θ_0 is any specified value, not necessarily zero. The test statistic, S, is now the number of observations (out of n) which are less than θ_0. Under H_0, $S \sim B(n, \frac{1}{2})$ (assuming that none of the n observations is exactly equal to θ_0).

Now suppose that the sign test of size α for H_0 against a two-sided alternative hypothesis has critical region of the form $\{S \leqslant s - 1 \text{ or } S \geqslant n - s + 1\}$, or equivalently an acceptance region $\{s \leqslant S \leqslant n - s\}$. (Note the symmetry of the acceptance region which follows from the symmetry of $B(n, \frac{1}{2})$.) We can invert this acceptance region to obtain a confidence interval for θ with confidence coefficient $1 - \alpha$. It can be seen that $\{s \leqslant S \leqslant n - s\}$ is equivalent to $\{X_{(s)} < \theta_0 < X_{(n-s+1)}\}$.

Now

$$\begin{aligned}\Pr[s \leqslant S \leqslant n - s \mid H_0] = 1 - \alpha & \\ &= \Pr[s \leqslant S \leqslant n - s \mid \theta = \theta_0] \\ &= \Pr[X_{(s)} < \theta_0 < X_{(n-s+1)} \mid \theta = \theta_0] \\ &= \Pr[X_{(s)} < \theta < X_{(n-s+1)}].\end{aligned}$$

By the definition of a confidence interval, the interval $(X_{(s)}, X_{(n-s+1)})$ is a confidence interval for θ, with confidence coefficient $1 - \alpha$.

A similar, though more complicated, argument leads to confidence intervals for θ based on the signed-rank test. Let W be the signed-rank test statistic. We can write W as

$$W = \sum_{i=1}^{n} \phi(D_i) r(|D_i|) = \sum_{i=1}^{n} \phi(D_i) + \sum_{\substack{i=1 \\ i \neq j}}^{n} \sum_{j=1}^{n} \phi(D_i)\phi(D_j - D_i) \tag{8.1}$$

where $D_i = (X_i - \theta_0)$, $r(|D_i|)$ is the rank of $|X_i - \theta_0|$,

$$\phi(D_i) = \begin{cases} 1, & D_i < 0 \\ 0, & \text{otherwise} \end{cases} \quad \text{and} \quad \phi(D_j - D_i) = \begin{cases} 1, & D_j < D_i \\ 0, & \text{otherwise} \end{cases};$$

see Exercise 8.6.

Consider the $\frac{1}{2}n(n + 1)$ pairwise averages $(X_i + X_j)/2$, $i, j = 1, 2, \ldots, n$, $i \leqslant j$, sometimes called *Walsh averages*. It can be shown that W is equal to the number of these Walsh averages which are less than θ_0.This can be seen by examining the positions of each possible pair X_i, X_j, $i \leqslant j$, relative to each other and to θ_0. For convenience assume that the X_i are labelled in ascending order, so $X_i \leqslant X_j$. There are six possibilities, ignoring ties:

(1) $i = j$, $X_i < \theta_0$
(2) $X_i < X_j < \theta_0$
(3) $X_i < \theta_0 < X_j$, $|X_i - \theta_0| > |X_j - \theta_0|$
(4) $i = j$, $X_i > \theta_0$
(5) $\theta_0 < X_i < X_j$
(6) $X_i < \theta_0 < X_j$, $|X_i - \theta_0| < |X_j - \theta_0|$.

Each of the first three possibilities has $(X_i + X_j)/2 < \theta_0$, and each provides a contribution of 1 to the second expression in equation (8.1). Conversely, $(X_i + X_j)/2 > \theta_0$ for possibilities (4)–(6), and none of these possibilities contributes anything to equation (8.1). Hence W equals the number of Walsh averages less than θ_0, as claimed.

Now suppose that C_α is a critical value for a test based on W, chosen so that

$$\Pr[C_\alpha \leqslant W \leqslant \tfrac{1}{2}n(n+1) - C_\alpha] = 1 - \alpha. \tag{8.2}$$

Then

$$\Pr[W_{(C_\alpha)} < \theta_0 < W_{(\frac{1}{2}n(n+1) - C_\alpha + 1)}] = 1 - \alpha \tag{8.3}$$

where $W_{(k)}$ is the kth smallest Walsh average. To see this, note that because W is the number of Walsh averages less than θ_0, the inequality $W \geqslant C_\alpha$ holds if and only if the C_αth smallest Walsh average is less than θ_0 (i.e. $W_{(C_\alpha)} < \theta_0$). A similar equivalence holds between the upper inequalities in (8.2) and (8.3). Thus, the interval $(W_{(C_\alpha)}, W_{(\frac{1}{2}n(n+1) - C_\alpha + 1)})$ is a confidence interval for θ, with confidence coefficient $1 - \alpha$. A corresponding point estimate for θ, sometimes known as the *Hodges–Lehmann estimator*, is the median of the $\frac{1}{2}n(n+1)$ Walsh averages. In the earlier numerical example of the signed-rank test we saw that $\Pr[W \leqslant 3] = 0.0049$. By symmetry $\Pr[W \geqslant 52] = 0.0049$ so that $\Pr[4 \leqslant W \leqslant 51] = 0.9902$. Hence $(W_{(4)}, W_{(52)})$ gives a 99.02% confidence interval for θ, and the Hodges–Lehmann estimate of θ is $W_{(28)}$ (there are $\frac{1}{2}n(n+1) = 55$ Walsh averages). For the given data, the estimate is 101.00 and the confidence interval is (100.05, 101.95).

For an interval based on the sign test, we have a very limited number of possible values for the confidence coefficient, because of the small number of values for S (smaller than for W). The interval $(X_{(2)}, X_{(9)}) = (99.9, 102.1)$ has confidence coefficient $0.989 = \Pr[2 \leqslant S \leqslant 8]$, whereas $(X_{(3)}, X_{(10)}) = (100.4, 101.6)$ has confidence coefficient 0.891. Note that the interval (99.9, 102.1) with confidence coefficient 0.989 is wider than the interval (100.05, 101.95) based on W although the latter has slightly higher confidence coefficient, 0.990. This reflects, in the confidence interval context, the fact that the sign test is often less efficient than the signed-rank test because it uses less information from the data. The corresponding point estimator of θ based on the sign test is, trivially, the sample median, 100.9.

Example 8.2 (*continued*)

Recall that the model in the randomized block design is $E[Y_{ij}] = \mu + \alpha_i + \beta_j$; the α_i are the parameters of interest, so we would like confidence intervals for $\alpha_i - \alpha_{i'}$, $i \neq i'$. Unfortunately, such intervals are difficult to construct. For example, attempting to invert Friedman's test will lead only to confidence intervals for differences between mean ranks, which is not particularly useful.

8.2.3 *Goodness-of-fit tests and related techniques*

In Sections 8.2.1 and 8.2.2 we outlined and illustrated many aspects of non-parametric inference in the case where interest centres on one or more parameters of an underlying distribution, even if the form of that distribution is unknown. Goodness-of-fit tests and related techniques are different: they seek to make inferences about the whole of a distribution.

The χ^2 goodness-of-fit test

The standard situation is where a random sample of observations $X_1, X_2, \ldots, X_n$ is taken from a distribution with p.d.f. $f(x; \theta)$. However, here we have no interest in any parameters θ, so we use the simplified notation $f(x)$ for the p.d.f. The null hypothesis is H_0: $f(x) \equiv f_0(x)$. In other words, we seek to establish whether or not the observations could have arisen from a distribution with a particular, fully specified, p.d.f. $f_0(x)$. The null hypothesis is often written in terms of the c.d.f. $F(x)$, that is H_0: $F(x) \equiv F_0(x)$.

The best-known procedure in this case is the χ^2 goodness-of-fit test, in which the range of values of X is divided into k classes, and a test statistic is computed as

$$C = \sum_{i=1}^{k} \frac{(O_i - E_i)^2}{E_i}.$$

Here O_i is the observed number of observations in the ith class, and E_i is the expected number of observations in the ith class under H_0, $i = 1, 2, \ldots, k$. Because this test is so well known, we say little further about it – see, for example, Gibbons (1985, Section 4.2) for further details. However, note that its asymptotic null distribution, $C \sim \chi^2_{(k-1)}$, can be derived from the large sample approximation to $-2 \ln \lambda$, where λ is the test statistic in the MLRT (Section 4.6.1). If an MLRT is constructed based on the multinomial distribution defined by the division of the range of X into k classes and H_0, then $-2 \ln \lambda = 2\Sigma_{i=1}^{k} O_i \ln(O_i/E_i)$, which in turn is approximated by C (Exercises 8.9, 8.10). In fact, C is the test statistic for a score test (Section 4.7.1) of the hypotheses in Exercise 8.9, and the closely related statistic $C' = \Sigma_{i=1}^{k} (O_i - E_i)^2/O_i$ arises in a Wald test (Section 4.7.2) of the same hypotheses – see Cox and Hinkley (1974, p. 316). Note also that the quantity $2\Sigma_{i=1}^{k} O_i \ln(O_i/E_i)$ is closely related to the concept of deviance, which plays a central role in generalized linear models – see Chapter 10.

The Kolmogorov–Smirnov test and a related confidence interval

An alternative approach to goodness-of-fit is to use a so-called *Kolmogorov–Smirnov test*. As well as providing a test of H_0: $F(x) \equiv F_0(x)$, this procedure can also be used to construct interval estimates for $F(x)$.

If $X_{(1)} \leqslant X_{(2)} \leqslant \cdots \leqslant X_{(n)}$ are the order statistics for our random sample from $F(x)$, the empirical distribution function for the sample is $F_n(x) = k/n$ for $X_{(k)} \leqslant x < X_{(k+1)}$, $k = 0, 1, 2, \ldots, n$, where $X_{(0)}, X_{(n+1)}$ are defined as $-\infty$, $+\infty$ respectively. $F_n(x)$ is simply a step function which approximates $F(x)$ and could be used as an estimate of $F(x)$. For a general alternative hypothesis, H_1: $F(x) \neq F_0(x)$ for some x, the test statistic in the Kolmogorov–Smirnov test is

$$D_n = \sup_x | F_n(x) - F_0(x) |.$$

It can be shown that the test based on D_n is, like the χ^2 goodness-of-fit test, distribution-free for any continuous distribution – see Gibbons (1985, Theorem 4.1). The exact null distribution of D_n can be computed (Gibbons, 1985, Theorem 4.2), and for large n there is a convenient simple approximation to the c.d.f. of D_n (Gibbons, 1985, Theorem 4.3).

Suppose that $D_{n;\alpha}$ is a critical value of D_n for a test of size α. We reject H_0 only for large values of D_n, so that

$$\Pr[D_n \geqslant D_{n;\alpha} \mid H_0] = \alpha.$$

It follows that

$$\Pr\left[\sup_x |F_n(x) - F_0(x)| < D_{n;\alpha} \mid H_0\right] = 1 - \alpha$$

leading to

$$\Pr[F_n(x) - D_{n;\alpha} < F(x) < F_n(x) + D_{n;\alpha}, \text{ for all } x] = 1 - \alpha.$$

Hence, the region

$$(F_n(x) - D_{n;\alpha}, F_n(x) + D_{n;\alpha})$$

gives a confidence band for $F(x)$, where the confidence coefficient $1 - \alpha$ refers to the probability of the band including $F(x)$ over the whole range of x. This situation, where we require simultaneous confidence intervals for a range of values of x, contrasts with the alternative set-up in finding interval estimates for a function, where confidence coefficients refer to a single value of x – see Section 5.4.

Comparisons

This section is not intended to give a comprehensive review of goodness-of-fit tests. There are many possibilities other than χ^2 or Kolmogorov–Smirnov, but we conclude with some comparisons between these two techniques which also indicate some points of more general interest. As well as having the advantage of a related interval estimation procedure, the Kolmogorov–Smirnov test can also be adapted, unlike χ^2, to deal with one-sided alternatives such as $H_1: F(x) \geqslant F_0(x)$ for all x. Because an exact distribution of D_n can be found, Kolmogorov–Smirnov tests can deal with smaller values of n than χ^2. The χ^2 test also has the disadvantage of needing an arbitrary division of the range of X into classes, which means that the test statistic is not uniquely defined.

Conversely, the Kolmogorov–Smirnov procedure makes the strong assumptions of continuous X and of a fully specified $F_0(x)$. The χ^2 test can cope easily with discrete data. It is also easily adapted to the case where $F_0(x)$ is not fully specified, but has some unknown parameters. For example, we may want to test whether a data set has come from a normal distribution, but are unable to specify the mean and variance of the distribution. The χ^2 test is simply modified by substituting MLEs of the unknown parameters when calculating the E_i and reducing the degrees of freedom by the number of estimated parameters (two, if we estimate mean and variance). Modifications of the Kolmogorov–Smirnov procedure to deal with the complications of discrete data and unknown parameters are much trickier.

Finally we note that there are extensions of the Kolmogorov–Smirnov technique to the comparison of two distributions – see, for example, Gibbons (1985, Section 7.3).

8.3 Semiparametric methods

In parametric inference we have a fully specified model, apart from values of unknown parameters, whereas non-parametric inference makes minimal assumptions about an underlying probability model. Semiparametric inference occupies an intermediate position, with some parts of the underlying model fully parameterized, whilst others are left as unknown functions. Generalized linear models (see Chapter 10) often have a semiparametric nature, with perhaps the best-known example being the proportional hazards regression model. If X is a survival time, with p.d.f. $f(x; \theta)$ and c.d.f. $F(x; \theta)$ then the hazard function is $h(x; \theta) = f(x; \theta)/(1 - F(x; \theta))$. The proportional hazards model expresses $h(x; \theta)$ as

$$h(x; \theta) = \exp(\eta)h_0(x)$$

where $h_0(x)$ is an unknown baseline hazard function, while $\eta = \theta_1 z_1 + \theta_2 z_2 + \cdots + \theta_p z_p$ is a linear function of explanatory variables $z_1, z_2, \ldots, z_p$. Usually interest centres on estimating the parameters $\boldsymbol{\theta} = (\theta_1, \theta_2, \ldots, \theta_p)$ in the first part of the model, with $h_0(x)$ treated as a nuisance function. In this example, and for some other semiparametric models, the nuisance function can be eliminated using the idea of marginal or partial likelihood – see Chapter 3.

8.4 Robust inference

In non-parametric inference we made few assumptions regarding $f(x; \theta)$ (or $f(x)$), the p.d.f. of the distribution from which our observations are drawn. The approach in robust inference is different. Here we have a working assumption about the form of $f(x; \theta)$, but we are not entirely convinced that the assumption is true. What is desired is an inference procedure which in some sense does almost as well as possible if the assumption is true, but does not perform much worse within a range of alternatives to the assumption. Such a procedure is *robust* to departures from the assumption on which it is based. Non-parametric or distribution-free procedures are, by their nature, often robust, but robust methods need not be distribution-free.

There are a number of different definitions of robustness. First, there is the question of from what deviations from assumptions we are seeking robustness. The most usual answer is deviations due to a different form of $f(x; \theta)$ from that assumed, $f_0(x; \theta)$, or deviations due to outliers. Robustness with respect to outliers is sometimes known as being *resistant* to outliers. In fact, resistance in this sense can be included within the more general concept of robustness over a range of distributions, because outliers can be generated by taking $f(x; \theta)$ as a mixture of $f_0(x; \theta)$ and an outlier-generating distribution. Frequently in discussions of robustness the assumed distribution, $f_0(x; \theta)$, is normal, so that the type of robustness of interest is 'robustness to non-normality'. We shall concentrate on deviations in distribution in this section but in some circumstances other types of robustness might be of interest, such as robustness to failure of independence assumptions, or robustness to the assumption of a particular prior distribution in a Bayesian analysis.

A second aspect of robustness is 'robustness with respect to what property?' For example, in a hypothesis testing context Tiku *et al.* (1986) distinguish between 'robustness of validity' and 'robustness of efficiency'. In this context 'robustness of validity' refers to keeping the Type I error of a test procedure stable over a range of distributions, whereas 'robustness of efficiency' means that the power function is stable over the same range of distributions and not much worse than the best power function assuming $f_0(x; \theta)$ is valid.

A third question in studies of robustness is the range of distributions over which a procedure should be robust. It is appealing to use a very broad range but, except with very small sample sizes, the general 'shape' of a data set should allow one to narrow the range considerably. Of course, it can be argued that looking at the data to determine the range is 'cheating' and will affect any inferences made, but we shall not consider such philosophical considerations here.

Much of what has been written about robust inference has been for point estimation, so we shall concentrate on this topic, with relatively brief discussions of hypothesis testing and interval estimation. As with some other topics in this text, we only scratch the surface of a rather large literature.

8.4.1 Point estimation

Breakdown points and influence functions

The best-known approach to robust estimation has its roots in Huber (1964). The book describing the so-called Princeton Study (Andrews *et al.*, 1972), which compared a large number of robust estimators, is another key reference, as is Hampel (1974). Technically, the approach involves some quite advanced ideas, and we shall only outline the main thrust of the arguments. For further details the reader may consult Huber (1981) or Hampel *et al.* (1986) or at a less technical level, Chapter 11 of Hoaglin *et al.* (1983). Suppose that $(x_1, x_2, \ldots, x_n)$ are a random sample from a distribution with c.d.f. $F(x; \theta)$, and $X_1, X_2, \ldots, X_n$ are the r.v.s. corresponding to $x_1, x_2, \ldots, x_n$. Let $\hat{\theta}_n(X_1, X_2, \ldots, X_n)$ be an estimator of θ. A qualitative definition of robustness is that small changes in $F(x; \theta)$ should cause only small changes in the probability distribution of $\hat{\theta}_n$. We shall not formally define what is meant by 'small' here (i.e. we shall not define a metric between probability distributions), but consider the case where the distribution of $X_1, X_2, \ldots, X_n$ is a mixture of an ideal distribution with c.d.f. $F_0(x; \theta)$, and a contaminant distribution with c.d.f. $F_1(x; \theta)$, so that the c.d.f. of the mixture distribution is

$$F(x; \theta) = (1 - \varepsilon)F_0(x; \theta) + \varepsilon F_1(x; \theta). \tag{8.4}$$

In (8.4) we can view ε as a measure of the distance between $F(x; \theta)$ and the ideal c.d.f., $F_0(x; \theta)$. Often there is a critical value of ε, say ε^*, below which the estimator $\hat{\theta}_n$ based on data from $F(x; \theta)$ remains close to that based on $F_0(x; \theta)$, and above which $\hat{\theta}_n$ for $F(x; \theta)$ and for $F_0(x; \theta)$ may be drastically different. This critical value, ε^*, is known as the *breakdown point* for the estimator. Roughly speaking, ε^* represents the proportion of contaminated data that can be present in a sample before the distribution of $\hat{\theta}_n$ changes drastically from its distribution when $F_0(x; \theta)$ is the correct c.d.f.

Example 8.3

When estimating the mean of a distribution with $F_0(x; \theta)$ corresponding to a normal distribution, we find that $\varepsilon^* = 0$ when the estimator is the sample mean, $\varepsilon^* = \frac{1}{2}$ for the sample median, and $\varepsilon^* = \alpha$ for a trimmed mean in which a proportion α of the observations is discarded and then the remainder averaged. These results are intuitively plausible, but our deliberate omission of much of the technical background means that we are unable to prove them – see Huber (1981), Hampel *et al.* (1986). However, the so-called *influence curve*, or *influence function*, provides an alternative way of viewing essentially the same phenomenon.

To define the influence curve, suppose that $T(F) = T[F(x; \theta)]$ is a functional defined on possible c.d.f.s F, such that $T(F) = \theta$. Consider a perturbation, F^*, to F which places a finite amount of probability at a particular value x of the r.v. X, so that $F^* = (1 - \varepsilon)F + \varepsilon\Delta_x$, where $0 < \varepsilon < 1$ and Δ_x represents the probability distribution which takes the value x with probability 1. The influence function is then defined as

$$IF(x; T, F) = \operatorname*{Lim}_{\varepsilon \to 0} \left[\frac{T(F^*) - T(F)}{\varepsilon}\right].$$

As might be suggested by this definition, working with the influence curve can be technically rather difficult, so we look instead at a sample analogue of the influence curve called the *sensitivity curve*. This is defined as

$$SC_{n-1}(x) = n[\hat{\theta}_n(x_1, x_2, \ldots, x_{n-1}, x) - \hat{\theta}_{n-1}(x_1, x_2, \ldots, x_{n-1})].$$

The sensitivity curve is a function of x, and is a measure of how much an estimate can change when an observation with value x is added to a data set consisting of $(n - 1)$ observations $x_1, x_2, \ldots, x_{n-1}$.

Example 8.3 (*continued*)

Consider again the problem of estimating the mean of a distribution. If $\hat{\theta}_n = \bar{x}_n$, then

$$SC_{n-1}(x) = n\left[\left(\sum_{i=1}^{n-1} x_i + x\right)\Big/ n - \left(\sum_{i=1}^{n-1} x_i\right)\Big/ (n - 1)\right]$$

$$= \left((n-1)x - \sum_{i=1}^{n-1} x_i\right) \Big/ (n-1)$$

$$= x - \bar{x}_{n-1},$$

so the estimate can change without bound as x varies.

Now suppose that the estimate is the sample median, and that $x_{(i)}$ is the ith order statistic among $x_1, x_2, \ldots, x_{n-1}$. Then, if n is odd and $n = 2k+1$, $\hat{\theta}_{n-1} = \frac{1}{2}[x_{(k)} + x_{(k+1)}]$, and

$$\hat{\theta}_n = \begin{cases} x_{(k)}, & \text{when } x < x_{(k)} \\ x, & \text{when } x_{(k)} \leqslant x \leqslant x_{(k+1)}. \\ x_{(k+1)}, & \text{when } x > x_{(k+1)} \end{cases}$$

Hence

$$SC_{n-1}(x) = \begin{cases} -\frac{n}{2}[x_{(k+1)} - x_{(k)}], & x < x_{(k)} \\ \frac{n}{2}[2x - x_{(k)} - x_{(k+1)}], & x_{(k)} \leqslant x \leqslant x_{(k+1)}. \\ \frac{n}{2}[x_{(k+1)} - x_{(k)}], & x > x_{(k+1)} \end{cases}$$

In contrast to the mean, however large or small x becomes, the median can only change within the limits $\pm\frac{1}{2}[x_{(k+1)} - x_{(k)}]$.

The corresponding influence function tells the same story – the median has a bounded influence function, the mean has unbounded influence. What this implies is that a small perturbation to the data (i.e. a change in only one observation of the data set) can have an unbounded effect on the mean, but not on the median. In constructing a robust estimator it is desirable to have bounded influence (an estimator cwith this property is said to be *B-robust*) and also to achieve a high breakdown point. Conversely, a high breakdown point is not the only consideration. In estimating a mean the sample median has the largest possible breakdown value, but if $F_0(x; \theta)$ corresponds to a normal distribution, the efficiency of the median is rather poor. A number of families of estimators have been suggested within which we can search for robust and efficient estimators. We now examine some of these families in turn.

M-estimates

Any estimate which is defined by minimizing $\Sigma_{i=1}^n \rho(x_i; \hat{\theta}_n)$ for some function $\rho(x_i; \hat{\theta}_n)$ of the observations and the estimate is called an *M-estimate*. The *M*-denotes 'maximum likelihood type' because maximum likelihood estimates are a special case in which $\rho(x; \theta) = -\log(f(x; \theta))$. Differentiation leads to the equation $\Sigma_{i=1}^n \psi(x_i; \hat{\theta}_n) = 0$, where

$$\psi(x; \hat{\theta}_n) = \left[\frac{\partial}{\partial\theta}\rho(x; \theta)\right]_{\theta = \hat{\theta}_n}$$

which must be solved to find $\hat{\theta}_n$. In the special case, where $\hat{\theta}_n$ is an estimate of location such as the mean, *M*-estimates minimize $\Sigma_{i=1}^n \rho(x_i - \hat{\theta}_n)$ and are obtained by solving

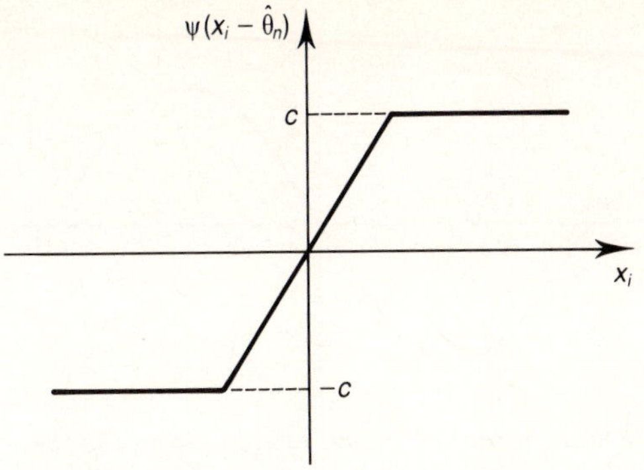

Fig. 8.1 Influence function for Huber *M*-estimator.

$$\sum_{i=1}^{n} \psi(x_i - \hat{\theta}_n) = 0. \tag{8.5}$$

There is a connection between *M*-estimates and influence functions. In particular, it can be shown (Hampel *et al.*, 1986, Section 2.3) that the influence function is proportional to ψ. Thus ψ can be chosen to have the properties of a desired influence function.

Example 8.3 (*continued*)

If we wish to estimate the mean of a distribution, then $\rho(x_i - \hat{\theta}_n) = (x_i - \hat{\theta}_n)^2$ will lead to $\hat{\theta}_n = \bar{x}$. The influence function in this case is $\partial\rho(x - \theta)/\partial\theta = -2(x - \theta) = \psi(x - \theta)$, which is unbounded. Note that $\psi(x - \theta)$ is usually quoted as $2(x - \theta)$, and indeed the change of sign is unimportant in solving (8.5). We shall adopt this standard modification of the definition in other examples which follow.

For $\rho(x_i - \hat{\theta}_n) = |x_i - \theta_n|$, $\hat{\theta}_n$ is the sample median and the influence function is

$$\frac{\partial}{\partial\theta}\rho(x - \theta) = \begin{cases} -1, & x > \theta \\ 0, & x = \theta. \\ 1, & x < \theta \end{cases}$$

A number of other *M*-estimates have been suggested which improve on $\bar{x}$ with respect to robustness whilst losing little power if normality holds. A popular choice, due to Huber, is illustrated in Figure 8.1, and has

$$\rho(x_i - \hat{\theta}_n) = \begin{cases} (x_i - \hat{\theta}_n)^2/2, & |x_i - \hat{\theta}_n| \leqslant c \\ c\,|x_i - \hat{\theta}_n| - c^2/2, & |x_i - \hat{\theta}_n| > c \end{cases}$$

where c is a suitably chosen constant. This leads to

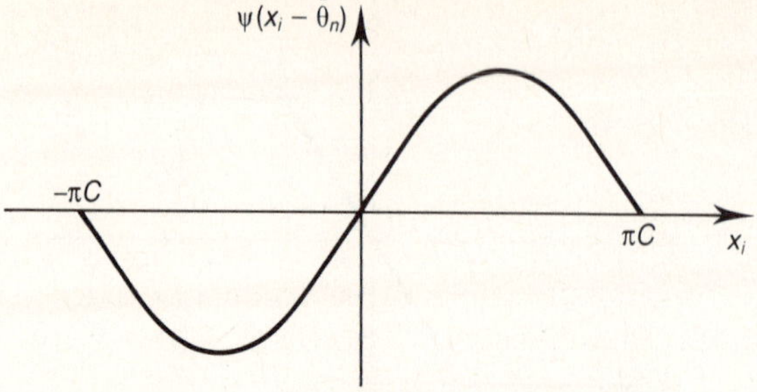

Fig. 8.2 Influence function for Andrews' sine function *M*-estimator.

$$\psi(x_i - \hat{\theta}_n) = \begin{cases} -c, & |x_i - \hat{\theta}_n| < -c \\ (x_i - \hat{\theta}_n), & -c \leqslant |x_i - \hat{\theta}_n| \leqslant c. \\ c, & |x_i - \hat{\theta}_n| > c. \end{cases}$$

Varying the tuning constant c changes the efficiency of the estimator relative to $\bar{X}$ for normally distributed data. For $c = 1$, the efficiency is 90%, for $c = 1.4$ it is 95% and for $c = 1.8$ it is 98%. Clearly there is a trade-off – smaller values of c do worse for near-normal data, but better if there are more substantial violations of the normality assumption.

Huber's *M*-estimate limits the influence function of extreme observations to a value equal to that of an observation which is distance c from the estimate. Other, so-called re-descending, estimates go further and progressively decrease the influence function beyond a threshold c, effectively discarding very extreme observations. One example, due to Andrews (see Hampel *et al.*, 1986, p. 151) has

$$\psi(x_i - \hat{\theta}_n) = \begin{cases} \sin\left(\dfrac{x_i - \hat{\theta}_n}{c}\right), & \text{for } |x_i - \hat{\theta}_n| \leqslant \pi c \\ 0, & \text{elsewhere} \end{cases}$$

where c is again a tuning constant – see Figure 8.2. Hoaglin *et al.* (1983) give some useful comparisons between the properties of various *M*-estimators, and Andrews *et al.* (1972) contains valuable information on the relative performance of many robust estimators.

Solving equation (8.5) must usually be done iteratively. One approach is to rewrite (8.5) as

$$\sum_{i=1}^{n} \frac{\psi(x_i - \hat{\theta}_n)}{(x_i - \hat{\theta}_n)} (x_i - \hat{\theta}_n) = 0 = \sum_{i=1}^{n} w_i(x_i - \hat{\theta}_n) \tag{8.6}$$

where

$$w_i = \frac{\psi(x_i - \hat{\theta}_n)}{(x_i - \hat{\theta}_n)}. \tag{8.7}$$

A solution to (8.6) is then

$$\hat{\theta}_n = \sum_{i=1}^{n} w_i x_i \Big/ \sum_{i=1}^{n} w_i \tag{8.8}$$

a weighted average of the observations.

But the weights w_i in (8.8) depend on $\hat{\theta}_n$, so we start with an initial estimate $\hat{\theta}_{n0}$ and then iterate between equations (8.7) and (8.8) until convergence. Similar iteratively weighted estimation schemes are discussed in other contexts in Sections 3.4.2 and 10.3.

One difficulty with M-estimates as defined in this section is that they are not usually invariant to changes in scale. To avoid this problem, equation (8.5) is replaced by

$$\sum_{i=1}^{n} \psi\left(\frac{x_i - \hat{\theta}_n}{S_n}\right) = 0$$

where S_n is a robust estimate of scale, with other equations modified in a similar manner.

L-estimates, R-estimates and other families of estimates

A number of other classes of potentially robust estimators have been suggested. We note briefly two which have connections with estimators already encountered in this chapter. L-estimators are linear combinations of order statistics. Clearly the median, which we have already met as an M-estimate, falls within this class, as does the family of trimmed means.

R-estimates are derived from rank tests; an example of such an estimate is the Hodges–Lehmann estimate introduced in Section 8.2.2. There are a number of other families of estimators – Hampel *et al.* (1986, Section 2.3d) discuss A-, D-, P-, S- and W-estimators.

Modified maximum likelihood estimators

Tiku *et al.* (1986) describe a different approach to robust estimation. The idea is that it is the smallest and largest observations which contribute to any non-robust behaviour, and it might be desirable to discard these extreme observations, as in a trimmed mean. However, discarding them entirely is wasteful of information, so they may instead be treated as censored observations in the following sense. Suppose that $X_{(1)}, X_{(2)}, \ldots, X_{(n)}$ are the order statistics of the observations $X_1, X_2, \ldots, X_n$, and we censor the first r_1 and last r_2 of them. If the p.d.f. of the observations is $(1/\sigma)f[(x-\theta)/\sigma]$, with corresponding c.d.f. $F[(x-\theta/\sigma]$, then the likelihood function for the censored data is

$$\frac{n!}{r_1!r_2!}\sigma^{-(n-r_1-r_2)}\left(F\left[\frac{x_{(r_1+1)}-\theta}{\sigma}\right]\right)^{r_1}\left(1-F\left[\frac{x_{(n-r_2)}-\theta}{\sigma}\right]\right)^{r_2}\prod_{i=r_1+1}^{n-r_2} f\left(\frac{x_{(i)}-\theta}{\sigma}\right).$$

Maximizing this likelihood gives possible estimators for the location and scale parameters θ, σ. However, for many distributions the equations whose solutions lead to these estimators are quite difficult to formulate and solve, so Tiku *et al.* (1986) propose simpler approximations, leading to their modified maximum likelihood estimators (MMLEs).

The amount of censoring will determine the properties of the estimators. Light censoring will preserve good properties at the assumed distribution, but will only provide protection against small deviations from that distribution. Heavy censoring will give protection against more substantial deviations, but will degrade performance at the assumed distribution. Tiku *et al.* (1986, Section 2.8) report a simulation study in which MMLEs performed well in comparison to a selection of *M*-estimators.

Compromise maximum likelihood estimators

Another suggestion for robustly estimating the location of a distribution is to use so-called compromise maximum likelihood estimators (CMLEs) (Easton, 1991). The idea behind CMLEs is to take a finite number of possible distributions for the data and find the MLE of a location parameter for each. The CMLE is then a weighted average of these individual MLEs. Simulation studies in Easton (1991) show good performance of CMLEs compared with *M*-estimators.

8.4.2 Hypothesis testing

Unlike non-parametric statistics, where much of the literature is concerned with hypothesis testing, robust inference is dominated by point estimation and less has been written on hypothesis testing or interval estimation. However, most robust approaches to estimation can be carried over to hypothesis testing. For example, in the one-sample location example (Example 8.1) we can modify the test statistic $T = \bar{X}/(S/\sqrt{n})$ to become $\hat{\mu}/(\hat{\sigma}/\sqrt{n})$, where $\hat{\mu}$, $\hat{\sigma}$ are MMLEs of μ and σ. Tiku *et al.* (1986, Theorem 4.2.1) show that if the underlying distribution is normal, then the null distribution of this modified t-statistic is still asymptotically a t-distribution, with $n - 2r$ degrees of freedom if r observations are censored from each tail of the distribution. They claim that the result is still true, to a close approximation, for n as small as 10.

Robust hypothesis testing may also be approached using influence functions and *M*-, *L*- or *R*-estimators. Hampel *et al.* (1986, Section 3.2) define a test influence function that is closely related to the influence function for estimators. As with estimators, a robust test procedure should have a bounded influence function. If $\hat{\theta}_n$ is an estimator of type M, L or R based on a sample of n observations, and we have a null hypothesis $H_0\colon \theta = \theta_0$, then a possible test statistic is $S_n = \sqrt{n}(\hat{\theta}_n - \theta_0)$. Such test statistics inherit the same influence functions as the estimators on which they are based, so that a choice of test statistic can be made on the same grounds as a choice of estimator.

8.4.3 *Interval estimation*

Little has been written on robust confidence intervals. However, it is clear that given a robust estimator, together with its distribution, a confidence interval can be constructed along the lines of Section 5.2.2. For example, if the test statistic, $\hat{\mu}(\hat{\sigma}/\sqrt{n})$, based on MMLEs, has an approximate t distribution with $n - 2r$ degrees of freedom, then $\hat{\mu} \pm t_{(n-2r);\,\alpha/2}\hat{\sigma}/\sqrt{n}$ gives endpoints of a confidence interval for μ with approximate confidence coefficient $1 - \alpha$, where $t_{(n-2r);\,\alpha/2}$ denotes a critical value, corresponding to tail probability $\alpha/2$, for $t_{(n-2r)}$.

Alternatively, a robust test of hypothesis may be inverted using the ideas of Section 5.2.3 to provide a robust confidence region.

Exercises

8.1 A random sample of six observations on an r.v. X takes the values $-1, 2, 4, 6, 7, 8$, where X is assumed to have a symmetric distribution centred at θ. Use
(a) the sign test;
(b) the signed rank test
to test $H_0\colon \theta = 0$ vs. $H_1\colon \theta > 0$.

Explain why the p-value for the signed-rank test is smaller than that for the sign test, but the two p-values would be identical if the -1 in the data set were relaced by $+1$.

8.2 For the data set in Exercise 8.1 carry out permutation tests for $H_0\colon \theta = 0$ vs. $H_1\colon \theta > 0$ based on
(a) the statistic $\bar{X}$;
(b) the statistic $T = \dfrac{\sqrt{n}\bar{X}}{S}$.
[*Hint*: By considering only permutations which lead to potentially extreme values of your statistic, you can avoid unnecessarily tedious calculations. Doing this is straightforward for (a), but less obvious for (b)].

8.3 Show that the signed-rank test for testing $H_0\colon \theta = \theta_0$ vs. $H_1\colon \theta \neq \theta_0$, based on a random sample of observations $X_1, X_2, \ldots, X_n$, is equivalent to a permutation test based on the statistic $T = \sqrt{n}\bar{Y}/S_y$, where $Y_1, Y_2, \ldots, Y_n$ are the ranks of $X_1, X_2, \ldots, X_n$ and $\bar{Y}$, S_y are the mean and standard deviation of $Y_1, Y_2, \ldots, Y_n$.

8.4 Using the formula given in Section 8.2.1, show that the asymptotic relative efficiency (ARE) of the sign test compared to the t-test is

(a) $\frac{2}{3}$, when $f(x;\theta) = \begin{cases} x - \theta + 1, & \theta - 1 \leqslant x \leqslant \theta \\ \theta + 1 - x, & \theta \leqslant x \leqslant \theta + 1 \\ 0, & \text{elsewhere} \end{cases}$

(b) 2, when $f(x;\theta) = \frac{1}{2}e^{-|x-\theta|}$.

8.5 Find the ARE of the signed-rank test compared to the t-test for the same two distributions as in Exercise 8.4.

8.6 Show that the signed-rank test statistic, W, can be written in the form given by equation (8.1).

8.7 For the data given in Exercise 8.1 a confidence interval for θ given by inverting the sign test is $(-1, 8)$. Find the confidence coefficient for this interval.

8.8 Given that W is the test statistic for a signed-rank test with a sample size $n = 6$, show that $\Pr[W \leqslant 3] = \frac{5}{64}$. Hence, find a confidence interval, with confidence coefficient $\frac{54}{64}$, for θ, using the data of Exercise 8.1.

Find also the Hodges–Lehmann estimate of θ for these data.

8.9 Consider a multinomial distribution with k classes. Let p_i be the probability of falling in the ith class, and let n_i be the number of observations, out of n, which fall in the ith class, $i = 1, 2, \ldots, k$. It is required to test H_0: $p_i = p_{i0}$, for some specified set of probabilities $p_{10}, p_{20}, \ldots, p_{k0}$, against the general alternative H_1: $p_i \neq p_{i0}$ for some i.

Find the test statistic, λ, for an MLRT of H_0 vs. H_1, and show that

$$-2 \ln \lambda = 2 \sum_{i=1}^{n} O_i \ln\left(\frac{O_i}{E_i}\right)$$

where $O_i = n_i$ and $E_i = np_{i0}$, $i = 1, 2, \ldots, k$.

8.10 By writing

$$\frac{O_i}{E_i} = \left(1 + \frac{(O_i - E_i)}{E_i}\right)$$

and expanding $\ln(O_i/E_i)$, show that $-2 \ln \lambda$ in Exercise 8.9 is approximately equal to $\Sigma_{i=1}^{k}(O_i - E_i)^2/E_i$.

8.11 A one-sided Kolmogorov–Smirnov test of H_0: $F(x) \equiv F_0(x)$ against the alternative H_1: $F(x) \geqslant F_0(x)$ for all x, uses the test statistic $D_n^+ = \mathrm{Sup}[F_n(x) - F_0(x)]$. For large n, $\Pr[D_n^+ < z/\sqrt{n}] \cong 1 - e^{-2z^2}$. Show that $4nD_n^{+2} \mathrel{\dot\sim} \chi_2^2$.

8.12 A 10% trimmed mean discards the largest 5% and the smallest 5% of a set of data and calculates the average of the remaining 90% of the data set. Find the form of the sensitivity curve for this estimator.

8.13 Investigate the form of the sensitivity curve for the Hodges–Lehmann estimate (Section 8.2.2).

8.14 Consider the Hampel M-estimator with

$$\rho(x_i, \hat\theta_n) = \begin{cases} (x_i - \hat\theta_n)^2/2, & |x_i - \hat\theta_n| \leqslant a \\ a[|x_i - \hat\theta_n| - a] + \dfrac{a^2}{2}, & a \leqslant |x_i - \hat\theta_n| \leqslant b \\ \dfrac{a}{2}[b + c - a] - \dfrac{a}{2(c-b)}[c - |x_i - \hat\theta_n|]^2, & b \leqslant |x_i - \hat\theta_n| \leqslant c \\ \dfrac{a}{2}[b + c - a], & |x_i - \hat\theta_n| \geqslant c \end{cases}$$

where $0 < a < b < c < \infty$.

Plot $\psi(x_i, \hat\theta_n)$ for this estimator.

8.15 Consider the M-estimator due to Andrews, with

$$\psi(x_i, \hat\theta_n) = \begin{cases} \sin\left(\dfrac{x_i - \hat\theta_n}{c}\right), & |x_i - \hat\theta_n| \leqslant \pi c \\ 0, & \text{elsewhere.} \end{cases}$$

Find, and sketch, the form of $\rho(x_i, \hat\theta_n)$ for this estimator.

Chapter Nine

Computationally intensive methods

9.1 Introduction

Computationally intensive methods are generally taken to be methods that repeat a relatively simple operation many times. Common features of the methods are that they

(1) require substantially more computer time as the number of data items increases,
(2) involve random number generation to assign values to random variables or choose a random sample.

Permutation tests exhibit trait (1), since the number of permutations increases dramatically with an increase in sample size. In Example 8.1, for instance, either a positive or negative sign is attached to each observation, giving 2^n 'permutations' when the sample size is n. Thus adding a single datum doubles the number of permutations that must be considered. Randomization tests illustrate (2) since they involve taking a *random* sample of the possible permutations.

Most of the methods we consider are highly versatile and well suited to the analysis of complex models. Further advantages vary. Some methods such as permutation tests, make few distributional assumptions. Other methods provide solutions to complex problems when alternative techniques do not exist or would yield only approximate answers. The disadvantage of the methods is that they are expensive in computer time, other than for comparatively small problems. However, this disadvantage has clearly become less important with the vast increase in cheap computer power that is now widely accessible.

9.2 Simulation and Monte Carlo methods

Simulation methods and Monte Carlo methods are versatile and they have wide application in statistics. The distinction between them is somewhat vague, although methods involving i.i.d. observations tend to be called *Monte Carlo methods* while those involving correlated observations from complex models are more likely to be called *simulation* methods. Here, space prevents us from doing more than outlining

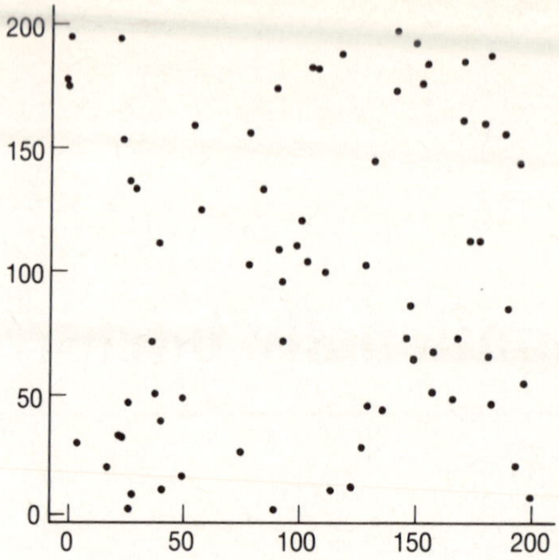

Fig. 9.1 The positions of 72 trees in a square with 200 metre sides.

two or three typical examples to show the use of simulation for inference and indicate some of its strengths and weaknesses.

The first example arises with spatial data. We suppose there are objects in some well-defined area and we are interested in the way they are dispersed over that area. The objects might be a species of wild-flower in a meadow, for example, in which case the flowers might cluster in those parts of the meadow where conditions suited them, or they might be trees in an orchard, where they would be very evenly spaced. Here we suppose the hypothesis to be tested is that the distribution of objects is random.

For illustration we examine data on pine trees given by Cressie (1993, pp. 581–5). The data are the location and diameters at breast height of longleaf pines in one section of a natural forest. The section was 200 m square and there were 72 large trees that had a breast height diameter of 50 cm or more. Their positions are mapped in Figure 9.1. (By eye, one might guess that they are either randomly distributed or loosely clustered.)

Cressie (1993, Chapter 8) describes several tests of randomness for spatial point patterns. We will apply one based on nearest-neighbour distances. For this, the distances from each tree to its nearest neighbour are determined and their average, d^* say, is calculated. Clearly d^* will be small if trees are clustered and relatively large if they tend to be evenly spaced. To assess whether d^* casts doubt on the assumption of randomness, a Monte Carlo test generates a large number of data sets. Each set consists of 72 points as this is the number of trees in the real data set. The simulated points must be randomly positioned in a 200 m square, which is achieved by generating independent values from a $U(0,200)$ distribution and using two values as the coordinates of a point.

For each simulated data set, the average distance from trees to their nearest neighbour is determined in exactly the same way as with the real data set. Denote these averages by $d_1, d_2, \ldots, d_n$, where n is the number of simulated data sets. If trees were randomly distributed, then d^* and $d_1, d_2, \ldots, d_n$ would all be observations from the same distribution. Thus we can determine the significance of d^* by comparing it with $d_1, d_2, \ldots, d_n$. For our example, 999 simulated data sets were generated and, of these, 369 gave values of d_i that were below d^* while the remaining 630 were above it. Hence we have 1000 observations (including d^*) and 370 of these are less than or equal to d^*. The hypothesis of randomness is consequently rejected at only the $370/1000 = 37\%$ significance level, providing little evidence against it.

Theoretical results have been derived that give the asymptotic distribution of d^*, but these ignore edge-effects at the boundary of the area. The Monte Carlo test thus seems preferable as it does not rely on asymptotic theory and also it takes edge-effects fully into account. Note that the test may be used with areas of any shape and may exclude well-defined sub-areas, such as a pond, in which trees cannot grow. Points would be generated at random within a rectangle that contains the region of interest. Only points within this region would be accepted, and points would be generated until the required number of acceptable points had been obtained. This type of flexibility in Monte Carlo methods is common – often changes of detail in a problem are easily accommodated by a Monte Carlo method.

Perhaps the most common use of simulation in statistics is to examine the accuracy of asymptotic results when samples are finite. To illustrate this application we will consider the Robbins–Monro process (Robbins and Monro, 1951), because this also leads to a Monte Carlo method for determining confidence intervals.

Suppose the result of a treatment is a binary response. Thus the response is one of two possible outcomes that might, for example, correspond to success–failure, zero–one, life–death or cured–not cured. Increasing the dose level of the treatment increases the probability of success and an estimate of the dose level that gives a specified proportion of successes is required. Denote this proportion by p and let D_p be the corresponding dose level. With the Robbins–Monro process a sequence of trials is conducted. At each trial, an item/patient/animal is given a dose whose level equals the current estimate of D_p. If the trial results in a success, the estimate of D_p is reduced; otherwise it is increased. Specifically, if x_j is the estimate of D_p before the jth trial, then

$$X_{j+1} = \begin{cases} x_j - \dfrac{c(1-p)}{j}, & \text{if the } j\text{th trial is a success} \\[2ex] x_j + \dfrac{cp}{j}, & \text{if the } j\text{th trial is a failure,} \end{cases} \tag{9.1}$$

where c is a chosen constant. (Notice that the revisions of the estimate get smaller as the sequence of trials progresses and j increases.) To see that the estimate of D_p is expected to improve at each trial, let y_j denote the probability of success when the dose level is x_j. Then

$$\begin{aligned} \mathrm{E}[X_{j+1}] &= y_j\{x_j - c(1-p)/j\} + (1-y_j)\{x_j + cp/j\} \\ &= x_j + c(p - y_j)/j. \end{aligned}$$

If $x_j > D_p$, then $y_j > p$ and X_{j+1} is expected to be smaller than x_j. Similarly, if $x_j < D_p$ then $y_j < p$ and X_{j+1} is expected to be bigger than x_j.

The asymptotic properties of the Robbins–Monro process are as follows. Let g be the value of $[dy/dx]$ evaluated at $x = D_p$, where $y = y(x)$ is the probability of success in a trial when the dose is x. As $j \to \infty$, $x_j - D_p$ is asymptotically normally distributed with a mean of 0 and a variance of $p(1-p)c^2/j(2gc-1)$. The value of the constant c that minimizes this variance is $c = g^{-1}$ and for this value the variance equals the Cramer–Rao lower bound to the variance of non-parametric estimates of D_p (Wetherill, 1975, Chapter 10). Hence asymptotic properties are very good for well-chosen values of c.

Simulation can be used to examine properties of the process for finite sample sizes. For each simulation run, four quantities must be specified: the relationship between y and x, c, D_p, and the first dose, x_1. For the dose x_j, the value of y_j (the probability of success) is calculated. A random variable from $U(0, 1)$ is generated and if its value is less than y_j, then the result of the jth simulated trial is a success; otherwise the result of the jth trial is a failure. After a specified number of trials, n say, the procedure is stopped and x_{n+1} is the estimate of D_p. The simulation is repeated many times so that $E(X_{n+1})$ and $\text{var}(X_{n+1})$ can be estimated. These can then be compared with D_p and the asymptotic estimate of the variance.

Wetherill (1963) conducted a simulation study of the Robbins–Monro process. He found the method performed very much worse than asymptotic theory predicted when n, the number of trials, equalled 15, 25 or 35. For example, in one set of simulations the task was to estimate the upper quartile of the logistic model,

$$y = \Pr(\text{success} \mid \text{dose} = x) = 1/(1 + e^{-x}).$$

Using the optimal value for c and starting each simulation run at the correct value (i.e. setting $x_1 = D_{0.75}$), the average estimate of $D_{0.75}$ after 35 trials was 1.296 and the variance of these estimates was 0.229. This average estimate shows marked bias, as $D_{0.75}$ actually equals 1.099, and the variance should be only 0.152 according to asymptotic theory. Many more than 35 trials are unlikely to be used in a sequential set of trials, where one trial must end before the next one starts, so the simulations suggest the Robbins–Monro process is unsuitable for most experiments involving a binary response.

The process can also be used to construct confidence intervals in one-parameter problems, provided the mechanism that gave the real data could be simulated if the parameter's value were known. Denote the parameter by θ and its estimate based on real data by $\hat{\theta}$. Suppose we assign θ some value and then generate a set of data using a mechanism similar to that which gave the real data. From the simulated data, estimate θ and if this estimate is greater than $\hat{\theta}$, call the result of the 'trial' a success, while if it is less than $\hat{\theta}$, call the result a failure. If the estimate equals $\hat{\theta}$, call the result a success when estimating the upper end point and a failure when estimating the lower end point. To obtain a $(1-2p)$ equal-tailed confidence interval for θ, use the Robbins–Monro process to estimate D_p, where $\Pr[\text{success}] = p$ when θ is given the value D_p. From the duality between hypothesis testing and confidence intervals described in Section 5.2.3, D_p is the lower end point of the required confidence interval. To obtain the upper end point, the Robbins–Monro process is used again, this time to estimate D_{1-p}. In determining a confidence interval in this way, it is practical to use thousands of simulated data sets (trials), so the asymptotic properties of the Robbins–Monro process hold much better. Strategies for choosing a starting value (x_1) and the constant c are suggested by Garthwaite and Buckland (1992), where some simulation results are also reported.

Example 9.1

As a simple illustration, suppose we have a random sample of 10 observations from $N(\theta,1)$ and we wish to use the Robbins–Monro process to estimate the lower end point of a 95% confidence interval for θ. Let $\hat{\theta}$ be the mean of the 10 observations and suppose x_j is the estimate of the lower end point after $j-1$ trials. Then for the jth trial we generate 10 observations from the distribution $N(x_j, 1)$. If the mean of the 10 simulated data is greater than $\hat{\theta}$, then the result of the trial is a success; otherwise it is a failure. If x_j is close to the lower end point, $D_{0.025}$, then the probability of success will be close to 0.025. The new estimate of $D_{0.025}$ is x_{j+1}, where

$$x_{j+1} = \begin{cases} x_j - 0.975c/j, & \text{if the } j\text{th trial was a success} \\ x_j + 0.025c/j, & \text{if the } j\text{th trial was a failure.} \end{cases}$$

If 1000 trials are simulated in total, then x_{1001} is taken as the estimate of the lower end point.

Another common use of simulation is to compare different inference procedures, the aim being to decide if one method is better than others or identify situations where particular methods are likely to give superior results. One important example that has attracted much attention is multiple regression. Many methods for estimating regression coefficients and forming prediction equations have been suggested: ordinary least squares, stepwise regression, ridge regression and regression on principal components, amongst others. These methods are described in many books on multiple regression (e.g. Gunst and Mason, 1980). The regression model has the form

$$Y = \beta_1 X_1 + \beta_2 X_2 + \ldots + \beta_k X_k + \varepsilon$$

and values or distributions for all the quantities on the right-hand side of the equation must be chosen in a simulation run. Clearly certain values would favour particular regression methods. For example, stepwise regression includes only a subset of the X-variables in a prediction equation and so should do well when many of the β-coefficients are very close to 0.

As particular values for the βs will favour some methods, it is quite common to give them different values in each run of a set of simulations. This is usually done by specifying a distribution for the βs and then, in each run, random values are generated from the distribution and assigned to the βs. This approach reduces bias between methods but does not eliminate it. Essentially, the distribution specified for the βs forms a prior distribution and Bayesian analysis indicates that some forms of prior distribution favour particular methods. For example, if the βs are i.i.d with normal distributions, then ridge regression is likely to be favoured. This is because the Bayesian posterior distribution and ridge regression with a particular choice of the ridge parameter give very similar estimators of the βs (Lindley and Smith, 1972). The conclusion to draw is that care should be taken in assigning values to parameters and other quantities in a simulation study. Ideally, one should have an idea of which quantities most influence the relative performances of methods, and vary these quantities systematically. Also, caution should be used in drawing general conclusions from simulations, as simulations are inevitably limited in their scope.

9.3 Permutation and randomization tests

Permutation and randomization tests were introduced earlier, in Section 8.2.1. The basic idea is to find a family of permutations of the data, such that the probability of each permutation is known under the null hypothesis, H_0. Typically, each permutation has equal probability. The probability distribution (under H_0) of a specified test statistic may then be evaluated by determining the statistic's value for each permutation. Alternatively, if the number of permutations is very large, the distribution is estimated using a random sample of the permutations. Comparison of the observed value of the test statistic with this probability distribution gives a measure of the evidence against H_0.

We have distinguished between permutation and randomization tests on the basis of whether all permutations are examined or just a randomly chosen subset of them. Quite commonly, however, both are referred to as randomization tests because their justification depends upon the randomization of treatments to experimental units. This randomization of treatments, together with H_0, determines the family of permutations that should be considered. Essentially, the relevant permutations are those that both (1) leave the responses of the experimental units unchanged if H_0 is true, and (2) might have occurred under the randomization used in the design of the experiment.

Example 9.2

In an experiment to examine two factors, A and B, the first factor had three levels (A_1–A_3) and the second had four levels (B_1–B_4). Thus there are 12 different factor-level combinations (A_1 with B_1, A_1 with B_2, ..., A_3 with B_4). These combinations were randomly allocated to 24 experimental units, subject to the restriction that each combination was used twice. Table 9.1 gives the observed responses from this 'two-factor experiment with replication'.

Suppose we wish to test the hypothesis that factor A had no effect on the response, i.e. the response of each experimental unit was not influenced by which level of factor A it received. Then a permutation of the experimental units does not change their

Table 9.1 Data from a two-factor experiment with replicate observations

	B_1	B_2	B_3	B_4	Totals
A_1	5 8	11 9	5 0	12 16	66
A_2	0 4	5 10	0 6	16 15	56
A_3	0 8	12 8	5 10	22 26	91
Totals	25	55	26	107	213

responses, provided each unit receives the same level of factor B that it received in the experiment. Thus we consider those permutations of the experimental units that might have occurred with the given experimental design and that also leave factor B unchanged. Hence the six units that received B_1 are permuted amongst themselves, as are the six that received B_2, and similarly for B_3 and B_4.

The number of ways of allocating levels A_1, A_2 and A_3 to six units, so that each level is used twice, is $6!/(2!2!2!) = 90$. Hence the number of permutations that leave the levels of B unchanged is $90^4 = 6561 \times 10^4$. This number is too large for each permutation to be examined so a random sample of 4999 permutations was used.

A test statistic is required that is sensitive to factor A. By analogy with the analysis of variance, a natural choice is the variance ratio, $F = M_A/M_W$, where M_A is the mean square for factor A and M_W is the mean square within replicates. For the observed data,

$$M_A = \{\tfrac{1}{8}(66^2 + 56^2 + 91^2) - \tfrac{1}{24}213^2\}/2 = 40.63$$

and, since the means of the replicates are 6.5, 10, 2.5, ..., 10, 7.5 and 24,

$$M_W = \{(5 - 6.5)^2 + (8 - 6.5)^2 + (11 - 10)^2 + (9 - 10)^2$$

$$+ \cdots + (22 - 24)^2 + (26 - 24)^2\}/12$$

$$= 10.54.$$

Thus, for the observed data $F = 40.63/10.54 = 3.854$. The values of F for each of 4999 permutations were also calculated. Of these, 273 exceeded 3.854, so 3.854 was one of the $274/5000 \times 100\% = 5.74\%$ largest F-values. Thus the hypothesis that factor A did not effect the responses is rejected at the 5.74% significance level.

A similar approach could be used to test whether factor B affected the responses, the main difference being that experimental units within each row of Table 9.1 would be permuted, so that the levels of factor A which units received would be unchanged.

The randomization test used in the above example is advocated by Edgington (1980, Chapter 6). A different form of randomization test for this type of data is suggested by Manly (1991, Chapter 5). He uses the factor mean square, M_A, as a test statistic, rather than $F = M_A/M_W$. This should have little effect on the conclusion drawn from the hypothesis test – a feature of randomization tests is that the choice of test statistic is somewhat arbitrary, but sensible choices tend to all give similar results. (Nevertheless, it is often useful to examine more than one test statistic.) Manly (1991) also suggests using unrestricted permutations in which each experimental unit can be allocated to any factor level combination. This has disadvantages for testing whether one factor affected the responses, since the influence of other factors may bias results. For instance, consider testing whether factor A affected responses in Example 9.1. Responses for level B_4 tended to be higher than for other levels of B. This is of little consequence when each level of factor A has the same number of experimental units that received B_4. With an unrestricted permutation, however, one level of A may receive most (or even all) of the experimental units that received B_4, and then M_A is likely to be larger than the value of M_A for the real data.

An important point about randomization tests is that, unlike most hypothesis tests, the experimental units need not be a random sample from a population. In an analysis of variance, for example, a randomization test has a null hypothesis of the form 'factor A *did* not affect the responses of the experimental units' and not 'factor A *does* not affect the responses of the population'. The validity of the test follows from the random allocation of factor levels to the experimental units. Sometimes, of course, the levels of a factor are a characteristic of an experimental unit and so cannot be randomly allocated e.g. if experimental units are people and one factor is a person's sex. Then for the randomization test to be valid for that factor, experimental units must be random samples from distributions that are identical under the null hypothesis. Note that distributional assumptions are now required.

There are situations where a randomization test is easily performed and there are no alternative methods of testing the hypothesis of interest. This is illustrated below in Example 9.3. First, however, we should mention that sometimes randomization tests are difficult to construct or can only be approximated, while standard parametric tests are straightforward to apply. For instance, consider a two-factor experiment with replication, as in Example 9.2, and suppose we wish to test for an interaction between the factors. A simple, widely used parametric test is through an analysis of variance and an F-test. However, there is no exact randomization test available (Edgington, 1980, Section 6.12). This is because observations cannot have their level of factor A changed, nor their level of factor B, so they cannot be rearranged to form permutations. As an approximate randomization test, Still and White (1981) suggest fitting a two-factor model with no interaction, estimating the residuals for this model, and then using unrestricted permutations of these residuals to test for interaction. (By *unrestricted* permutation we mean the residuals' levels of both factor A and factor B can be changed.) A drawback of this approach is that the sum of residuals for each factor level should be zero, but this is unlikely to be the case *after* permutation. Thus permutation of the residuals does not retain all the deterministic characteristics of their original arrangement. Despite this, simulation results reported by Still and White indicate the method should work well in practice. (Similar problems can obviously also occur in multiple regression, since factorial models can be expressed as regression models.)

Example 9.3

Suppose we have n individuals and the 'distances' between all possible pairs of individuals are measured in two different ways or at two different times. We might wish to test if the two sets of distances are associated. For example, to examine whether a particular disease is contagious we might consider the geographic distances between n cases of the disease and their distance apart in time. If there is contagion the geographical distances and time differences should be associated. Another example, considered by Besag and Diggle (1977), relates to the location of 84 blackbirds ringed in the British Isles during the winter months and recovered in northern Europe during a subsequent summer. The question was whether there was any pattern transference between the winter and summer locations.

Performing a randomization test for association between two sets of distances is straightforward. In each set there are $m = n(n-1)/2$ distances. Denote the first set by $x_1, x_2, \ldots, x_m$ and the second set by $y_1, y_2, \ldots, y_m$, where the ordering is such that x_i and y_i relate to the same pair of individuals ($i = 1, 2, \ldots, m$). Some measure of association between the xs and ys is required, and one reasonable choice is the correlation.

$$r = \frac{\Sigma x_i y_i - \Sigma x_i \Sigma y_i / m}{\sqrt{\Sigma(x_i - \bar{x})^2 \Sigma(y_i - \bar{y})^2}}. \tag{9.2}$$

(An equivalent choice as a test statistic is $\Sigma x_i y_i$ – see Exercise 9.7).

One set of distances, the ys say, is permuted perhaps 4999 times and the correlation between x and y for each permutation is determined. These correlations are compared with the observed correlation between the two sets of distances. If the latter is one of the k% largest correlations, then the hypothesis of no association between the sets of distances is rejected at the k% significance level.

Note that standard parametric tests of correlation cannot be used here because of the complex interrelationships within a set of distances. For example, if the distances from A to B and from A to C are 20 and 5, then the distance from B to C must be between 15 and 25. With more individuals, the constraints relating the distances between them would be more numerous and complex. Hence it seems essential that any test of association between sets of distances should be based on data permutation.

It should be mentioned that randomization tests will usually be more powerful than non-parametric tests that are based on ranks, such as the signed-rank test or Friedman's test (cf. Section 8.2.1). This is because the latter tests are equivalent to ranking the data and then performing a randomization test on the ranks. Using the actual data, rather than the ranks, retains more information and hence should give a more powerful test.

For further information on randomization tests, useful books are Edgington (1980) and Manly (1991). Edgington concentrates on analysis of variance and correlation. Manly's book is much broader and, as well as analysis of variance, considers regression, time series and analyses for spatial data.

9.4 Cross-validation

Validation and cross-validation are methods of assessing the predictive ability of a statistical model and choosing between alternative models. The idea is that models should be judged by their ability to predict future data, rather than the degree to which they fit the data from which their parameters were estimated. We let Y denote the quantity to be predicted, so Y would be the dependent variable in the context of regression, or the category to which an item belongs in a discriminant analysis, or the response in a designed experiment. In these cases, predictions conditional on information $\mathbf{x}$ would be made, where $\mathbf{x}$ is typically a vector of independent or explanatory variables, or might indicate factor levels. We shall assume there is information $\mathbf{x}$ about each datum, although validation and cross-validation can also be used in scalar problems where Y is the only variable.

In its simplest form, validation involves dividing a sample into two subsamples, referred to as the training set and the test set. The parameters of a model are estimated from data in the training set and the model is then used to predict values of $Y \mid \mathbf{x}$ for items in the test set. Comparison of the predictions with the observed values in the test set, under some suitable loss function, gives a measure of the predictive accuracy of the model. If the purpose is to choose between models, then the one with the best predictive accuracy would be selected.

Cross-validation is an extension of validation that aims to use the data more efficiently. Instead of dividing the data into two subsets, it is divided into k subsets whose sizes are as nearly equal as possible. One subset is selected as the test set and the union of the remaining $k - 1$ subsets forms the training set. As in simple validation, model parameters are estimated with the training set and values of $Y \mid \mathbf{x}$ for the test set are predicted. The loss for prediction errors is determined and the process is then repeated k times, each time using a different subset as the test set. Thus each datum in the full sample is predicted once. The sum of the k losses is taken as the measure of the model's predictive accuracy. Once a model has been selected, its parameters would be estimated using the complete sample.

In practice, values of k between 3 and 20 are often used (Hjorth, 1993) although ideally k would be set equal to the sample size so that each subset contained just one datum. Then the parameter estimates used for assessing accuracy would be as similar as possible to the estimates finally adopted. The disadvantage, of course, is that more computer time is needed as k increases.

Example 9.4

Measurements x and y were made on ten items in a random sample and their values are given in Table 9.2. Each item was omitted in turn and the remaining nine data used to estimate the parameters in the simple linear regression model

$$E[Y] = \alpha + \beta x.$$

The least squares estimates are denoted by $\hat{\alpha}_{\setminus i}$ and $\hat{\beta}_{\setminus i}$ for the case where item i was omitted ($i = 1, 2, \ldots, 10$) and $\hat{y}_i = \hat{\alpha}_{\setminus i} + \hat{\beta}_{\setminus i} x_i$ is the resultant prediction of y_i. These

Table 9.2 Data and cross-validation estimates of the parameters in a simple linear model with predictions and their squared errors

i	x_i	y_i	$\hat{\alpha}_{\backslash i}$	$\hat{\beta}_{\backslash i}$	$\hat{y}_i$	$(y_i - \hat{y}_i)^2$
1	22.0	96	34.7	2.564	91.1	24.3
2	23.4	88	47.8	2.190	99.0	121.1
3	24.9	105	34.9	2.550	98.4	43.2
4	28.5	111	38.3	2.453	108.2	7.8
5	29.8	107	40.1	2.419	112.2	27.3
6	31.6	113	39.2	2.443	116.4	11.7
7	34.2	132	41.6	2.318	120.9	123.5
8	36.4	122	36.1	2.556	129.2	51.2
9	37.7	135	42.1	2.320	129.6	29.3
10	39.0	131	36.2	2.544	135.4	19.7

quantities are also tabulated. Using prediction squared error as the loss function, so that

$$\text{loss} = \sum_{i=1}^{10} (y_i - \hat{y}_i)^2 \tag{9.3}$$

gives a loss of $24.3 + 121.1 + \cdots + 19.7 = 459.0$ for the model. If a quadratic model, $E[Y] = \alpha + \beta_1 x + \beta_2 x^2$, is treated in similar fashion, a bigger loss of 569.2 is obtained. Hence cross-validation here chooses the simple linear model in preference to the quadratic model. (If the residual sum of squares from fitting models to the whole data set were used as the loss criteria then, of course, the quadratic model would inevitably be preferred.)

As well as cross-validation, another widely applicable procedure for choosing between models is Akaike's information criterion. While discussing model selection, this important criterion should be defined. Suppose a model is parameterized by a vector θ of dimension p. Let $\hat{\theta}$ be the maximum likelihood estimate of θ (based on all data, not a training sample) and let $l(\hat{\theta})$ denote the logarithm of the likelihood at its maximum. In comparing models some account should be taken of their sizes. If a model has too many parameters it will follow random variation in the data, rather than merely capturing the data's underlying structure. Nevertheless, for the data used to estimate its parameters it will give a better fit than a model that omits the redundant parameters. Hence, in choosing a model some balance must be struck between its ability to fit the data and its size. Akaike's criterion is to determine

$$l(\hat{\theta}) - p \tag{9.4}$$

for the available models and choose the model for which it is maximized. Thus the criterion uses $l(\hat{\theta})$ to measure the fit of a model to the data and penalizes larger models through the term $-p$. Stone (1977) shows that choosing a model by cross-validation is asymptotically equivalent to using Akaike's criterion provided cross-validation assesses a model's predictive accuracy by the loss function,

$$\text{loss} = -\sum_i \log f_{\setminus i}(y_i \mid \mathbf{x}_i).$$

Here $f_{\setminus i}(. \mid \mathbf{x}_i)$ is the predictive distribution of the ith observation, given all data except y_i.

Further uses and examples of cross-validation are given by Hjorth (1993), Stone (1974) and Geisser (1993). Amongst the topics they cover, Hjorth (Section 3.6) suggests a strategy for variable selection in multiple regression that combines stepwise methods with cross-validation. Stone gives a method of using cross-validation to simultaneously select a model and evaluate its predictive accuracy. The method first uses cross-validation to select a model for each possible subset of $n - 1$ data (where n is the total number of data) and then predicts the omitted datum with the selected model. Geisser considers applications of *predictive sample reuse* (another term for cross-validation) in Bayesian statistics. All three authors also consider the use of cross-validation for parameter estimation. We illustrate this last use in the following example.

Example 9.5

Suppose we have observations $(x_1, y_1), (x_2, y_2), \ldots, (x_n, y_n)$ from the regression model $E(Y \mid x) = \alpha + \beta x$. Using least squares estimates, the predicted value of y is

$$\hat{y} = \hat{\alpha} + \hat{\beta}x = \bar{y} + \hat{\beta}(x - \bar{x}).$$

Under some circumstances, $\hat{y}$ is likely to overestimate large values of y and underestimate small values of y. This is the case, for example, if there is a choice of several explanatory variables for the regression model and X is selected because it fitted the data best. (Biases caused by variable selection in regression are discussed by Copas (1983), who also gives examples.) Hence we might shrink the prediction towards the mean, $\bar{y}$, by forming a revised prediction, $\tilde{y}$ say, as a weighted average of $\bar{y}$ and $\hat{y}$. This gives the shrunken estimator,

$$\tilde{y} = \omega\bar{y} + (1 - \omega)\hat{y} = \bar{y} + \omega\hat{\beta}(x - \bar{x})$$

where $\omega \in [0, 1]$.

The parameter ω can be estimated by cross-validation. Let $\bar{y}_{\setminus i}$ and $\bar{x}_{\setminus i}$ denote the sample means of x and y when the ith datum is omitted. Using the notation of Example 9.4 and again using prediction squared error as the loss, $\omega \in [0, 1]$ must be chosen to minimize

$$\text{loss} = \sum_{i=1}^{n} \{y_i - \bar{y}_{\setminus i} - \omega\hat{\beta}_{\setminus i}(x_i - \bar{x}_{\setminus i})\}^2.$$

Differentiation with respect to ω leads to

$$\hat{\omega} = \Sigma\hat{\beta}_{\setminus i}(y_i - \bar{y}_{\setminus i})(x_i - \bar{x}_{\setminus i})/\Sigma\hat{\beta}_{\setminus i}(x_i - \bar{x}_{\setminus i})^2. \tag{9.5}$$

If $\hat{\omega} < 0$ or $\hat{\omega} > 1$, then ω would be estimated by 0 or 1, respectively.

Properties of shrunken estimators are examined in several papers by Stein (e.g. Stein, 1960) and estimators derived from his work are often referred to as Stein estimators. He has shown that in various multivariate problems a Stein estimator dominates the least squares estimator, so that the latter is inadmissible. Stone (1974) indicates that cross-validation can give shrunken estimators that are related to Stein estimators.

9.5 Jackknife and bootstrap methods

Suppose inferences about a parameter θ are to be made from a sample of independent and identically distributed random variables $Y_1, Y_2, \ldots, Y_n$. For this task, both the jackknife and bootstrap methods construct or generate additional sample sets based on the observed values of the Ys. From each sample set an estimate of θ is determined and inferences are drawn from these estimates without making further use of the Ys. Although the methods can be computationally demanding, their mechanics are straightforward and they are simple to use in a wide variety of problems. In particular cases they perform less well than parametric alternatives if the parametric assumptions hold, but in some complex situations they can be the only viable means of analyzing the data.

9.5.1 *The jackknife*

Let $\hat{\theta}$ be an estimator of θ based on all the variables, $Y_1, Y_2, \ldots, Y_n$. With the jackknife, as in cross-validation, subsamples are considered in which each of the Y_i is omitted in turn. For $i = 1, 2, \ldots, n$, let $\hat{\theta}_{\backslash i}$ denote the estimator of θ that is based on all the Ys except Y_i. Define

$$\tilde{\theta}_i = n\hat{\theta} - (n-1)\hat{\theta}_{\backslash i}, \qquad i = 1, 2, \ldots, n. \tag{9.6}$$

The $\tilde{\theta}_i$ are referred to as *pseudo-values* and the jackknife estimator of θ is defined to be

$$\tilde{\theta}_J = \frac{1}{n}\sum_{i=1}^{n} \tilde{\theta}_i. \tag{9.7}$$

Clearly $\tilde{\theta}_J$ is unbiased if $\hat{\theta}$ and $\hat{\theta}_{\backslash i}$ $(i = 1, 2, \ldots, n)$ are unbiased.

Example 9.6

For the data in Table 9.2, $\hat{\beta} = 2.434$ is the least squares estimate of β in the model $\hat{y} = \alpha + \beta x$. Table 9.2 also gives the values of $\beta_{\backslash i}$ $(i = 1, 2, \ldots, 10)$ and these lead to the pseudo-values $\tilde{\beta}_1 = 10(2.434) - 9(2.564) = 1.264$, $\tilde{\beta}_2 = 10(2.434) - 9(2.190) = 4.630, \ldots, \tilde{\beta}_{10} = 10(2.434) - 9(2.544) = 1.444$. Hence the jackknife estimate of β is $\tilde{\beta}_J = (1.264 + 4.630 + \cdots + 1.444)/10 = 2.419$. Similar computations give $\tilde{\alpha}_J = 39.7$ as the jackknife estimate of α, so $\tilde{\theta}_J = (39.7, 2.419)$ is the jackknife estimate of $\theta = (\alpha, \beta)$.

The jackknife was initially introduced by Quenouille (1949) as a technique for bias reduction. That is, $\tilde{\theta}_J$ will often have smaller bias than the original estimator, $\hat{\theta}$, if the latter is biased. This follows from Theorem 9.1, whose conditions hold for many estimators.

Theorem 9.1

Suppose the expectation of $\hat{\theta}$ can be expressed as

$$\mathrm{E}[\hat{\theta}] = \theta + \sum_{i=1}^{\infty} c_j/n^j \tag{9.8}$$

where the c_j may be functions of θ but not of n. Then the bias of $\tilde{\theta}_J$ is of order n^{-2}.

Proof

By assumption, $\mathrm{E}[\hat{\theta}_{\backslash i}] = \theta + \Sigma_{j=1}^{\infty} c_j/(n-1)^j$, so

$$\mathrm{E}[\tilde{\theta}_i] = n\left[\theta + c_1/n + \sum_{j=2}^{\infty} c_j/n^j\right] - (n-1)\left[\theta + c_1/(n-1) + \sum_{j=2}^{\infty} c_j/(n-1)^j\right]$$

$$= \theta + \sum_{j=2}^{\infty} (c_j/n^{j-1} - c_j/(n-1)^{j-1}).$$

Hence $\mathrm{E}[\theta_i] = \theta + 0(n^{-2})$, so we also have that $\mathrm{E}[\tilde{\theta}_J] = \theta + 0(n^{-2})$, as $\tilde{\theta}_J$ is the average of $\tilde{\theta}_1, \tilde{\theta}_2, \ldots, \tilde{\theta}_n$. □

The theorem states that if bias in $\hat{\theta}$ has the form

$$\mathrm{E}[\hat{\theta}] - \theta = c_1/n + 0(1/n^2)$$

then the jackknife eliminates the term of order $1/n$. Obviously, when $E(\hat{\theta}) = \theta + c_1/n$, the jackknife eliminates the bias completely.

Example 9.7

Suppose an estimate of $\sigma^2 = \mathrm{var}(Y)$ is required and we choose

$$\hat{\theta} = \frac{1}{n}\sum_{i=1}^{n} (Y_i - \bar{Y})^2.$$

If the Y_i were normally distributed, then $\hat{\theta}$ would be the maximum likelihood estimate. Since $\mathrm{E}[\Sigma(Y_i - \bar{Y})^2/(n-1)] = \sigma^2$, we have that $\mathrm{E}[\hat{\theta}] = \sigma^2 - \sigma^2/n$. Thus the bias in $\hat{\theta}$ is of the form c_1/n, so the jackknife must be unbiased. This is easily verified directly, as

$$\mathrm{E}[\tilde{\theta}_i] = nE(\hat{\theta}) - (n-1)E(\hat{\theta}_{\backslash i})$$

$$= n[\sigma^2 - \sigma^2/n] - (n-1)[\sigma^2 - \sigma^2/(n-1)] = \sigma^2.$$

Thus $\tilde{\theta}_i$ and, consequently, $\tilde{\theta}_J = \Sigma\tilde{\theta}_i/n$ are unbiased.

Quenouille (1956) shows that a 'second-order jackknife' can be used to also eliminate the bias term of order $1/n^2$. The technique involves forming subsamples with two data omitted and jackknifing the jackknife estimator.

As well as point estimation, the jackknife is also used in practice for interval estimation. The critical idea is to treat the pseudo-values $\tilde{\theta}_1, \tilde{\theta}_2, \ldots, \tilde{\theta}_n$ as though they were i.i.d. random observations from a distribution whose mean is θ. To illustrate that it can be reasonable to treat the pseudo-values as observations, suppose $\hat{\theta} = (1/n)\Sigma Y_i$. Then

$$\begin{aligned}\tilde{\theta}_i &= n\hat{\theta} - (n-1)\hat{\theta}_{\backslash i} \\ &= \sum_{j=1}^{n} Y_j - \sum_{j \neq i} Y_j \\ &= Y_i\end{aligned}$$

and the pseudo-values are the original data.

Treating the pseudo-values as approximately i.i.d. random variables, $\hat{\sigma}^2 = \Sigma_i(\tilde{\theta}_i - \tilde{\theta}_J)^2/(n-1)$ is their estimated variance and the statistic

$$\frac{\tilde{\theta}_J - \theta}{\sqrt{\hat{\sigma}^2/n}} \tag{9.9}$$

is presumed to have an approximate t distribution on $n - 1$ degrees of freedom. Then an approximate $100(1-\alpha)\%$ confidence interval for θ is $\tilde{\theta}_J \pm t_{(n-1);\alpha/2}\sqrt{\hat{\sigma}^2/n}$.

Example 9.6 (*continued*)

Suppose a 95% confidence interval for β is required. We have $\tilde{\beta}_J = 2.419$ and pseudo-values 1.264, 4.630, 1.390, 2.263, 2.569, 2.353, 3.478, 1.336, 3.460, and 1.444 for the $\tilde{\beta}_i$. Hence $\hat{\sigma}^2 = \{(1.264 - 2.419)^2 + (4.630 - 2.419)^2 + \cdots + (1.444 - 2.419)^2\}/9 = 1.296$. As $t_{9;\,.025} = 2.262$, an approximate 95% confidence interval for β is $2.419 \pm 2.262\sqrt{1.296/10} = 2.419 \pm 0.814$. For comparison, if errors are assumed to be normally distributed, the 'usual' 95% confidence interval given by standard methods is 2.434 ± 0.766, which is similar to the approximate confidence interval given by the jackknife.

The jackknife is a useful way of setting reasonable confidence intervals in a wide variety of complex situations although, in particular cases, there can be specialized methods that set more accurate intervals. It is because of these features that the jackknife was given its name – after a Boy Scout's pocket knife with pull-out tools that make it useful for a multitude of tasks, although purpose-made tools might do individual tasks better.

In many situations, the pseudo-values are not completely independent of each other, violating one of the assumptions made in forming jackknife confidence intervals. However, Mosteller and Tukey (1977, p. 135) assert that the t-distribution still performs well in many circumstances where the pseudo-values deviate substantially from independence, so that inaccuracy in confidence intervals is often small. Mosteller and Tukey (1977) also give some heuristic rules for improving the behaviour of jackknife intervals. They suggest that, ideally, the distribution of pseudo-values should be close to symmetrical with tails that are neither very straggling nor truncated. Transformation of θ is advocated if these requirements are not satisfied.

9.5.2 The Bootstrap

Efron (1979) gave the bootstrap its name because in using the method one seems to be *pulling oneself up by one's own bootstraps.* Efron (1979, p. 25) mentions that he thought about calling the method the *shotgun* because it '... can blow the head off any problem if the statistician can stand the resulting mess.' The method is indeed capable of wide application and the 'mess' presumably relates to the volume of numbers that may be produced.

We have i.i.d. random variables $Y_1, Y_2, \ldots, Y_n$ that give an estimate $\hat{\theta}$ of θ. Let F denote the cumulative distribution function of the Ys and let $y_1, y_2, \ldots, y_n$ be the observed data. The basic idea of the bootstrap is to construct an empirical distribution function, $\hat{F}_n$, that puts probability $1/n$ on each of the values $y_1, y_2, \ldots, y_n$. Hence

$$\hat{F}_n(y) = \frac{1}{n}\{\text{Number of } y_i \leqslant y\}. \tag{9.10}$$

From this distribution, a random sample of size n is drawn. This resample is called a *bootstrap sample.* Let $Y_1^*, Y_2^*, \ldots, Y_n^*$ denote random variables in the resample. Thus the values of the Y^*s are selected *with replacement* from $y_1, y_2, \ldots, y_n$. From the bootstrap sample an estimate of θ, say $\hat{\theta}^*$, is constructed in the same way as $\hat{\theta}$ was constructed from $y_1, y_2, \ldots, y_n$. The distribution of $\hat{\theta}^*$ is referred to as a *bootstrap distribution* because it is determined by $\hat{F}_n$.

Standard statistical methods usually relate $\hat{\theta}$ to θ by considering the distribution of $\hat{\theta} - \theta$, or $\hat{\theta}/\theta$, or some other convenient expression such as a pivotal quantity. Here we will suppose that (perhaps after transformation of the original parameter) $\hat{\theta} - \theta$ is the quantity of interest. We shall also suppose θ is a quantity that is well defined for a class of distributions which include $\hat{F}_n$. For example, θ might be the mean or variance of a distribution. Let $\tilde{\theta}$ be the parameter value when $\hat{F}_n$ is the distribution. Often $\hat{\theta}$ and $\tilde{\theta}$ will be equal but they may differ. For example, $\hat{\theta}$ might be the trimmed mean of the data $y_1, y_2, \ldots, y_n$ while $\tilde{\theta}$ is the mean of the distribution $\hat{F}_n(y)$. Simple bootstrap methods make one or other of the following big assumptions.

Assumption A $\hat{F}_n$ is a good approximation of F (so that the distribution of $\hat{\theta}^*$ is similar to that of $\hat{\theta}$).

Assumption B The distribution of $\hat{\theta}^* - \tilde{\theta}$ is similar to that of $\hat{\theta} - \theta$.

Under either assumption, the task of making inferences about θ largely reduces to learning about the bootstrap distribution of $\hat{\theta}^*$. Sometimes the relevant features of the bootstrap distribution can be determined mathematically. In most non-trivial problems, however, the distribution must be estimated using Monte Carlo methods. Although computationally intensive, this is straightforward as $\hat{F}_n$ is known. A large number of bootstrap samples, each of size n, is generated from $\hat{F}_n$. From each of these an estimate of θ is determined. Denote these estimates by $\hat{\theta}_1^*, \hat{\theta}_2^*, \ldots, \hat{\theta}_B^*$, where B is the number of bootstrap samples. The histogram of these estimates is taken as an approximation to the bootstrap probability distribution of $\hat{\theta}^*$.

The mean and variance of the approximate bootstrap distribution are

$$\bar{\theta}^* = \frac{1}{B}\sum_{i=1}^{B} \hat{\theta}_i^* \tag{9.11}$$

and

$$\hat{\sigma}^2(\hat{\theta}^*) = \frac{1}{B-1}\sum_{i=1}^{B} (\hat{\theta}_i^* - \bar{\theta}^*)^2. \tag{9.12}$$

These immediately give estimates of the variance and bias of $\hat{\theta}$. Since $\mathrm{Var}[\hat{\theta}] = \mathrm{Var}[\hat{\theta} - \theta]$ and $\mathrm{Var}[\hat{\theta}^*] = \mathrm{Var}[\hat{\theta}^* - \tilde{\theta}]$, assumptions A and B lead to the same estimate of $\mathrm{Var}(\hat{\theta})$,

$$\begin{aligned}\mathrm{Var}[\hat{\theta}] &= \mathrm{Var}[\hat{\theta}^*] \\ &\approx \hat{\sigma}^2(\hat{\theta}^*).\end{aligned}$$

Under assumption B, the bias of $\hat{\theta}$ is

$$\begin{aligned}\mathrm{E}[\hat{\theta} - \theta] &= \mathrm{E}[\hat{\theta}^* - \tilde{\theta}] \\ &= \mathrm{E}[\hat{\theta}^*] - \tilde{\theta} \approx \bar{\theta}^* - \tilde{\theta},\end{aligned}$$

and a bias adjusted estimate of θ is

$$\hat{\theta} - (\bar{\theta}^* - \tilde{\theta}). \tag{9.13}$$

This adjusted estimator can be motivated as follows. Suppose that, when sampling from both $\hat{F}_n$ and F, the estimator is expected to exceed the true parameter value by an amount k. Then $\mathrm{E}[\bar{\theta}^* - \tilde{\theta}] = k$ for sampling from $\hat{F}_n$ and $\theta = E(\hat{\theta}) - k$ for sampling from F, so $\mathrm{E}[\hat{\theta} - (\bar{\theta}^* - \tilde{\theta})] = \theta$.

Under assumption A, $\theta = \tilde{\theta}$ and $\mathrm{E}[\hat{\theta}] = \mathrm{E}[\hat{\theta}^*]$, so one might estimate the bias as

$$\begin{aligned}\mathrm{E}[\hat{\theta} - \theta] &= \mathrm{E}[\hat{\theta}^*] - \tilde{\theta} \\ &\approx \bar{\theta}^* - \tilde{\theta}.\end{aligned}$$

This is the same as under assumption B. However, since $\theta = \tilde{\theta}$ was used to derive this expression, $\tilde{\theta}$ should presumably be taken as the best estimate of θ under assumption A, and $\hat{\theta} - \tilde{\theta}$ as the error in $\hat{\theta}$. However, this is rarely done in practice, perhaps because $\bar{\theta}^* - \hat{\theta}$ has some justification as an estimate of bias under both assumption A and assumption B.

Example 9.8

The following eight values are a random sample from a population with unknown mean θ:

7.0 19.8 12.8 6.0 15.2 5.1 15.0 7.6.

To illustrate that the complexity of an estimator has little effect on the simplicity of the bootstrap, suppose the 25% trimmed mean is used to estimate θ. Then, omitting the smallest and largest data,

Table 9.3 Ten bootstrap samples and their 25% trimmed means ($\hat{\theta}_i^*$)

	1	2	3	4	5	6	7	8	9	10
	7.6	12.8	7.6	12.8	15.0	7.6	7.0	12.8	7.0	19.8
	15.0	7.6	12.8	19.8	15.2	19.8	15.2	7.0	7.0	12.8
	15.0	12.8	7.6	15.2	7.6	15.2	5.1	7.6	7.6	7.6
	19.8	15.2	7.6	15.2	12.8	12.8	7.6	6.0	7.6	15.0
	15.0	7.0	19.8	15.2	7.6	7.0	15.2	7.0	15.0	5.1
	15.2	5.1	12.8	12.8	7.6	5.1	7.6	5.1	19.8	19.8
	12.8	12.8	7.6	6.0	7.0	7.6	15.2	5.1	15.0	7.0
	7.0	5.1	15.0	15.0	15.2	15.2	15.0	6.0	6.0	7.0
$\hat{\theta}_i^*$	13.43	9.68	10.57	14.37	10.97	10.90	11.27	6.45	9.87	11.53

$$\hat{\theta} = \tfrac{1}{6}(7.0 + 12.8 + 6.0 + 15.2 + 15.0 + 7.6) = 10.600$$

while $\tilde{\theta} = \frac{1}{8}(7.0 + 19.8 + \cdots + 7.6) = 11.063$. To form a bootstrap sample, 8 random draws were made with replacement from $\{7.0, 19.8, \ldots, 7.6\}$, giving the values in the first column of Table 9.3. Nine further bootstrap samples were formed in a similar way, giving the subsequent columns of the table. The 25% trimmed mean was determined for each bootstrap sample, resulting in estimates $\hat{\theta}_1^* = 13.43$, $\hat{\theta}_2^* = 9.68$, ..., $\hat{\theta}_{10}^* = 11.53$. The mean and variance of these estimates are $\bar{\theta}^* = 10.904$ and $\hat{\sigma}^2(\hat{\theta}^*) = 4.604$. Under assumption A, the best estimate of θ is $\tilde{\theta} = 11.063$ and the error in $\hat{\theta}$ is $10.600 - 11.063 = -0.463$. Under assumption B, the estimated bias is $\bar{\theta}^* - \tilde{\theta} = 10.904 - 11.063 = -0.159$ and a bias adjusted estimate of θ is $\hat{\theta} - (\bar{\theta}^* - \tilde{\theta}) = 10.600 + 0.159 = 10.759$. In practice, a computer would be used to generate, say, 1000 bootstrap samples which would give greater accuracy.

Quantiles of the bootstrap distribution of $\hat{\theta}^*$ are used to form confidence intervals for θ. Let $\theta^*(\alpha)$ denote the α-level quantile of the bootstrap distribution of $\hat{\theta}^*$, i.e. $\Pr^*[\hat{\theta}^* \leqslant \theta^*(\alpha)] = \alpha$, where $\Pr^*$ denotes probability when $\hat{F}_n^*$ is the distribution of Y. Then under assumption A, a $100(1 - 2\alpha)$% equal-tailed confidence interval for θ is $(\theta^*(\alpha), \theta^*(1 - \alpha))$. This method of constructing confidence intervals is usually called the *percentile method* and the intervals are called *percentile intervals*. Under assumption B,

$$\alpha = \Pr^*[\hat{\theta}^* - \tilde{\theta} < \theta^*(\alpha) - \tilde{\theta}] = \Pr[\hat{\theta} - \theta < \theta^*(\alpha) - \tilde{\theta}]$$

so $1 - \alpha = \Pr[\theta \leqslant \hat{\theta} + \{\tilde{\theta} - \theta^*(\alpha)\}]$. Similarly, $\alpha = \Pr[\theta \leqslant \hat{\theta} - \{\theta^*(1 - \alpha) - \tilde{\theta}\}]$. Hence, under assumption B, a $100(1 - 2\alpha)$% confidence interval for θ is $(\hat{\theta} - \{\theta^*(1 - \alpha) - \tilde{\theta}\}, \hat{\theta} + \{\tilde{\theta} - \theta^*(\alpha)\})$. These will be referred to as bootstrap B intervals.

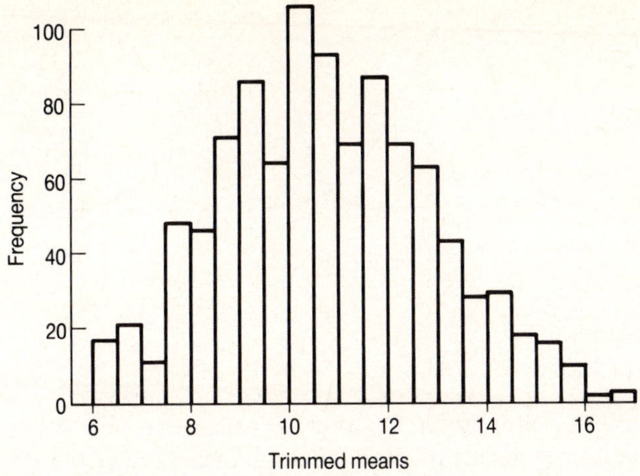

Fig. 9.2 Trimmed means in 1000 bootstrap samples.

Example 9.8 (*continued*)

A thousand bootstrap samples were generated by computer and the 25% trimmed mean ($\hat{\theta}^*$) calculated for each. The histogram in Figure 9.2 shows their distribution. The 50th smallest and 50th largest $\hat{\theta}^*$ were 7.517 and 14.500, so these are θ^* (0.05) and θ^* (0.95), respectively. Hence a 90% percentile confidence interval is (7.52, 14.50) and the 90% bootstrap B interval is $(10.600 - \{14.500 - 11.063\},\ 10.600 + \{11.063 - 7.517\}) = (7.16, 14.15)$.

We next consider when the methods give confidence intervals with the correct coverage probabilities. Suppose $g(\cdot)$ is a monotonic increasing function and let

$$\phi = g(\theta), \qquad \hat{\phi} = g(\hat{\theta}), \qquad \hat{\phi}^* = g(\hat{\theta}^*).$$

Then

$$\begin{aligned}\Pr[\theta \leqslant \theta^*(\alpha)] &= \Pr[g(\theta) \leqslant g\{\theta^*(\alpha)\}] \\ &= \Pr[\phi \leqslant g\{\theta^*(\alpha)\}] \qquad (9.14)\end{aligned}$$

Suppose, now, that $\hat{\phi} - \phi$ and $\phi^* - \hat{\phi}$ have the same distribution, symmetric about the origin. Then, as $\phi - \hat{\phi}$ and $\hat{\phi}^* - \hat{\phi}$ must have the same distribution (from the symmetry assumption)

$$\begin{aligned}\Pr[\phi \leqslant g\{\theta^*(\alpha)\}] &= \Pr{}^*[\hat{\phi}^* \leqslant g\{\theta^*(\alpha)\}] \\ &= \Pr{}^*[\hat{\theta}^* \leqslant \theta^*(\alpha)] \\ &= \alpha.\end{aligned}$$

Similarly $\Pr[\theta \geq \theta^*(1 - \alpha)] = \alpha$, so we have the following.

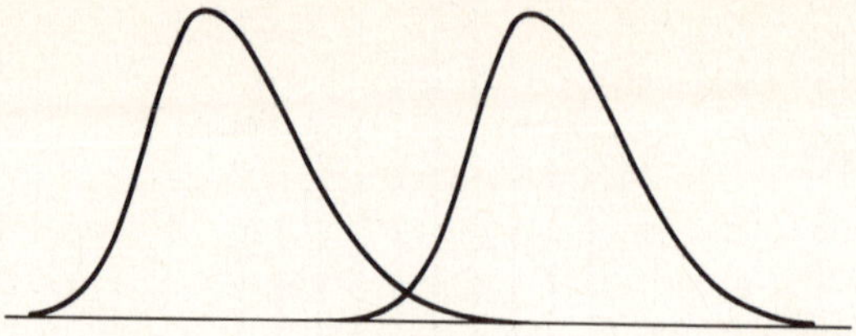

Fig. 9.3 Distributions from a location family.

Theorem 9.2

Suppose there exists a monotonic function $g(\cdot)$ such that $\hat{\phi} - \phi$ and $\hat{\phi}^* - \hat{\phi}$ have the same distribution, symmetric about the origin, where ϕ, $\hat{\phi}$ and $\hat{\phi}^*$ are defined as above. Then the percentile interval $(\theta^*(\alpha), \theta^*(1 - \alpha))$ has the correct coverage probability.

An important point is that for the theorem to hold the function $g(\cdot)$ must merely exist; we do not need to know what the function is.

When percentile intervals have the correct coverage probability for all confidence levels, then assumption A must hold. Hence it follows that the conditions of the theorem are not more general than assumption A. However, known theory will sometimes indicate that the conditions of the theorem hold when one would not otherwise realize that assumption A was satisfied.

Turning to bootstrap B intervals, suppose the distributions of $\hat{\theta}$ and $\hat{\theta}^*$ are members of a location family (cf Figure 9.3). That is, $\hat{\theta} = b_1 + X_1$ and $\hat{\theta}^* = b_2 + X_2$ where b_1 and b_2 are constants and X_1 and X_2 are identically distributed random variables. Then $b_1 - \theta = b_2 - \hat{\theta} = c$, for some constant c, so $\hat{\theta} - \theta = c + X_1$ and $\hat{\theta}^* - \hat{\theta} = c + X_2$. Thus assumption B is satisfied and we have the following result.

Theorem 9.3

If $\hat{\theta}$ and $\hat{\theta}^*$ are members of a location family then bootstrap B confidence intervals have the correct coverage probabilities.

Example 9.9

Let $\hat{\theta}$ be the sample correlation for a sample of n data from a bivariate normal population with population correlation θ. It is well known that the $\tanh^{-1}$ transformation gives approximate normality. Specifically, if $\phi = \tanh^{-1}(\theta)$, $\hat{\phi} = \tanh^{-1}(\hat{\theta})$ and $\hat{\phi}^* = \tanh^{-1}(\hat{\theta}^*)$, then

$$\hat{\phi} - \phi \doteq N(0, 1/\sqrt{n-3}) \qquad \text{and} \qquad \hat{\phi}^* - \hat{\phi} \doteq N(0, 1/\sqrt{n-3}). \tag{9.15}$$

Thus the conditions for Theorem 9.2 are satisfied so percentile intervals for θ have the correct coverage probability.

Suppose the bootstrap B method is applied to $\phi = \tanh^{-1}(\theta)$. Then $\hat{\phi} = b_1 + X_1$ and $\hat{\phi}^* = b_2 + X_2$ with $b_1 = \phi$, $b_2 = \hat{\phi}$, $X_i \sim N(0, 1/\sqrt{n-3})$ for $i = 1, 2$. Hence the bootstrap B interval for ϕ will have the correct coverage probability, from Theorem 9.3. Denoting the interval by (a_1, a_2) and transforming back, $[\tanh(a_1), \tanh(a_2)]$ is a confidence interval for θ that also has the correct coverage probability. This illustrates the potential benefit of transformations with the bootstrap B method, but requires knowledge of the appropriate transformation.

9.5.3 *Comparisons and extensions*

The jackknife was proposed as a means of reducing bias in point estimation and it is clearly somewhat crude in its approach to interval estimation. Taking $\tilde{\theta}_J \pm t_{(n-1);\alpha/2}\sqrt{\sigma^2/n}$ as a confidence interval may be poor for small sample sizes if $\tilde{\theta}_J$ is not normally distributed, or if pseudo-values are highly correlated. The bootstrap, in contrast, is primarily concerned with approximating the distribution of estimators ($\hat{\theta}$ or $\hat{\theta} - \theta$) and so should be better suited to interval estimation. Indeed, simulation work tends to support the superiority of bootstrap confidence intervals over jackknife intervals (see, for example, Efron 1982), although for asymptotic justification of the accuracy of intervals, both the jackknife and the bootstrap generally rely on the asymptotic normality of $\hat{\theta}$.

The percentile method is invariant to monotonic transformations. If $\phi = g(\theta)$ is a monotonic increasing transformation and (a_1, a_2) is a $(1 - 2\alpha)$ confidence interval for ϕ, then $(g^{-1}(a_1), g^{-1}(a_2))$ is a $(1 - 2\alpha)$ confidence interval for θ. (The end points of the interval must be interchanged if $g(\cdot)$ is a monotonic decreasing transformation.) The bootstrap B method and the jackknife do not have this property. With these methods it is usually beneficial, when possible, to transform to a pivotal quantity that has an approximately normal distribution. Example 9.9 illustrated this for the bootstrap B method.

In some problems the way bootstrap resamples should be constructed is not clear cut. As an example, consider the important problem of estimating the parameters of the regression model.

$$y_i = \beta_1 x_{1i} + \beta_2 x_{2i} + \cdots + \beta_k x_{ki} + \varepsilon_i = \mathbf{x}_i^{\mathrm{T}}\beta + \varepsilon_i \tag{9.16}$$

from n data points, $(\mathbf{x}_1, y_1), (\mathbf{x}_2, y_2), \ldots, (\mathbf{x}_n, y_n)$. Putting $\mathbf{X} = (\mathbf{x}_1, \mathbf{x}_2, \ldots, \mathbf{x}_n)^{\mathrm{T}}$ and $\mathbf{y} = (y_1, y_2, \ldots, y_n)^{\mathrm{T}}$, the last squares estimate of $\boldsymbol{\beta}$ is $\hat{\boldsymbol{\beta}} = (\mathbf{X}^{\mathrm{T}}\mathbf{X})^{-1}\mathbf{X}^{\mathrm{T}}\mathbf{y}$.

In this problem there are two main approaches to bootstrap resampling. The most natural is to select n data points with replacement from $(\mathbf{x}_1, y_1), \ldots, (\mathbf{x}_n, y_n)$. This retains the relationship between the $\mathbf{x}$ and y values so that, for example, outliers with high leverage will not lose this characteristic in resampling (An outlier has high leverage if its deletion from the data set would markedly change the estimated regression equation. This depends on both the values $\mathbf{x}$ and y of the outlier.) Denoting the resample by $(\mathbf{x}_1^*, y_1^*), \ldots, (\mathbf{x}_n^*, y_n^*)$, the bootstrap estimate of $\boldsymbol{\beta}$ is

$\hat{\boldsymbol{\beta}}^* = (\mathbf{X}^{*T}\mathbf{X}^*)^{-1}\mathbf{X}^{*T}\mathbf{y}^*$ where $\mathbf{X}^* = (\mathbf{x}_1^*, \ldots, \mathbf{x}_n^*)^T$ and $\mathbf{y}^* = (y_1^*, \ldots, y_n^*)^T$. A drawback of this approach is that $\text{Var}[\hat{\boldsymbol{\beta}}^*] = \sigma^2(\mathbf{X}^{*T}\mathbf{X}^*)^{-1}$ so, as $\mathbf{X}^{*T}\mathbf{X}^*$ and $\mathbf{X}^T\mathbf{X}$ will seldom be equal, $\hat{\boldsymbol{\beta}}^*$ and $\hat{\boldsymbol{\beta}}$ typically have different variances. For a similar reason, bootstrap estimates of $\boldsymbol{\beta}$ from different resamples will not be identically distributed. If the x-values are fixed, as in a designed experiment, then differences between $\mathbf{X}$ and $\mathbf{X}^*$ might be particularly undesirable, especially if $\mathbf{X}$ was designed to have special properties (e.g. orthogonality) not held by $\mathbf{X}^*$.

The second approach to forming bootstrap resamples is based on the residuals

$$\hat{\varepsilon}_i = y_i - \mathbf{x}_i^T\hat{\boldsymbol{\beta}} \tag{9.17}$$

for $i = 1, 2, \ldots, n$. A resample of n values is drawn with replacement from these residuals. Denoting the ith value by ε_i^*, we put

$$y_i^* = \mathbf{x}_i^T\hat{\boldsymbol{\beta}} + \varepsilon_i^* \tag{9.18}$$

and take $(\mathbf{x}_i, y_i^*)$ as the ith datum in the resample. Then in every resample $\mathbf{X}^* = (\mathbf{x}_1, \mathbf{x}_2, \ldots, \mathbf{x}_n)^T = \mathbf{X}$, so the drawbacks of the first resampling method do not arise. Instead of course, there are different drawbacks. In particular, bootstrap resamples will seldom display either (1) non-linearity between the dependent and independent variables, (2) heterogeneous error variance that is dependent on $\mathbf{x}$, or (3) correlated errors, even when these traits are present in the original data. Obviously this means the bootstrap resamples have little value as a diagnostic tool. It has also been shown that the bootstrap estimate of $\text{var}(\hat{\boldsymbol{\beta}})$ will be biased and inconsistent when the original data are heterogeneous.

Many refinements of bootstrap and jackknife methods have been proposed. Several designed specifically for regression models are reviewed or suggested in a paper by Wu (1986) and in a lengthy discussion which accompanies that paper. Other variations of the bootstrap are applicable to a broad class of models. Three that have received much attention are the bias-corrected percentile method, the accelerated bias-corrected method, and the bootstrap t method. We describe these briefly in turn. They each aim to give more accurate confidence intervals than simpler bootstrap methods when assumptions A or B hold only approximately.

Bias-corrected percentile method (BC method)

This method assumes there is a monotonically increasing transformation $\phi = g(\theta)$ such that

$$\hat{\phi} - \phi \sim N(-z_0\sigma, \sigma^2) \qquad \text{and} \qquad \hat{\phi}^* - \phi \sim N(-z_0\sigma, \sigma^2).$$

Both z_0 and σ^2 are unknown parameters but only an estimate of z_0 will be needed. To obtain this estimate, determine the proportion of bootstrap samples, p, for which $\hat{\theta}^* \leqslant \hat{\theta}$. Then

$$\begin{aligned} p = \Pr{}^*[\hat{\theta}^* \leqslant \hat{\theta}] &= \Pr[\hat{\phi}^* \leqslant \hat{\phi}] \\ &= \Pr[(\hat{\phi}^* - \hat{\phi} + z_0\sigma)/\sigma \leqslant z_0]. \end{aligned}$$

By assumption, $(\hat{\phi}^* - \hat{\phi} + z_0\sigma)/\sigma \sim N(0, 1)$, so an estimate of z_0 is $\Phi^{-1}(p)$ where $\Phi(.)$ is the standard normal c.d.f.

As $\hat{\phi} - \phi \sim N(-z_0\sigma, \sigma^2)$, a $1 - 2\alpha$ confidence interval for θ is

$$[g^{-1}\{\hat{\phi} + z_0\sigma - z_\alpha\sigma\}, g^{-1}\{\hat{\phi} + z_0\sigma + z_\alpha\sigma\}] \tag{9.19}$$

where z_α is the $1 - \alpha$ quantile of a standard normal distribution. To find the lower limit, consider

$$\begin{aligned}\Pr{}^*[\hat{\phi}^* \leqslant \hat{\phi} + z_0\sigma - z_\alpha\sigma] &= \Pr[(\hat{\phi}^* - \hat{\phi} + z_0\sigma)/\sigma \leqslant 2z_0 - z_\alpha] \\ &= \Phi(2z_0 - z_\alpha)\end{aligned}$$

where the latter equality follows because $(\hat{\phi}^* - \hat{\phi} + z_0\sigma)/\sigma \sim N(0, 1)$. Using the estimate of z_0, determine $\beta_1 = \Phi(2z_0 - z_\alpha)$ and let $\theta^*(\beta_1)$ be the β_1-level quantile of the bootstrap distribution of $\hat{\theta}^*$. Then $\beta_1 = \Pr^*[g(\hat{\theta}^*) \leqslant g(\theta^*(\beta_1))]$, so

$$g(\theta^*(\beta_1)) = \hat{\phi} + z_0\sigma - z_\alpha\sigma. \tag{9.20}$$

Thus a $1 - 2\alpha$ confidence interval for θ is $[\theta^*(\beta_1), \theta^*(\beta_2)]$, where $\beta_2 = \Phi(2z_0 + z_\alpha)$. The following example illustrates that determining a bias-corrected percentile interval is not difficult.

Example 9.10

Suppose a 95 % confidence interval for θ is required and, in 10 000 bootstrap resamples, $\hat{\theta}^*$ exceeded $\hat{\theta}$ 5635 times. Then $z_\alpha = 1.96$ and $z_0 = \Phi^{-1}(5635/10000) = 0.16$. Thus $\beta_1 = \Phi(2\{0.16\} - 1.96) = 0.0506$ and $\beta_2 = \Phi(2\{0.16\} + 1.96) = 0.9886$. Hence the end points of the interval for θ are the $10\,000 \times 0.0506 = 506$th and 9886th largest of the $\hat{\theta}^*$s.

Accelerated bias-corrected method (ABC method)

This method assumes there is a monotonically increasing transformation $\phi = g(\theta)$ and constants τ, a and z_0 such that for all values of ϕ,

$$\hat{\phi} - \phi \sim N(-z_0\sigma_\phi, \sigma_\phi^2), \qquad \sigma_\phi = \tau(1 + a\phi).$$

The constants z_0 and a (the latter called the acceleration constant) must be estimated. As with the bias-corrected percentile method, we put $z_0 = \Phi^{-1}(p)$. Estimating a is quite complex and the reader is referred to Efron (1987) for details. The $1 - 2\alpha$ confidence interval for θ is then given by $[\theta^*(\gamma_1), \theta^*(\gamma_2)]$, where

$$\gamma_1 = \Phi\left[z_0 + \frac{(z_0 - z_\alpha)}{1 - a(z_0 - z_\alpha)}\right] \quad \text{and} \quad \gamma_2 = \Phi\left[z_0 + \frac{(z_0 + z_\alpha)}{1 - a(z_0 + z_\alpha)}\right].$$

From Theorem 9.2, the percentile method gives intervals with the correct coverage probability if the distribution of $\hat{\phi} - \phi$ is symmetric with $E[\hat{\phi} - \phi] = 0$. The BC method is designed to correct for bias if $E[\hat{\phi} - \phi] \neq 0$, while the ABC method in addition corrects for skewness in the distribution of $\hat{\phi} - \phi$. Obviously, when $a = 0$

the ABC method reduces to the BC method. When $a = 0$ and $z_0 = 0$, it reduces to the ordinary percentile method. Like the percentile method, the BC and ABC methods give confidence intervals that are invariant under transformation.

Bootstrap t method

This method assumes that $(\hat{\theta} - \theta)/\hat{\sigma}(\hat{\theta})$ and $(\hat{\theta}^* - \hat{\theta})/\hat{\sigma}(\hat{\theta}^*)$ have similar distributions, where $\hat{\sigma}(\cdot)$ denotes the estimated standard error of the estimator. Much work suggests that $(\hat{\theta} - \theta)/\hat{\sigma}(\theta)$ is generally more stable than $\hat{\theta} - \theta$, so this assumption is likely to hold better than assumption B. The method treats $(\hat{\theta} - \theta)/\hat{\sigma}(\hat{\theta})$ as a pivotal quantity and its name derives from the fact that when $\hat{\theta} \sim N(\theta, \sigma^2)$, then $(\hat{\theta} - \theta)/\hat{\sigma}(\hat{\theta})$ has a t-distribution.

The mechanics of the method are straightforward in theory. For the original sample $\hat{T} = (\hat{\theta} - \theta)/\hat{\sigma}(\hat{\theta})$ is determined and, for each bootstrap resample, $\hat{T}^* = (\hat{\theta}^* - \hat{\theta})/\hat{\sigma}(\hat{\theta}^*)$ is determined. Let $T^*(\alpha)$ denote the α-level quantile of $\hat{T}^*$. Then

$$\alpha \approx \Pr[(\hat{\theta} - \theta)/\hat{\sigma}(\hat{\theta}) < T^*(\alpha)] = \Pr[\theta > \hat{\theta} - \hat{\sigma}(\hat{\theta})T^*(\alpha)]$$

so $[\hat{\theta} - \hat{\sigma}(\hat{\theta})T^*(1 - \alpha), \hat{\theta} - \hat{\sigma}(\hat{\theta})T^*(\alpha)]$ is taken as the $1 - 2\alpha$ confidence interval for θ.

The method requires estimates $\hat{\sigma}(\hat{\theta})$ and $\hat{\sigma}(\hat{\theta}^*)$. The former can be obtained from the bootstrap estimates $\hat{\theta}^*$, as described in Section 9.5.2. The latter might be estimated using the jackknife. An alternative is to use *second-level bootstrapping*. Let $y_1^*, y_2^*, \ldots, y_n^*$ be a bootstrap resample and let θ_y^* denote the estimate of θ it gives. We can randomly sample, with replacement, nB^* items from $y_1^*, y_2^*, \ldots, y_n^*$ to form B^* second-level resamples, each of size n. Each second-level resample yields an estimate of θ, say $\hat{\theta}^{**}$, and the empirical variance of the $\hat{\theta}^{**}$ is an estimate of $\{\hat{\sigma}(\hat{\theta}_y^*)\}^2$. The obvious drawback of second-level bootstrapping is that it is computationally very intensive, from each bootstrap resample one should take perhaps 50 or more second-level resamples.

Theoretical considerations favour the bootstrap t and ABC methods over the other bootstrap methods described here. This is because inaccuracy in estimating the end point of a confidence interval is of order $O(n^{-3/2})$ for the bootstrap t and ABC methods, while it is of order $O(n^{-1})$ for the other methods. Choosing between the bootstrap t and ABC methods on theoretical grounds is difficult. The ABC method is somewhat ungainly in that it is formed by adding correction terms to the percentile method. In contrast, the bootstrap t method seems far more natural, but the confidence intervals it gives are not invariant under transformation, unlike those of the ABC method.

Choosing between methods on the basis of simulation studies is also difficult. The studies show that neither one of the bootstrap t and ABC methods is consistently better than the other and, indeed, they are quite commonly both outperformed by the BC method, percentile method or bootstrap B method. A further complication is that the percentile and bootstrap B methods are by far the most commonly used, which may suggest that practitioners are satisfied with the results they produce. Thus recommending a method is hard and the best method to use will depend on the task at hand. For further information on bootstrap methods the reader is referred to Efron and Tibshirani (1993) and Hall (1992).

9.6 Gibbs sampling and related methodology

In implementing Bayesian methods, the need to evaluate integrals occurs at many stages. For example, the posterior distribution is

$$q(\theta; \mathbf{x}) = L(\theta; \mathbf{x})p(\theta)/h(\mathbf{x})$$

where $L(\theta; \mathbf{x})$ is the likelihood, $p(\theta)$ is the prior distribution and $h(\mathbf{x})$ is the normalizing constant. The expression giving $h(\mathbf{x})$ is the *integral*

$$h(\mathbf{x}) = \int L(\theta; \mathbf{x})p(\theta)\, d\theta.$$

The risk function for the loss $L(\theta, \hat{\theta})$ is the *integral*

$$R(\theta, \hat{\theta}) = \int L_S(\theta, \hat{\theta})L(\theta; \mathbf{x})\, d\mathbf{x}.$$

Calculating expectations, forming marginal distributions and predictive distributions, eliminating nuisance parameters, and so on, all require integrals to be evaluated (cf. Chapters 6 and 7).

For relatively simple models, the required integrals can frequently be determined analytically if prior information corresponds to a natural conjugate prior distribution. For more complex models, however, or if a conjugate distribution is inappropriate, numerical approximation methods are usually needed to estimate the integrals. One such method that is straightforward to implement and which has wide application is Gibbs sampling.

9.6.1 *Gibbs sampling*

The usefulness of Gibbs sampling is not restricted to applications in Bayesian inference, so we describe it in a more general setting. Suppose $X_1, X_2, \ldots, X_k$ have the joint probability density $f(\mathbf{X}) = f(X_1, X_2, \ldots, X_k)$, where the X_i are scalar or vector variables. In Bayesian inference the X_i may be parameters, as Bayesian inference (unlike classical statistics) treats unknown parameters as variables that have probability distributions. This is the reason Gibbs sampling is so useful in Bayesian inference.

Let $f_i(X_i | \mathbf{X}_{\setminus i})$ denote the conditional probability density of X_i, given values of all the other Xs: $X_1, X_2, \ldots, X_{i-1}, X_{i+1}, \ldots, X_k$. We require that, for each $i = 1, 2, \ldots, k$, a random observation can be generated from the conditional density $f_i(X_i | \mathbf{X}_{\setminus i})$. It is assumed that together these conditional densities uniquely determine the joint density, $f(\mathbf{X})$. Gibbs sampling is an iterative algorithm that aims to obtain random observations from $f(\mathbf{X})$.

The algorithm starts by arbitrarily choosing starting values $\mathbf{x}^{(0)} = (x_1^{(0)}, x_2^{(0)}, \ldots, x_k^{(0)})$. To give its general step, suppose that after j iterations we have the values $\mathbf{x}^{(j)} = (x_1^{(j)}, x_2^{(j)}, \ldots, x_k^{(j)})$. Then $\mathbf{x}^{(j+1)}$ is obtained as follows.

- Generate a random observation from $f_1(X_1 \mid \mathbf{X}_{\backslash 1} = (x_2^{(j)}, x_3^{(j)}, \ldots, x_k^{(j)}))$. Denote the observation by $x_1^{(j+1)}$.
- Generate a random observation from $f_2(X_2 \mid \mathbf{X}_{\backslash 2} = (x_1^{(j+1)}, x_3^{(j)}, x_4^{(j)}, \ldots, x_k^{(j)}))$. Denote the observation by $x_2^{(j+1)}$.
- Generate $x_3^{(j+1)}$ from $f_3(X_3 \mid \mathbf{X}_{\backslash 3} = (x_1^{(j+1)}, x_2^{(j+1)}, x_4^{(j)}, x_5^{(j)}, \ldots, x_k^{(j)}))$.

⋮

- Generate $x_k^{(j+1)}$ from $f_k(X_k \mid \mathbf{X}_{\backslash k} = (x_1^{(j+1)}, x_2^{(j+1)}, \ldots, x_{k-1}^{(j+1)}))$.
- Put $\mathbf{x}^{(j+1)} = (x_1^{(j+1)}, x_2^{(j+1)}, \ldots, x_k^{(j+1)})$.

As $j \to \infty$, intuitively one might expect the distribution of $\mathbf{x}^{(j)}$ to approach some limiting distribution. What distribution? Perhaps unsurprisingly (there seem no other candidates), under suitable regularity conditions the limiting distribution is $f(\mathbf{X})$, i.e. for large j, $\mathbf{x}^{(j)}$ is a random observation from $f(\mathbf{X})$.

The sequence $\mathbf{x}^{(0)}, \mathbf{x}^{(1)}, \mathbf{x}^{(2)}, \ldots$ is one realization of a Markov chain, since the probability distribution of $\mathbf{x}^{(j+1)}$ is dependent on $\mathbf{x}^{(j)}$ but, given $\mathbf{x}^{(j)}$, it is independent of $\mathbf{x}^{(0)}, \mathbf{x}^{(1)}, \ldots, \mathbf{x}^{(j-1)}$. For this reason Gibbs sampling is referred to as a Markov chain Monte Carlo (MCMC) method. Consecutive observations $\mathbf{x}^{(j)}$ and $\mathbf{x}^{(j+1)}$ will usually be correlated so, if a random sample of approximately *independent* observations is required, one might sample intermittently from the Markov chain, perhaps taking observations $\mathbf{x}^{(j)}, \mathbf{x}^{(j+t)}, \mathbf{x}^{(j+2t)}, \ldots, \mathbf{x}^{(j+nt)}$ as a sample of size n, where j and t are large, and ignore the other observations generated. Alternatively, one might choose n different starting values $(\mathbf{x}^{(0)})$, follow the iterative procedure to obtain an $\mathbf{x}^{(j)}$ from each and take these as the sample of observations.

Suppose, now, that we have generated a large random sample of observations $\mathbf{x}^{[1]}, \mathbf{x}^{[2]}, \ldots, \mathbf{x}^{[n]}$ from $f(\mathbf{X})$ and we wish to make inferences about the distribution of one component, X_i. Let $x_i^{[j]}$ denote the ith component of $\mathbf{x}^{[j]}$ and let $\mathbf{x}_{\backslash i}^{[j]}$ denote $\mathbf{x}^{[j]}$ with the ith component deleted ($j = 1, 2, \ldots, n$). Inferences about X_i may be based on $x_i^{[1]}, x_i^{[2]}, \ldots, x_i^{[n]}$ or $\mathbf{x}_{\backslash i}^{[1]}, \mathbf{x}_{\backslash i}^{[2]}, \ldots, \mathbf{x}_{\backslash i}^{[n]}$. Using the former is more obvious. An estimate of the mean, say $\mathrm{E}[X_i]$, would be $(1/n)\sum_j x_i^{[j]}$ and the probability density of X_i would be approximated by smoothing the empirical distribution of $x_i^{[1]}, \ldots, x_i^{[n]}$. [For methods of density estimation, the reader is referred to Silverman (1986).] However, it is more efficient to base inferences on $\mathbf{x}_{\backslash i}^{[1]}, \mathbf{x}_{\backslash i}^{[2]}, \ldots, \mathbf{x}_{\backslash i}^{[n]}$. The conditional density $f_i(X_i \mid \mathbf{X}_{\backslash i})$ is obviously known as we have drawn samples from it, so $f(X_i)$ may be estimated by

$$\hat{f}(X_i) = \frac{1}{n} \sum_{j=1}^{n} f_i(X_i \mid \mathbf{X}_{\backslash i} = \mathbf{x}_{\backslash i}^{[j]}) \tag{9.21}$$

and

$$\mathrm{E}[X_i] = \frac{1}{n} \sum_{j=1}^{n} E(X_i \mid \mathbf{X}_{\backslash i} = \mathbf{x}_{\backslash i}^{[j]}). \tag{9.22}$$

Obtaining estimates in this way is sometimes referred to as 'Rao–Blackwellization' because the Rao–Blackwell theorem underpins the proof that these estimators are preferable to those based on $x_i^{[1]}, x_i^{[2]}, \ldots, x_i^{[n]}$. Note that to estimate densities or expected values of functions of $X_1, \ldots, X_k$, the observations $\mathbf{x}^{[1]}, \mathbf{x}^{[2]}, \ldots, \mathbf{x}^{[n]}$ need not be independent of each other. However, the greater the correlation between them, the larger n should be.

Similar methods can be used for inferences about a variable V that is a function $g(X_1, \ldots, X_k)$ of $X_1, \ldots, X_k$. For each observation $\mathbf{x}^{[j]}$ we can put $v^{[j]} = g(\mathbf{x}^{[j]})$ $(j = 1, 2, \ldots, n)$. Then $(1/n)\sum v^{[j]}$ is an estimate of $\mathrm{E}[V]$ and the marginal probability density of V can be approximated by smoothing the empirical distribution of $v^{[1]}, v^{[2]}, \ldots, v^{[n]}$. Alternatively, we could choose any X_i that is in the argument of g and by transformation of variables determine the density $f(V\,|\,\mathbf{X}_{\backslash i})$ from $f(X_i\,|\,\mathbf{X}_{\backslash i})$. 'Rao–Blackwellized' estimators could then be based on the $f(V\,|\,\mathbf{X}_{\backslash i} = \mathbf{x}_{\backslash i}^{[j]})$, $j = 1, 2, \ldots, n$.

There are many problems that are easily handled by the combination of Bayesian methods and Gibbs sampling (or some other MCMC method) but are difficult to handle by other means. The following is an example.

Example 9.11

We use an example given by Gelfand *et al.* (1990) that is similar to Example 7.7 (cf. Section 7.7). The data are observations Y_{ij}; $i = 1, 2, \ldots, k$; $j = 1, 2, \ldots, r$; where $Y_{ij}\,|\,\theta_i, \tau_1 \sim N(\theta_i, \tau_1^{-1})$ and $\theta_i\,|\,\phi, \tau_2 \sim N(\phi, \tau_2^{-1})$. Thus the data consist of r observations from each of k populations whose means are from a common distribution, $N(\phi, \tau_2^{-1})$. Suppose ϕ, τ_1 and τ_2 are independent with priors $\phi \sim N(\mu_0, \tau_0^{-1})$, $\tau_1 \sim \text{gamma}(\alpha_1, \beta_1)$ and $\tau_2 \sim \text{gamma}(\alpha_2, \beta_2)$. Let $\theta = (\theta_1, \ldots, \theta_k)$ and let $\mathbf{Y} = (Y_{11}, Y_{12}, \ldots, Y_{kr})$, so that $\mathbf{Y}$ denotes all the data.

The unknown parameters are τ_1, τ_2, ϕ and θ. Their joint posterior distribution is

$$q(\tau_1, \tau_2, \phi, \theta\,|\,\mathbf{Y}) \propto p_1(\tau_1)p_2(\tau_2)p_3(\phi)p_4(\theta\,|\,\phi, \tau_2)L(\tau_1, \tau_2, \phi, \theta; \mathbf{Y})$$

$$\propto \frac{\beta_1^{\alpha_1}\tau_1^{\alpha_1 - 1}\exp(-\beta_1\tau_1)}{\Gamma(\alpha_1)} \cdot \frac{\beta_2^{\alpha_2}\tau_2^{\alpha_2 - 1}\exp(-\beta_2\tau_2)}{\Gamma(\alpha_2)} \cdot \frac{\tau_0^{1/2}}{\sqrt{2\pi}}\exp\left[-\frac{\tau_0}{2}(\phi - \mu_0)^2\right]$$

$$\cdot \prod_{i=1}^{k} \frac{\tau_2^{1/2}}{\sqrt{2\pi}}\exp\left[-\frac{\tau_2}{2}(\theta_i - \phi)^2\right] \cdot \prod_{i=1}^{k}\prod_{j=1}^{r} \frac{\tau_1^{1/2}}{\sqrt{2\pi}}\exp\left[-\frac{\tau_1}{2}(Y_{ij} - \theta_i)^2\right].$$

To apply the Gibbs sampler we require formulae for (1) $f_1(\tau_1\,|\,\mathbf{Y}, \tau_2, \phi, \theta)$, (2) $f_2(\tau_2\,|\,\mathbf{Y}, \tau_1, \phi, \theta)$, (3) $f_3(\phi\,|\,\mathbf{Y}, \tau_1, \tau_2, \theta)$ and (4) $f_4(\theta\,|\,\mathbf{Y}, \tau_1, \tau_2, \phi)$.

If we ignore terms in the posterior joint distribution that do not involve τ_1, then

$$f_1(\tau_1\,|\,\mathbf{Y}, \tau_2, \phi, \theta) \propto \tau_1^{\alpha_1 + (kr/2) - 1}\exp\{-\tau_1[\beta_1 + \tfrac{1}{2}\Sigma\Sigma(Y_{ij} - \theta_i)^2]\}.$$

Thus, for (1),

$$\tau_1\,|\,\mathbf{Y}, \tau_2, \phi, \theta \sim \Gamma(\alpha_1 + kr/2, \beta_1 + \tfrac{1}{2}\Sigma\Sigma(Y_{ij} - \theta_i)^2). \tag{9.23}$$

A similar procedure for τ_2 gives, for (2),

$$\tau_2 \mid \mathbf{Y}, \tau_1, \phi, \theta \sim \Gamma\left(\alpha_2 + k/2, \beta_2 + \tfrac{1}{2}\sum_i (\theta_i - \phi)^2\right). \tag{9.24}$$

Ignoring terms in the posterior distribution that do not involve ϕ gives

$$f_3(\phi \mid \mathbf{Y}, \tau_1, \tau_2, \theta) \propto \exp\left[-\frac{1}{2}\left\{\tau_0(\phi - \mu_0)^2 + \tau_2 \sum_i (\theta_i - \phi)^2\right\}\right].$$

Using results derived in Example 6.3 leads (for (3)) to

$$\phi \mid \mathbf{Y}, \tau_1, \tau_2, \theta \sim N\left(\frac{\mu_0\tau_0 + \tau_2\Sigma\theta_i}{\tau_0 + k\tau_2}, \frac{1}{\tau_0 + k\tau_2}\right). \tag{9.25}$$

For (4), similar calculations show that the components θ, given $\mathbf{X}$, τ_1, τ_2 and ϕ, are independently distributed with

$$\theta_i \mid \mathbf{Y}, \tau_1, \tau_2, \phi \sim N\left(\frac{\phi\tau_2 + \tau_1 \sum_{j=1}^{r} Y_{ij}}{\tau_2 + r\tau_1}, \frac{1}{\tau_2 + r\tau_1}\right). \tag{9.26}$$

To run the Gibbs sampler, values of the sample data and prior distribution parameters must be inserted in equations (9.23)–(9.26). Gelfand *et al.* (1990) addressed a problem in which the available data were Y_{ij}; $i = 1, 2, \ldots, 6 = k; j = 1, 2, \ldots, 5 = r$. For prior parameters they chose $\mu_0 = 0$, $\tau_0 = 10^{-12}$, $\alpha_1 = \beta_1 = \alpha_2 = \beta_2 = 0$. After specifying starting values $\tau_1^{(0)}, \tau_2^{(0)}, \phi^{(0)}$ and $\theta^{(0)}$, the observations $\tau_1^{(1)}, \tau_2^{(1)}, \phi^{(1)}$ and $\theta^{(1)}$ are successively sampled from $f_1(\tau_1 \mid \mathbf{Y}, \tau_2^{(0)}, \phi^{(0)}, \theta^{(0)})$, $f_2(\tau_2 \mid \mathbf{Y}, \tau_1^{(1)}, \phi^{(0)}, \theta^{(0)})$, $f_3(\phi \mid \mathbf{Y}, \tau_1^{(1)}, \tau_2^{(1)}, \theta^{(0)})$ and $f_4(\theta \mid \mathbf{Y}, \tau_1^{(1)}, \tau_2^{(1)}, \phi^{(1)})$; then $\tau_1^{(2)}, \tau_2^{(2)}, \phi^{(2)}$ and $\theta^{(2)}$ are sampled from $f_1(\tau_1 \mid \mathbf{Y}, \tau_2^{(1)}, \phi^{(1)}, \theta^{(1)})$, $f_2(\tau_2 \mid \mathbf{Y}, \tau_1^{(2)}, \phi^{(1)}, \theta^{(1)})$, $f_3(\phi \mid \mathbf{Y}, \tau_1^{(2)}, \tau_2^{(2)}, \theta^{(1)})$ and $f_4(\theta \mid \mathbf{Y}, \tau_1^{(2)}, \tau_2^{(2)}, \phi^{(2)})$; and so on. Initially, observations are influenced by the starting values but after a 'burn-in' phase, $(\tau_1^{(t)}, \tau_2^{(t)}, \phi^{(t)}, \theta^{(t)})$ is a random observation from the posterior joint distribution. Gelfand *et al.* found 19 iterations sufficient for the burn-in phase and used $(\tau_1^{(20)}, \tau_2^{(20)}, \phi^{(20)}, \theta^{(20)})$ as an observation from $q(\tau_1, \tau_2, \phi, \theta \mid \mathbf{Y})$. They repeated the procedure 100 times to obtain 100 random independent observations, from which they estimated marginal densities of the parameters using Rao–Blackwellization.

The most difficult problem in applying Gibbs sampling and other MCMC methods is deciding whether the Markov chain has converged to $f(\mathbf{X})$. Some theoretical results are available (e.g. Smith and Roberts, 1993) but they are not easy to implement and, in practice, monitoring the progress of the chain through output diagnostics is generally used to decide convergence. A common approach is to plot the averages of selected scalar quantities, such as the first two moments of the marginal densities, and visually assess whether convergence has occurred. Figure 9.4 gives an example. It shows convergence of the mean had not occurred fully until after 50 000 observations, although the mean stayed within about 5% of its correct value after only 1000 iterations.

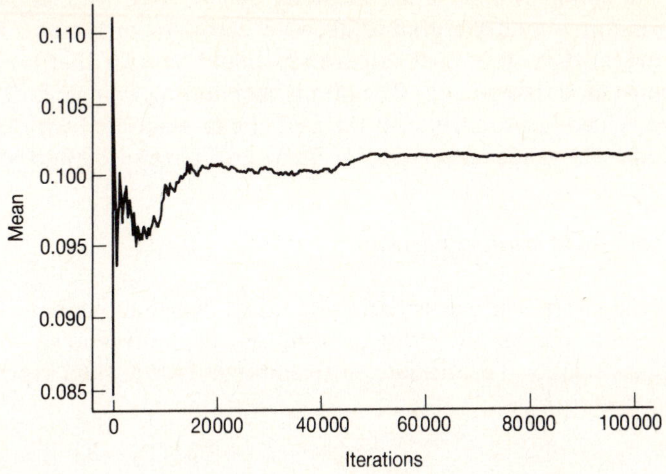

Fig. 9.4 Monitoring convergence of the mean of a marginal distribution.

It is also helpful to run the Markov chain more than once, using varied starting points, and monitor variation within each sequence and between the different sequences. Obviously convergence has not occurred while the sequences differ, but it must be stressed that the converse does not necessarily hold, i.e. sequences from different runs may overlap even though convergence has not occurred.

As mentioned in Example 9.11, Gelfand *et al.* used a burn-in of 19 iterations. This is shorter than is generally used and, quite commonly, upwards of the first 100 iterations would be discarded. (The example in Figure 9.4 requires a burn-in period of several thousand iterations but this is longer than usual.) One factor that influences the number of iterations before convergence is the correlation between the different components of $\mathbf{X}$. If correlations are high, then convergence can be slow, and reducing the correlations through a variable transformation (or a reparameterization in a Bayesian context) is desirable.

Another factor that may slow down convergence substantially is if the joint density has multiple modes with regions of low probability between some of them. For instance, suppose there are two regions of high probability that are well separated by a region of very low probability. Then the Markov chain may stay within one of the high regions for a very long period, unless the location of the other high region differs from it in only a single coordinate of $\mathbf{X}$. Note that if one runs a single long Markov chain and it stays within one of the regions of high probability, then the other high region will not be evident and the chain will appear to converge. Consequently, if multiple modes are suspected, then several separate runs from dispersed starting points should be carried out, so that the problem of multiple modes, if it exists, can be detected.

In simple problems, multimodality is seldom a problem – otherwise maximum likelihood, for example, would be a poor method of estimation. However MCMC methods, in conjunction with Bayesian methods, can be used to analyze complex data sets when multiple modes are more likely to occur. In such cases, Besag and Green (1993) suggest that an MCMC algorithm should be used that is specifically designed to change modes frequently. The idea is that, during a single long run, each mode should be visited in proportion to the probability associated with the density around that mode. The reader is referred to Besag and Green (1993) for details.

9.6.2 Metropolis–Hastings algorithm

A drawback of the Gibbs sampler is that conditional distributions must be known completely but, quite commonly, these distributions are only known to within a constant of proportionality. To elucidate, in Example 9.11 we formed expressions for conditional densities such as

$$f_1(\tau_1 \mid \mathbf{Y}, \tau_2, \phi, \theta) \propto \tau_1^{\alpha_1 + (kr/2) - 1} \exp\{-\tau_1[\beta_1 + \tfrac{1}{2}\Sigma\Sigma(Y_{ij} - \theta_i)^2]\}.$$

We then *recognized* this as a gamma distribution and hence knew the constant of proportionality. However, there will obviously be occasions when the conditional density is non-standard and then we will not know this constant. This prevents use of the Gibbs sampler, but there are other MCMC methods that can still be used. The best known of these is the Metropolis–Hastings algorithm, which we briefly describe next.

We have a probability density $f(\mathbf{X})$ that is known up to a constant of proportionality, so $f(\mathbf{X}) = cf^*(\mathbf{X})$ where $f^*(\mathbf{X})$ is known. In Bayesian inference, $f(\mathbf{X})$ might be the joint density of all unknown parameters, for example, or perhaps the conditional density of one parameter. To generate a random observation from $f(\mathbf{X})$, we choose a (so far arbitrary) probability density $u(\mathbf{x}' \mid \mathbf{x})$. For given $\mathbf{x}$, let $A_\mathbf{x}$ be the set of points $\{\mathbf{x}' : f^*(\mathbf{x}')u(\mathbf{x} \mid \mathbf{x}') < f^*(\mathbf{x})u(\mathbf{x}' \mid \mathbf{x})\}$ and define the probability density $p(\mathbf{x}' \mid \mathbf{x})$ by

$$p(\mathbf{x}' \mid \mathbf{x}') = \begin{cases} u(\mathbf{x}' \mid \mathbf{x}), & \text{if } \mathbf{x}' \neq \mathbf{x} \text{ and } \mathbf{x}' \notin A_\mathbf{x} \\ u(\mathbf{x} \mid \mathbf{x}')f^*(\mathbf{x}')/f^*(\mathbf{x}), & \text{if } \mathbf{x}' \neq \mathbf{x} \text{ and } \mathbf{x}' \in A_\mathbf{x} \\ 1 - \sum_{x' \notin A_\mathbf{x}} u(\mathbf{x}' \mid \mathbf{x}) - \sum_{x' \in A_\mathbf{x}} u(\mathbf{x} \mid \mathbf{x}')f^*(\mathbf{x}')/f^*(\mathbf{x}), & \text{if } \mathbf{x}' = \mathbf{x}. \end{cases} \tag{9.27}$$

Consider the Markov chain $\mathbf{X}^{(1)}, \mathbf{X}^{(2)}, \mathbf{X}^{(3)}, \ldots$ with transition probability function $p(\mathbf{x}' \mid \mathbf{x})$. i.e. $p(\mathbf{x}' \mid \mathbf{x}) = \Pr[\mathbf{X}^{(j+1)} = \mathbf{x}' \mid \mathbf{X}^{(j)} = \mathbf{x}]$ for all j, $\mathbf{x}$, $\mathbf{x}'$. It is readily checked that

$$f(\mathbf{x})p(\mathbf{x}' \mid \mathbf{x}) = f(\mathbf{x}')p(\mathbf{x} \mid \mathbf{x}') \tag{9.28}$$

so $f(\mathbf{X})$ is the equilibrium distribution of this chain (cf. Exercise (9.26)). Hence for large j, $\mathbf{X}^{(j)}$ is a random observation from $f(\mathbf{X})$. To indicate how a sequence from the chain is produced, suppose $\mathbf{X}^{(t)} = \mathbf{x}$ and a value for $\mathbf{X}^{(t+1)}$ is required. A value $\mathbf{x}'$ is randomly generated from $u(\mathbf{X}' \mid \mathbf{x})$. If $\mathbf{x}' \notin A_\mathbf{x}$, we put $\mathbf{X}^{(t+1)} = \mathbf{x}'$. If $\mathbf{x}' \in A_\mathbf{x}$, we perform a further randomization and accept $\mathbf{x}'$ with probability $u(\mathbf{x} \mid \mathbf{x}')f^*(\mathbf{x}')/\{u(\mathbf{x}' \mid \mathbf{x})f^*(\mathbf{x})\}$. If it is accepted, we again put $\mathbf{X}^{(t+1)} = \mathbf{x}'$. Otherwise we put $\mathbf{X}^{(t+1)} = \mathbf{x}$. This procedure gives the required transition probability from $\mathbf{x}$ to $\mathbf{x}'$.

In theory, the function $u(\mathbf{x}' \mid \mathbf{x})$ must satisfy very mild conditions for the algorithm to work. (For those familiar with Markov chains, $u(\mathbf{x}' \mid \mathbf{x})$ must be irreducible and aperiodic when regarded as a transition probability function on a suitable state space.) However, the choice of $u(\mathbf{x}' \mid \mathbf{x})$ will markedly influence the computer time needed to run the algorithm, so it should be selected carefully. Tierney (1991) considers some possible choices of $u(\mathbf{x}' \mid \mathbf{x})$.

9.6.3 Further points

The potential power of MCMC methods for performing complex Bayesian computations was realized towards the end of the 1980s. Since then the methods have been applied to a variety of problems. To mention but a few, Gelfand *et al.* (1990) apply them to variance component models, hierarchical models and crossover trials with missing data; Smith and Roberts (1993) consider constrained parameter problems, changepoint problems (where the distribution changes at some unknown point in time) and generalized linear models; Gilks *et al.* (1993) use the methods for a variety of applications in medicine.

A topic we have not touched on is methods for generating random variates from specified distributions. This is outside the scope of this book and the reader is referred to Devroye (1986) and Press (1992), which are good sources of information on such methods. Interest in MCMC has itself motivated development of some new techniques for generating random variates and, in particular, Gilks and Wild (1992) and Wakefield *et al.* (1991) give some useful methods.

Exercises

9.1 Suppose X has an exponential distribution with mean 1, $Y \sim N(\mu, 1)$ and $Z = Y/(X + 1)$.
(a) Outline a method of determining the cumulative distribution function of Z using simulation when it is known that $\mu = 1$.
(b) Given sample data $z_1, z_2, \ldots, z_n$, outline a simulation method for testing the hypothesis H_0: $\mu = 1$ against H_1: $\mu > 1$.
You may assume you have routines for generating random observations from an $N(0, 1)$ distribution and from an exponential distribution with mean 1.)

9.2 In one type of calibration problem there are two measurement methods, one that is accurate but expensive and the other is cheap but somewhat inaccurate. Measurements are made using both methods on each of n items, giving paired observations (x_i, y_i) for $i = 1, 2, \ldots, n$, where x is the accurate measurement and y is the cheap measurement. It is assumed that $y_i = \alpha + \beta x_i + \varepsilon_i$ where $\varepsilon_i \sim N(0, \sigma^2)$ and an equation is required for estimating a future x_0 from a measurement y_0 (i.e. the cheap inaccurate measurement is to be used to estimate the accurate value). One approach to the problem is to regress y on x and estimate $\hat{x}_0 = \bar{x} + (y_0 - \bar{y})/\hat{\beta}$, where $\hat{\beta}$ is the estimated regression coefficient and $\bar{x}$ and $\bar{y}$ are the sample means. Another approach is to regress x on y and put $\hat{x}_0 = \bar{x} + b(y_0 - \bar{y})$, where b is the estimated regression coefficient in this latter regression. Discuss how simulation might be used to compare these two approaches.

9.3 Let X have p.d.f. $f_X(x)$ and a *continuous* cumulative distribution function $F_X(x)$. Computers usually have routines for generating random observations from the uniform distribution, $U(0, 1)$. Given that y_0 is such an observation, show that $x_0 = F_X^{-1}(y_0)$ is a random observation of X.

Assuming y_0 is a random observation from $U(0, 1)$, show that

(a) $x_0 = -\ln(1 - y_0)/\theta$ is a random observation from the exponential distribution $f(x) = \theta e^{-\theta x}$, $x > 0$.

(b) $x_0 = \ln(y_0/(1 - y_0))$ is a random observation from the logistic distribution $f(x) = e^{-x}(1 + e^{-x})^{-2}$, $-\infty < x < \infty$.

9.4 Write a program to estimate by simulation the cumulative distribution function of Z in Exercise 9.1(a). (To obtain random observations from $N(0, 1)$, generate two independent observations y_1 and y_2 from $U(0, 1)$ and put

$$z_1 = \cos(2\pi y_1)\sqrt{-2 \ln y_2}, \qquad z_2 = \sin(2\pi y_1)\sqrt{-2 \ln y_2}.$$

Then z_1 and z_2 are independent observations from $N(0, 1)$. This is the Box–Muller transformation.)

9.5 Suppose we have m groups of observations with n_i observations in the ith group and we wish to test whether the variance differs across groups. One parametric method is based on Bartlett's test statistic

$$B = \sum_{i=1}^{m} (n_i - 1) \ln(s^2/s_i^2)$$

where s_i^2 is the sample variance for the ith group and $s^2 = \{\Sigma(n_i - 1)s_i^2\}/(n - m)$. Asymptotically $B \sim \chi^2_{m-1}$ but a randomization test may be preferred if sample sizes are small. Consider a test based on the random permutations of all observations (ensuring only that the number of observations in each group does not change) where B is determined for each permutation as well as for the actual data. The hypothesis of constant variance across groups is rejected at the α% significance level if the actual data gives one of the α% largest values of B.

(a) Criticize this randomization test.

(b) Suggest a preferable test based on residuals $x_{ij} - \bar{x}_i$, where x_{ij} is the jth observation in the ith group and $\bar{x}_i$ is the mean of observations in the ith group. Outline any obvious criticisms of your suggested test.

9.6 By writing a computer program, perform a randomization test to evaluate the evidence that factor B affected the results in Table 9.1.

9.7 Explain why r and $\Sigma x_i y_i$ are equivalent test statiftics in Example 9.3. Would r^2 be equivalent to r as a test statistic?

9.8 For the data on x and y in Table 9.2, determine the values of Akaike's information criterion for the models (a) $\hat{y} = \alpha + \beta x$ and (b) $\hat{y} = \alpha$. Under this criterion, which model is preferred?

9.9 Suppose we have observations $x_1, x_2, \ldots, x_n$ and define

$$\bar{x}_{\setminus i} = \sum_{j \neq i} x_i/(n - 1).$$

Show that the cross-validation squared error $\Sigma(x_i - \bar{x}_{\setminus i})^2$ is equal to $n^2 s^2/(n - 1)$, where s^2 is the sample variance.

9.10 Express equation (9.5) in terms of $\bar{x}$, $\bar{y}$, $x_{\setminus i}$, $y_{\setminus i}$, and $\hat{\beta}_{\setminus i}$ $(i = 1, 2, \ldots, n)$ and calculate $\hat{\omega}$ for the data in Table 9.2.

9.11 Let $Y_1, Y_2, \ldots, Y_n$ be i.i.d. random variables from the Bernoulli process, where $\Pr[Y_i = 1] = p$ and $\Pr[Y_i = 0] = 1 - p$. Suppose $\hat{\theta} = (\Sigma Y_i/n)^2$ is taken as an estimate of p^2. (This is plausible as $\Sigma Y_i/n$ is an unbiased estimator of p.) Show that the jackknife estimator to which this leads is unbiased both by

(a) determining $E(\hat{\theta})$;

(b) first showing that

$$\tilde{\theta}_J = \frac{X(X-1)}{n(n-1)} \quad \text{where} \quad X = \sum_{i=1}^{n} Y_i$$

and then determining $E[X(X-1)]$.

9.12 In one type of sampling problem (cf. Exercise 9.13), paired observations (X_i, Y_i) are obtained where the ratio Y_i/X_i has smaller variance than Y_i, and $E[X_i] = \mu$ is known while $E[Y_i] = \eta$ is unknown. One approach to estimating η is to first estimate $\theta = \eta/\mu$ by $\hat{\theta} = \bar{Y}/\bar{X}$ and to then put $\hat{\eta} = \mu\hat{\theta}$. However, $\bar{Y}/\bar{X}$ is a biased estimator of $\hat{\theta}$, so a jackknife estimator of $\hat{\theta}$ is sometimes preferred. For the following data

(a) calculate the pseudo-values $\tilde{\theta}_i$ $(i = 1, 2, \ldots, 6)$.

(b) determine the jackknife estimator of θ and the approximate 95% confidence interval for θ that the pseudo-values yield.

X	3.4	7.2	4.6	1.3	5.5	2.4
Y	14	32	23	3	31	14

9.13 Cochran (1963, p. 156) gives the population sizes in 1920 and 1930 of 49 randomly selected large US cities. These data are given below, where X and Y are the population sizes (in thousands of people) of a city in 1920 and 1930, respectively. The total population living in large US cities in 1920 was known and was to be used to estimate the corresponding figure for 1930, using the ratio estimate:

$$1930 \text{ population size} \approx 1920 \text{ population size} \times E[Y]/E[X].$$

Let $\theta = E[Y]/E[X]$ and write a computer program to

(a) calculate the jackknife pseudo-values $\tilde{\theta}_i$ $(i = 1, 2, \ldots, 49)$

(b) determine the jackknife estimator of θ and the approximate 95% confidence interval for θ that the pseudo-values yield.

X	Y	X	Y	X	Y	X	Y	X	Y
76	80	93	104	50	64	116	130	43	50
138	143	172	183	44	58	46	53	298	317
67	67	78	106	77	89	243	291	36	46
29	50	66	86	64	63	87	105	161	232
381	464	60	57	64	77	30	111	74	93
23	48	46	65	56	142	71	79	45	53
37	63	2	50	40	60	256	288	36	54
120	115	507	634	40	64	43	61	50	58
61	69	179	260	38	52	25	57	48	75
387	459	121	113	136	139	94	85		

9.14 A random sample of n independent observations, $Y_1, Y_2, \ldots, Y_n$, is taken from the uniform distribution, $U(0, \theta)$. Suppose jackknife and bootstrap estimators are to be based upon the maximum likelihood estimator.
(a) The MLE of θ is $\hat{\theta} = Y_{(n)}$, the largest observation. Show that $E[\hat{\theta}] = n\theta/(n+1)$.
(b) Show that $E[Y_{(n-1)}] = (n-1)\theta(n+1)$ where $Y_{(n-1)}$ is the second largest observation. Hence show that the bias of the jackknife estimator is $-\theta/(n^2+n)$.
(c) If $n = 4$, show that the bootstrap bias-adjusted estimator of θ given by equation (9.13) approximately equals

$$1.3164Y_{(4)} - 0.2539Y_{(3)} - 0.0586Y_{(2)} - 0.0039Y_{(1)}.$$

Calculate the bias of this estimator and compare it with the bias of the jackknife estimator when $n = 4$.

9.15 A random sample of six items is drawn from a population with unknown mean θ and gives values 14, 32, 23, 3, 31, 14. Using random number tables, generate five bootstrap samples and set $\hat{\theta}_i^*$, the estimate of θ from the ith sample, equal to the mean of the ith resample ($i = 1, 2, \ldots, 5$). Determine the bias-adjusted estimate of θ that these give and its variance. Compare them with the sample mean and the usual estimate of its variance.

9.16 Suppose items from one distribution equal either 1, with probability θ, or 0, with probability $1-\theta$. A random sample of size n is taken and let $\bar{y}$ denote the sample mean. If a bootstrap sample is taken and its mean is denoted by $\hat{\theta}^*$, find the expectation and variance of $\hat{\theta}^*$ in terms of $\bar{y}$ and n.

9.17 The estimate of θ given by a random sample was $\hat{\theta} = 77.6$. From this random sample, 50 bootstrap samples were generated and these gave the following estimates of θ (the estimates have been ordered by size).
32.9, 42.8, 50.2, 52.0, 56.2, 58.4, 59.9, 62.0, 64.4, 64.6, 65.2, 65.8, 66.2, 67.9, 69.6, 71.6, 71.9, 71.9, 72.2, 74.6, 75.3, 75.6, 76.6, 76.9, 78.1, 78.8, 78.8, 79.5, 81.2, 81.2, 81.9, 83.0, 83.3, 83.7, 84.1, 84.2, 85.3, 85.8, 85.9, 87.0, 87.5, 88.3, 89.4, 89.9, 92.3, 93.1, 93.4, 93.9, 95.3, 96.7.
Assuming $\tilde{\theta} = \hat{\theta}$ determine (a) a 90% percentile interval and (b) a 90% bootstrap B interval for θ.

9.18 Let θ be the correlation between the sizes of large US cities in 1920 (X) and 1930 (Y). The data in Exercise 9.13 are to be used to determine a 90% confidence interval for θ. Treating each (x, y) pair as an item, write a program to generate 1000 bootstrap samples and calculate a 90% percentile interval and a 90% bootstrap B interval for θ
(a) using no transformation of the bootstrap sample correlations $\hat{\theta}^*$
(b) using the transformation $\hat{\phi}^* = \tanh^{-1}(\hat{\theta}^*)$.
Compare the four intervals you obtain.

9.19 (a) A random sample gives observations $y_1, y_2, \ldots, y_n$ from a distribution with unknown mean, θ, and unknown variance, σ^2. Denote the sample variance by $s^2 = \Sigma(y_i - \bar{y})^2/(n-1)$. A large number of bootstrap resamples are generated and the mean of each sample is taken as an estimate of θ. Show that the variance of the bias-adjusted bootstrap estimate of θ is $(n-1)s^2/n^2$.
(b) Consider the regression model given by equation (9.16) and assume $\sigma^2 = \text{var}(\varepsilon_i)$ is unknown. Suppose bootstrap resamples are based on residuals as in equations (9.17) and (9.18). If the least squares estimator of $\boldsymbol{\beta}$ is $\hat{\boldsymbol{\beta}}$ and the bootstrap estimator (from a large number of resamples) is $\bar{\boldsymbol{\beta}}^*$, show that

$$\text{Var}[\bar{\boldsymbol{\beta}}^*] = \left(1 - \frac{k}{n}\right)\text{Var}[\hat{\boldsymbol{\beta}}].$$

9.20 The following bivariate data have one influential outlier.

X	7.7	13.8	14.2	14.8	16.7	17.6	22.7	23.2	27.1	29.1	29.3	34.6	41.6	41.5	49.1
Y	5.1	7.0	5.1	4.7	6.2	3.9	4.1	5.5	3.0	8.1	5.5	4.0	2.5	4.8	25.7

Because of the outlier, the model $\mathrm{E}[Y_i | x_i] = \alpha + \beta x_i + \varepsilon_i$ with $\varepsilon_i \sim N(0, \sigma^2)$ is probably inappropriate. However, assume this model holds and write a computer program to generate 1000 bootstrap estimates of β using
(a) resamples in which each pair (x_i, y_i) is treated as a single item;
(b) resamples based on residuals as in equations (9.17) and (9.18).
Plot histograms of the estimates of β obtained by each method and comment on their shapes.

9.21 For the data in Exercise 9.17, determine a 90% confidence interval for θ using (a) the bias-corrected percentile method and (b) the bootstrap t method.

9.22 Consider the situation in Example 9.11, where the data are observations $Y_{ij}, i = 1, 2, \ldots, k$; $j = 1, 2, \ldots, r$; with $Y_{ij} | \theta_i, \tau_1 \sim N(\theta_i, \tau_1^{-1})$ and $\theta_i | \phi, \tau_2 \sim N(\phi, \tau_2^{-1})$. As before, suppose $\phi \sim N(\mu_0, \tau_0^{-1})$ and $\tau_2 \sim \text{gamma}(\alpha_2, \beta_2)$, but suppose now that τ_1 is known.
(a) Describe how the Gibbs sampler would be used to estimate the marginal distribution of the θ_i. In your description, include details of the distributions from which observations would be sampled.
(b) Indicate how your analysis would change if you wished to obtain a 'Rao–Blackwellized' estimate of the distribution of $V = \Sigma_{i=1}^k \theta_i$.

9.23 Suppose we have independent counts, Y_i, over different lengths of time, t_i, and Y_i is an observation from a Poisson process with rate θ_i $(i = 1, 2, \ldots, k)$. Thus $Y_i | \theta_i \sim \text{Poisson}(\theta_i t_i)$ where the t_i are known. Assume the θ_i are independently and identically distributed from the gamma distribution with density $\theta_i^{\alpha-1} \mathrm{e}^{-\beta\theta_i} \beta^\alpha / \Gamma(\alpha)$, where α is known and β has the tprior distribution

$$p(\beta) = \beta^{\gamma-1} \mathrm{e}^{-\delta\beta} \delta^\gamma / \Gamma(\gamma).$$

Derive the appropriate conditional distributions and explain how the Gibbs sampler would be used to estimate the marginal distributions of the θ_i and β, given data $\mathbf{y} = (y_1, y_2, \ldots, y_k)$.

9.24 For the situation considered in Example 9.11 and using its notation, suppose we have four populations and five observations from each population (i.e. $r = 4$ and $k = 5$). Let the parameters of the prior distribution be $\mu_0 = 0$, $\tau_0 = 10^{-5}$, $\alpha_1 = \alpha_2 = \beta_1 = \beta_2 = 0$ and suppose the data are as follows for the four populations:

(1) 3.1, −0.7, 15.0, −1.0, 0.2 (2) 14.9, −4.5, 7.8, −19.1, −8.9

(3) 4.1, −1.2, 14.9, −1.4, 1.9 (4) 6.8, 7.5, 8.7, 4.6, 1.3.

Write a computer program that implements the Gibbs sampler and use it to obtain point estimates of $\theta = (\theta_1, \theta_2, \theta_3, \theta_4)$, ϕ, τ_1 and τ_2. Check for convergence by monitoring the parameter estimates during the Gibbs sampling.

9.25 Suppose X_1 and X_2 are discrete random variables that can each take any integer value between 1 and 20, and that their bivariate probability distribution is

$$p(x_1, x_2) \propto \exp[-\{(x_1 - 15)^2 + (x_2 - 15)^2\}/(2 \times 0.048^2)]$$
$$+ \exp[-\{(x_1 - 6)^2 + (x_2 - 6)^2\}/(2 \times 0.048^2)]$$

$x_1 = 1, 2, \ldots, 20$; $x_2 = 1, 2, \ldots, 20$.
To generate a random sample from this distribution we might consider using an MCMC method such as the Metropolis–Hastings algorithm (or the Gibbs sampler if we first determined the constant of proportionality in the definition of $p(x_1, x_2)$). Explain why convergence problems would occur if this were attempted.

9.26 Verify equation (9.28).

Chapter Ten

Generalized linear models

10.1 Introduction

In Chapter 2, Example 2.12, we described the general linear model for normally distributed data $Y_1, Y_2, \ldots, Y_n$. This model stated that

$$E[\boldsymbol{Y}] = \boldsymbol{X\beta}, \qquad \text{Var}[\boldsymbol{Y}] = \sigma^2\mathbf{I}$$

where $\boldsymbol{Y} = (Y_1, Y_2, \ldots, Y_n)^T$ is the $n \times 1$ vector of independent responses, $\boldsymbol{\beta}$ is the $q \times 1$ vector of unknown parameters, $\boldsymbol{X}$ is the $n \times q$ *model matrix* of explanatory variables, σ^2 is the constant variance and $\mathbf{I}$ is the $n \times n$ identity matrix. We assume that $\boldsymbol{X}$ is of full rank q. For convenience we are now using q to denote the number of parameters, not p as done in Chapter 2.

Whilst this model is important for normally distributed data, it is less useful for other distributions such as the binomial, Poisson and gamma. The context in which such distributions are used often means that we need to model $E[\boldsymbol{Y}]$ as a non-linear function of $\boldsymbol{X\beta}$. In a *generalized linear model* (*glm*) we are able to work with both non-normal data and non-linear functions of $\boldsymbol{X\beta}$. The distributions for which glms are applicable belong to the exponential family. Maximum likelihood estimation of the parameters is conveniently undertaken by using an iteratively weighted least squares procedure. The relative goodness-of-fit of nested models can be tested using something called the *deviance* or *scaled deviance*.

The assumptions made about a particular model may be checked by calculating various types of *residuals* and other diagnostics.

Sometimes it is not possible to fully specify the likelihood and so estimation and testing based on full maximum likelihood is not possible. However, when the first two moments of the distribution of $\boldsymbol{Y}$ can be specified, then *quasi-likelihood* methods can be used.

In this chapter we explain how to specify a glm and describe the iterative fitting procedure. We define various types of residual and model-checking diagnostics and some useful plots are described. The use of quasi-likelihood as a way of dealing with overdispersed data is illustrated. Three data sets are used to illustrate the techniques.

Much of the material on which this chapter is based is taken from the following sources: McCullagh and Nelder (1989), Firth (1991, 1993), McCullagh (1991), Collett (1991) and Jorgensen (1992). A non-parametric approach to glms is described by Green and Silverman (1994).

10.2 Specifying the model

Let $Y_1, Y_2, \ldots, Y_n$ be a set of independent random variables such that the probability density function, or probability function, can be written as

$$f(y_i; \theta_i, \phi) = \exp\{[y_i\theta_i - b(\theta_i)]/a_i(\phi) + c(y_i, \phi)\} \tag{10.1}$$

for some specific functions $a_i(.)$, $b(.)$ and $c(.)$. If ϕ is known this is a (linear) *exponential family model* (see Section 2.6) with *canonical* or *natural parameters* θ_i, $i = 1, 2, \ldots, n$. If ϕ is unknown the model belongs to the class of *exponential dispersion models* (Jorgensen, 1987, 1992, Cordeiro *et al.*, 1994). Quite often $a_i(\phi) = a_i\phi$ for known weight a_i and ϕ is called the *scale* or *dispersion parameter*. For example, when each Y_i is the mean of n_i independent normally distributed random variables with constant variance σ^2, $a_i(\phi) = \sigma^2/n_i$, i.e. $\phi = \sigma^2$ and $a_i = 1/n_i$.

Thus the likelihood of $Y_1, Y_2, \ldots, Y_n$ is

$$l(\boldsymbol{\theta}; \boldsymbol{y}) = \exp\left[\sum_{i=1}^{n} [y_i\theta_i - b(\theta_i)]/a_i(\phi) + \sum_{i=1}^{n} c(y_i, \phi)\right]. \tag{10.2}$$

Expressions for $a_i(\phi)$, $b(\theta_i)$ and $c(y_i, \phi)$ for some well-known distributions are given in Table 10.1, where we are using π to denote the probability of success in the binomial distribution. The expressions for the gamma distribution are derived in Example 10.1.

We note that the term linear exponential family derives from the fact that y_i and θ_i occur together as a linear product. If θ_i was a non-linear function involving other parameters we would refer to the family as the *curved* exponential family (Lindsey, 1995, p. 32).

Table 10.1 Distributions expressed in standard form

Distribution	$\mu = E[Y]$	$a(\phi)$	$b(\theta)$	$c(y, \phi)$
Normal, $N(\mu, \sigma^2)$	$\theta = \mu$	σ^2	$\theta^2/2$	$-\frac{1}{2}\left[\frac{y^2}{\sigma^2} + \ln(2\pi\sigma^2)\right]$
Binomial, $B(n, \pi)$	$n\left[\frac{e^\theta}{1+e^\theta}\right] = n\pi$	1	$n \ln(1 + e^\theta)$	$\ln\left[\binom{n}{y}\right]$
Poisson, $P(\mu)$	$e^\theta = \mu$	1	e^θ	$-\ln(y!)$
Gamma (ν, λ)	$-\frac{1}{\theta} = \frac{\nu}{\lambda}$	$\frac{1}{\nu}$	$-\ln(-\theta)$	$\nu \ln(\nu y) - \ln(y) - \ln(\Gamma(\nu))$

Example 10.1

Suppose Y has the gamma distribution with parameters v and λ, i.e.

$$f(y; v, \lambda) = \frac{\lambda^v}{\Gamma(v)} y^{v-1} \exp[-\lambda y], \qquad 0 < y < \infty, \quad v > 0, \lambda > 0.$$

To obtain the standard form we first write the above p.d.f. as

$$\begin{aligned} f(y; v, \lambda) &= \frac{1}{y\Gamma(v)} (\lambda y)^v \exp[-\lambda y] \\ &= \exp[v \ln(\lambda y) - \lambda y - \ln(y) - \ln(\Gamma(v))] \\ &= \exp[v \ln(\lambda) + v \ln(y) - \lambda y - \ln(y) - \ln(\Gamma(v))] \\ &= \exp\left[-y\lambda + v \ln\left(\frac{\lambda}{v}\right) + v \ln(v) + v \ln(y) - \ln(y) - \ln(\Gamma(v))\right] \\ &= \exp\left[\frac{y(-\lambda/v) + \ln(\lambda/v)}{1/v} + v \ln(v) + v \ln(y) - \ln(y) - \ln(\Gamma(v))\right] \end{aligned}$$

and then identify

$$\theta = -\frac{\lambda}{v}, \qquad b(\theta) = -\ln(-\theta), \qquad a(\phi) = \frac{1}{v}$$

and

$$c(y, \phi) = v \ln(vy) - \ln(y) - \ln(\Gamma(v)).$$

We easily find that

$$\mathrm{E}[Y] = -\frac{1}{\theta} = \frac{v}{\lambda}$$

and note that when $v = 1$ this is the exponential distribution.

In general we will not be interested in fitting a model with as many parameters as there are observations. We will want to model $\mathrm{E}[Y_i]$ or some function of $\mathrm{E}[Y_i]$ in terms of a set of p explanatory variables, which for convenience we will denote as $x_{i2}, x_{i3}, \ldots, x_{iq}$, where $q = p + 1$ and $q < n$. Including the intercept term β_1 and corresponding dummy variable $x_{i1} = 1$, our standard notation will be that

$$\boldsymbol{X} = \begin{bmatrix} \boldsymbol{x}_1^{\mathrm{T}} \\ \boldsymbol{x}_2^{\mathrm{T}} \\ \vdots \\ \boldsymbol{x}_n^{\mathrm{T}} \end{bmatrix}$$

$$\boldsymbol{x}_i^{\mathrm{T}} = [1, x_{i2}, x_{i3}, \ldots, x_{iq}]$$

and

$$\boldsymbol{\beta}^{\mathrm{T}} = [\beta_1, \beta_2, \ldots, \beta_q].$$

In terms of this notation the glm consists of three components:

(1) Independent response variables $Y_1, Y_2, \ldots, Y_n$ which share the same form of parametric distribution from the exponential family.
(2) A $q \times 1$ vector of parameters $\boldsymbol{\beta}$ and an $n \times q$ model matrix $\boldsymbol{X}$.
(3) A monotone and differentiable link function $g(\cdot)$ that defines the relationship between $\mu_i = \mathrm{E}[Y_i]$ and $\boldsymbol{x}_i^{\mathrm{T}}\boldsymbol{\beta}$, i.e. $g(\mu_i) = \boldsymbol{x}_i^{\mathrm{T}}\boldsymbol{\beta}$.

In other words, the link function takes the form

$$g(\mathrm{E}(Y_i)) = \beta_1 + \beta_2 x_{i2} + \beta_3 x_{i3} + \cdots + \beta_q x_{iq}.$$

We have already met an example of a glm in Chapters 2 and 3. The linear regression model given in Example 2.12 and Section 3.4.2 is such that $Y_1, Y_2, \ldots, Y_n$ are independent, normally distributed random variables with the identity as the link function.

It is perhaps worth pointing out at this early stage in the chapter that glms are not 'generalized' in the widest possible sense. We assume that ϕ is held constant and do not permit $\boldsymbol{X\beta}$ to be replaced by a non-linear function of $\boldsymbol{\beta}$. However, even with these limitations glms form a very useful and versatile class of models.

Example 10.2

Suppose $Y_1, Y_2, \ldots, Y_n$ are independent binomial random variables such that $Y_i \sim B(n_i, \pi_i)$, i.e.

$$\Pr[Y_i = y_i] = \binom{n_i}{y_i}\pi_i^{y_i}(1-\pi_i)^{n_i - y_i}, \qquad i = 1, 2, \ldots, n$$

and

$$\mathrm{E}(Y_i) = n_i \pi_i.$$

Such random variables might be observed, for example, in a set of trials to assess the effectiveness of an insecticide at n increasing doses $d_1, d_2, \ldots, d_n$. In the ith trial n_i insects are exposed to dose d_i and the number of insects responding, y_i, is recorded. The probability of observing a response is π_i. The link function often used with such data is the *logit* of μ_i, i.e.

$$g(\mu_i) = \ln\left[\frac{\mu_i}{n_i - \mu_i}\right] = \ln\left[\frac{\pi_i}{1-\pi_i}\right]$$

which corresponds to the cumulative distribution function of a logistic distribution. One reason for the popularity of this link with the binomial distribution is that it is the only one that admits sufficient statistics for the parameters in the linear model.

Other link functions that could have been considered are the *probit* and *complementary-log-log* (*c-l-l*) functions.

Expressed in terms of π_i the probit link is

$$g(\pi_i) = \Phi^{-1}(\pi_i)$$

where $\Phi(\pi_i)$ is the cumulative distribution function of the standard normal distribution, and the c-l-l link is

$$g(\pi_i) = \ln[-\ln(1 - \pi_i)]$$

which is derived from an extreme value distribution.

One feature that distinguishes the c-l-l function is that it is not symmetric about $\pi_i = 0.5$. We will illustrate the use of the c-l-l link in Example 10.6.

The binomial distribution is a member of the exponential family and so we have the necessary ingredients for a glm. Using the logit, the simplest model we might consider fitting to the increasing doses is a straight line

$$g(\mu_i) = g(\pi_i) = \ln\left(\frac{\pi_i}{1 - \pi_i}\right) = \beta_1 + \beta_2 d_i.$$

A more complicated model might define a polynomial regression on d_i. If we define $x_{i1} = 1$, $x_{i2} = d_i$, $x_{i3} = d_i^2, \ldots, x_{iq} = d_i^p$, then this model can be written as

$$g(\pi_i) = \sum_{k=1}^{q} \beta_k x_{ik} = \boldsymbol{x}_i^{\mathrm{T}} \boldsymbol{\beta}.$$

Staying with the simple straight-line model, the inverse transformation gives

$$\pi_i = \frac{\exp(\beta_1 + \beta_2 d_i)}{1 + \exp(\beta_1 + \beta_2 d_i)}.$$

When the link function for a given distribution equals the function that defines the canonical parameter for that distribution, the link is called a *canonical link* (cf. Section 2.6.1 and Table 2.1). The link used in Example 10.2 is a canonical link because $g(\mu) = \ln[\pi/(1 - \pi)]$, the canonical parameter for the binomial distribution. Another canonical link is given in Example 10.3, where we consider the Poisson distribution.

Example 10.3

Suppose $Y_1, Y_2, \ldots, Y_n$ are independent Poisson random variables with means $\mu_1, \mu_2, \ldots, \mu_n$, respectively. The canonical link function is

$$g(\mu_i) = \ln(\mu_i), \qquad i = 1, 2, \ldots, n.$$

To see this we note that for the Poisson distribution

$$P(Y = y) = \frac{e^{-\mu}\mu^y}{y!}$$

$$= \exp[y \ln(\mu) - \mu - \ln(y!)].$$

The canonical parameter is

$$\theta = \ln(\mu)$$

and

$$b(\theta) = \mu = e^{\theta}.$$

10.3 Fitting a generalized linear model using maximum likelihood

The log-likelihood function for a single observation y, written in canonical form, is

$$l(\theta;\, y) = \frac{[y\theta - b(\theta)]}{a(\phi)} + c(y,\, \phi).$$

It will be recalled from Chapter 2, Section 2.4.1, that $U = \partial l(\theta;\, y)/\partial\theta$ is such that

$$\mathrm{E}[U] = 0 \text{ and } \mathrm{E}[U^2] = -\mathrm{E}[U'], \quad \text{where} \quad U' = \frac{\partial^2 l(\theta;\, y)}{\partial\theta^2}.$$

Here, and in the following, we will denote the derivative of a function f, say, by f' and its second derivative by f''.

Applying the above results to the glm gives

$$U = \frac{y - b'(\theta)}{a(\phi)}$$

and

$$U' = -\frac{b''(\theta)}{a(\phi)}.$$

Hence

$$\mu = \mathrm{E}[Y] = b'(\theta)$$

and

$$\mathrm{E}[U^2] = \frac{1}{a^2(\phi)}\,\mathrm{E}[Y-\mu]^2 = \frac{\mathrm{Var}[Y]}{a^2(\phi)}.$$

Setting

$$\mathrm{E}[U^2] = -\mathrm{E}[U'] \qquad \text{gives} \qquad \mathrm{Var}[Y] = a(\phi)b''(\theta).$$

Since $b''(\theta)$ depends on μ via $b'(\theta)$ and $a(\phi)$ is independent of μ, we often write $\mathrm{Var}[Y]$ as

$$\mathrm{Var}[Y] = a(\phi)V(\mu),$$

where $V(\mu)$ is the *variance function*.

If $g(\mu)$ is the canonical link a simplification occurs in that $V(\mu) = [g'(\mu)]^{-1}$. To see this we note that for the canonical link

$$g(\mu_i) = \theta_i.$$

This implies that

$$\frac{\partial g(\mu_i)}{\partial \mu_i} = \frac{\partial \theta_i}{\partial \mu_i}$$

and

$$\frac{\partial \mu_i}{\partial \theta_i} = \frac{\partial b'(\theta_i)}{\partial \theta_i} = b''(\theta_i) = V(\mu_i).$$

That is

$$V(\mu_i) = [g'(\mu_i)]^{-1}.$$

Example 10.3 (*continued*)

Noting that $a(\phi) = 1$ for the Poisson distribution, we have

$$\mathrm{E}[Y] = b'(\theta) = \mathrm{e}^{\theta} = \mu$$

and

$$\mathrm{Var}[Y] = b''(\theta) = \mathrm{e}^{\theta} = \mu = V(\mu).$$

The canonical link is

$$g(\mu) = \ln(\mu)$$

and

$$g'(\mu) = \frac{1}{\mu} = \frac{1}{V(\mu)}.$$

Returning to our set of n random variables $Y_1, Y_2, \ldots, Y_n$, the log-likelihood function is

$$l(\boldsymbol{\theta}; \boldsymbol{y}) = \sum_{i=1}^{n} \frac{[y_i\theta_i - b(\theta_i)]}{a_i(\phi)} + \sum_{i=1}^{n} c(y_i, \phi). \tag{10.3}$$

To obtain the maximum likelihood estimators we must maximize this function. This is most easily accomplished by expressing the maximization in terms of a weighted least squares regression, and how this is done is explained next.

To obtain the maximum likelihood estimators of β_j, $j = 1, 2, \ldots, q$, we need to solve

$$\frac{\partial l(\boldsymbol{\theta}; \boldsymbol{y})}{\partial \beta_j} = 0, \qquad j = 1, 2, \ldots, q.$$

If we define

$$l_i = \frac{y_i\theta_i - b(\theta_i)}{a_i(\phi)} + c(y_i, \phi)$$

then

$$l(\boldsymbol{\theta}; \boldsymbol{y}) = \sum_{i=1}^{n} l_i.$$

Also we note that $\partial l_i/\partial \beta_j$ can be written as

$$\frac{\partial l_i}{\partial \beta_j} = \frac{\partial l_i}{\partial \theta_i}\frac{\partial \theta_i}{\partial \mu_i}\frac{\partial \mu_i}{\partial \eta_i}\frac{\partial \eta_i}{\partial \beta_j},$$

where $\eta_i = g(\mu_i) = \boldsymbol{x}_i^{\mathrm{T}}\boldsymbol{\beta}$ and is called the *linear predictor*. Further we have that

(a) $$\frac{\partial l_i}{\partial \theta_i} = \frac{y_i - b'(\theta_i)}{a_i(\phi)} = \frac{(y_i - \mu_i)}{a_i(\phi)}$$

(b) $$\frac{\partial \mu_i}{\partial \theta_i} = b''(\theta_i) = \frac{\mathrm{Var}(Y_i)}{a_i(\phi)} = V(\mu_i)$$

(c) $$\frac{\partial \eta_i}{\partial \beta_j} = x_{ij},$$ where x_{ij} is the (i, j)th element of $\boldsymbol{X}$,

and hence

$$\frac{\partial l_i}{\partial \beta_j} = \frac{(y_i - \mu_i)}{a_i(\phi)}\frac{1}{V(\mu_i)}\frac{1}{g'(\mu_i)}x_{ij}$$

where

$$g'(\mu_i) = \frac{\partial \eta_i}{\partial \mu_i}.$$

The required *estimating equations* for maximum likelihood are then

$$\sum_{i=1}^{n}\frac{\partial l_i}{\partial \beta_j} = 0, \qquad j = 1, 2, \ldots, q$$

that is

$$\sum_{i=1}^{n}\frac{(y_i - \mu_i)}{a_i(\phi)V(\mu_i)}\frac{x_{ij}}{g'(\mu_i)} = 0, \qquad j = 1, 2, \ldots, q. \tag{10.4}$$

Before solving these equations in general we consider some special cases.

If $g(\mu_i)$ is the canonical link, so that $V(\mu_i) = [g'(\mu_i)]^{-1}$, then the estimating equations simplify to

$$\sum_{i=1}^{n}\frac{y_i x_{ij}}{a_i(\phi)} = \sum_{i=1}^{n}\frac{\mu_i x_{ij}}{a_i(\phi)}, \qquad j = 1, 2, \ldots, q. \tag{10.5}$$

If $a_i(\phi)$, $i = 1, 2, \ldots, n$, are taken as known then

$$\left(\sum_{i=1}^{n} \frac{y_i x_{ij}}{a_i(\phi)}, \qquad j = 1, 2, \ldots, q\right)$$

is a set of minimal sufficient statistics for $(\beta_j, j = 1, 2, \ldots, q)$. This follows from considering the general form of the likelihood equation (10.3). If $g(\mu_i)$ is the canonical link, then the log-likelihood can be written as

$$l(\boldsymbol{\theta}; \boldsymbol{y}) = \sum_{i=1}^{n} \left(\frac{y_i(\sum_{j=1}^{q} \beta_j x_{ij})}{a_i(\phi)}\right) - \sum_{i=1}^{n} \frac{b(\theta_i)}{a_i(\phi)} + \sum_{i=1}^{n} c(y_i, \phi).$$

The first term on the right-hand side can be written as

$$\sum_{j=1}^{q} \left(\beta_j \frac{\sum_{i=1}^{n} y_i x_{ij}}{a_i(\phi)}\right)$$

and from Lemmas 2.4 and 2.5 we see that if the $a_i(\phi)$ are taken as known then

$$\left(\sum_{i=1}^{n} \frac{y_i x_{ij}}{a_i(\phi)}, \qquad j = 1, 2, \ldots, q\right)$$

is a set of minimal sufficient statistics for $(\beta_j, j = 1, 2, \ldots, q)$.

For the special case where $a_i(\phi) = a_i \phi$ and the a_i are known, equations (10.4) simplify to

$$\sum_{i=1}^{n} \frac{(y_i - \mu_i)}{a_i V(\mu_i)} \frac{x_{ij}}{g'(\mu_i)} = 0, \qquad j = 1, 2, \ldots, q$$

and are independent of the unknown parameter ϕ (and $V(\mu_i)$ for a canonical link).

Returning to the general case, the equations we need to solve are

$$U_j = \frac{\partial l(\boldsymbol{\theta}; \boldsymbol{y})}{\partial \beta_j} = 0, \qquad j = 1, 2, \ldots, q$$

where

$$U_j = \sum_{i=1}^{n} \frac{\partial l_i}{\partial \beta_j}$$

$$= \sum_{i=1}^{n} \frac{(y_i - \mu_i)}{\mathrm{Var}[Y_i]} x_{ij} \left(\frac{\partial \mu_i}{\partial \eta_i}\right), \tag{10.6}$$

where we have written $\mathrm{Var}[Y_i]$ for $a_i(\phi)V(\mu_i)$.

As described in Chapter 3, Section 3.2, this can be done using either the Newton–Raphson procedure or Fisher's Method of Scoring. We will now describe both of these and show that the Method of Scoring can be expressed as a weighted least squares procedure.

The methods we will describe begin by taking an initial estimate of $\boldsymbol{\beta}$, $\boldsymbol{b}_0$, say, and iteratively improving it to give estimates $\boldsymbol{b}_1, \boldsymbol{b}_2, \ldots$, that converge to $\hat{\boldsymbol{\beta}}$, the maximum likelihood estimate of $\boldsymbol{\beta}$.

Let $\mathbf{U}(\boldsymbol{b}_r)$ denote the vector whose elements are $[\partial l(\boldsymbol{\theta}; \boldsymbol{y})]/\partial\beta_j$, evaluated at $\boldsymbol{\beta} = \boldsymbol{b}_r$ and let $\boldsymbol{H}(b_r)$ be the $q \times q$ matrix whose (j, k)th element is $[\partial^2 l(\boldsymbol{\theta}; \boldsymbol{y})]/\partial\beta_j\partial\beta_k$, evaluated at $\boldsymbol{\beta} = \boldsymbol{b}_r$.

In the Newton–Raphson procedure the rth iteration is given by

$$\boldsymbol{b}_r = \boldsymbol{b}_{r-1} - \boldsymbol{H}^{-1}(\boldsymbol{b}_{r-1})\mathbf{U}(\boldsymbol{b}_{r-1}). \tag{10.7}$$

(Note that previously in Chapter 3 we used $\mathbf{g}$ to denote the vector of first derivatives of $l(\boldsymbol{\theta}; \boldsymbol{y})$.)

In the Method of Scoring, $\boldsymbol{H}(\boldsymbol{b}_{r-1})$ is replaced by the corresponding matrix of expected second derivatives, whose (j, k)th element is

$$\mathrm{E}\left[\frac{\partial^2 l(\boldsymbol{\theta}; \boldsymbol{y})}{\partial\beta_j\partial\beta_k}\right].$$

This matrix is the negative of the information matrix (see Lemma 2.1) and so we can replace $\boldsymbol{H}(\boldsymbol{b}_{r-1})$ in formula (10.7) by $-\mathbf{I}(\boldsymbol{b}_{r-1})$, where $\mathbf{I}(\boldsymbol{b}_{r-1})$ is the information matrix evaluated at $\boldsymbol{\beta} = \boldsymbol{b}_{r-1}$. The iterations required for the Method of Scoring are then

$$\boldsymbol{b}_r = \boldsymbol{b}_{r-1} + \mathbf{I}^{-1}(\boldsymbol{b}_{r-1})\mathbf{U}(\boldsymbol{b}_{r-1}). \tag{10.8}$$

For glms the (j, k)th element of $\mathbf{I}(\boldsymbol{b}_{r-1})$ is

$$\begin{aligned}
-\mathrm{E}\left[\frac{\partial^2 l(\boldsymbol{\theta}; \boldsymbol{y})}{\partial\beta_j\partial\beta_k}\right] &= -\mathrm{E}\left[\sum_{i=1}^{n}\frac{\partial l_i^2}{\partial\beta_j\partial\beta_k}\right] = \mathrm{E}\left[\sum_{i=1}^{n}\frac{\partial l_i}{\partial\beta_j}\frac{\partial l_i}{\partial\beta_k}\right] \\
&= \mathrm{E}\left[\sum_{i=1}^{n}\frac{(Y_i - \mu_i)^2 x_{ij}x_{ik}(\partial\mu_i/\partial\eta_i)^2}{\mathrm{Var}^2[Y_i]}\right] \\
&= \sum_{i=1}^{n}\frac{x_{ij}x_{ik}(\partial\mu_i/\partial\eta_i)^2}{\mathrm{Var}[Y_i]}
\end{aligned} \tag{10.9}$$

where this expression is evaluated at $\boldsymbol{\beta} = \boldsymbol{b}_{r-1}$.

If we multiply both sides of equation (10.8) by $\mathbf{I}(\boldsymbol{b}_{r-1})$ we obtain

$$\mathbf{I}(\boldsymbol{b}_{r-1})\boldsymbol{b}_r = \mathbf{I}(\boldsymbol{b}_{r-1})\boldsymbol{b}_{r-1} + \mathbf{U}(\boldsymbol{b}_{r-1}). \tag{10.10}$$

The jth element of the vector on the right-hand side of this equation can be expressed as

$$\sum_{k=1}^{q}\sum_{i=1}^{n}\frac{x_{ij}x_{ik}}{\mathrm{Var}[Y_i]}\left(\frac{\partial\mu_i}{\partial\eta_i}\right)^2 b_{k,r-1} + \sum_{i=1}^{n}\frac{(y_i - \mu_i)x_{ij}}{\mathrm{Var}[Y_i]}\left(\frac{\partial\mu_i}{\partial\eta_i}\right) \tag{10.11}$$

where $b_{k,r-1}$ is the kth element of $\boldsymbol{b}_{r-1}$. Both this expression and the one given in equation (10.9) can be written in matrix notation if we define $\boldsymbol{W}(\boldsymbol{b}_{r-1})$ to be the $(n \times n)$ diagonal matrix whose ith diagonal element is

$$W_{ii} = \left(\frac{\partial\mu_i}{\partial\eta_i}\right)^2\frac{1}{\mathrm{Var}[Y_i]}$$

evaluated at $\boldsymbol{\beta} = \boldsymbol{b}_{r-1}$ and let $\boldsymbol{Z}(\boldsymbol{b}_{r-1})$ be the $n \times 1$ vector whose ith element is

$$Z_i = \sum_{k=1}^{q} x_{ik} b_{k,r-1} + (y_i - \mu_i)\left(\frac{\partial \eta_i}{\partial \mu_i}\right)$$

$i = 1, 2, \ldots, n$. The variable Z_i is referred to as the *adjusted dependent variable*. The quantity in (10.9) is the element of a matrix that can be written as $\boldsymbol{X}^{\mathrm{T}}\boldsymbol{W}(\boldsymbol{b}_{r-1})\boldsymbol{X}$ and (10.11) is the element of a matrix that can be written as $\boldsymbol{X}^{\mathrm{T}}\boldsymbol{W}(\boldsymbol{b}_{r-1})\boldsymbol{Z}(\boldsymbol{b}_{r-1})$. The iterative equation given in (10.10) for the Method of Scoring can then be expressed as

$$(\boldsymbol{X}^{\mathrm{T}}\boldsymbol{W}(\boldsymbol{b}_{r-1})\boldsymbol{X})\boldsymbol{b}_r = \boldsymbol{X}^{\mathrm{T}}\boldsymbol{W}(\boldsymbol{b}_{r-1})\boldsymbol{Z}(\boldsymbol{b}_{r-1}). \tag{10.12}$$

These equations are of the same form as those for a weighted least squares regression (see Section 3.4.2) with model matrix $\boldsymbol{X}$, vector of responses $\boldsymbol{Z}(\boldsymbol{b}_{r-1})$ and weight matrix $\boldsymbol{W}(\boldsymbol{b}_{r-1})$. However, it can be seen that both the weights and the responses depend on $\boldsymbol{b}_{r-1}$, the current estimate of the parameter vector. By specifying an initial estimate $\boldsymbol{b}_0$ for $\boldsymbol{\beta}$, the equations can be solved iteratively to give a sequence of estimates $\boldsymbol{b}_1$, $\boldsymbol{b}_2, \ldots$, that converge to the maximum likelihood estimate $\hat{\boldsymbol{\beta}}$. This method of obtaining $\hat{\boldsymbol{\beta}}$ is known as iteratively weighted least squares (IWLS – see Section 3.4.2). The initial estimate $\boldsymbol{b}_0$ for $\boldsymbol{\beta}$ can be obtained by setting $\hat{\mu}_i = y_i$, adjusted if necessary. For example, for the log-link we might set y_i equal to the maximum of y_i or ε, where ε is a small positive value. If $g(\mu_i)$ is a canonical link, then the matrix of second derivatives, $\boldsymbol{H}$, does not depend on $\boldsymbol{y}$, the observed vector of data. This means that $\boldsymbol{H}(\boldsymbol{\beta}) = -\mathbf{I}(\boldsymbol{\beta})$ and the Newton–Raphson method and the Method of Scoring are identical.

When a solution vector $\hat{\boldsymbol{\beta}}$ can be found, it is consistent, asymptotically normal and asymptotically efficient. The mean of the asymptotic normal distribution is $\boldsymbol{\beta}$ and the variance–covariance matrix is $\mathbf{I}^{-1}$, where the (j, k)th element of $\mathbf{I}$ is

$$I_{jk} = \sum_{i=1}^{n} \left[\frac{x_{ij}x_{ik}}{a_i(\phi)V(\mu_i)[g'(\mu_i)]^2}\right].$$

If $a_i(\phi) = a_i\phi$, then

$$\mathbf{I} = \frac{1}{\phi}\boldsymbol{X}^{\mathrm{T}}\boldsymbol{W}\boldsymbol{X}$$

where $\boldsymbol{W}$ is a diagonal matrix whose ith diagonal element is

$$W_{ii} = \frac{1}{a_i V(\mu_i)[g'(\mu_i)]^2}.$$

To estimate the variance–covariance matrix of $\hat{\boldsymbol{\beta}}$ we use

$$\hat{V}(\hat{\boldsymbol{\beta}}) = \phi(\boldsymbol{X}^{\mathrm{T}}\hat{\boldsymbol{W}}\boldsymbol{X})^{-1}$$

where $\hat{\boldsymbol{W}}$ is the matrix $\boldsymbol{W}$ evaluated at $\hat{\boldsymbol{\beta}}$.

Unfortunately unless ϕ is known, as it is for the binomial and Poisson distributions for example, we will need to estimate it. A consistent estimator of

$$\phi = \frac{\mathrm{Var}[Y_i]}{a_i V(\mu_i)}$$

that also allows for the estimation of $\beta_1, \beta_2, \ldots, \beta_q$ is the 'degrees-of-freedom' corrected estimate

$$\tilde{\phi} = \frac{1}{(n-q)} \sum_{i=1}^{n} \frac{(y_i - \hat{\mu}_i)^2}{a_i V(\hat{\mu}_i)}. \tag{10.13}$$

We see that (10.13) can be written as $X^2(\boldsymbol{y}; \hat{\boldsymbol{\mu}})/(n-q)$, where $X^2(\boldsymbol{y}; \hat{\boldsymbol{\mu}})$ is known as the generalized Pearson statistic (Firth, 1991, p. 69; McCullagh and Nelder, 1989, p. 34). For normally distributed random variables $Y_1, Y_2, \ldots, Y_n$ with $\mathrm{E}[Y_i] = \mu_i$ and $\mathrm{Var}[Y_i] = \sigma^2$, this gives the familiar estimator

$$\tilde{\phi} = \hat{\sigma}^2 = \frac{1}{(n-q)} \sum_{i=1}^{n} (y_i - \hat{y}_i)^2.$$

Example 10.2 *(continued)*

The log-likelihood function is

$$l(\pi; \boldsymbol{y}) = \sum_{i=1}^{n} \left[\ln\binom{n_i}{y_i} + y_i \ln(\pi_i) + (n_i - y_i) \ln(1 - \pi_i) \right]$$

where

$$\mu_i = \mathrm{E}[Y_i] = n_i \pi_i, \quad i = 1, 2, \ldots, n$$

and the link function is $g(\mu_i) = \boldsymbol{x}_i^{\mathrm{T}} \boldsymbol{\beta}$ where

$$g(\mu_i) = \ln\left(\frac{\mu_i}{n_i - \mu_i}\right) = \ln\left(\frac{\pi_i}{1 - \pi_i}\right).$$

To obtain the IWLS equations we need to derive expressions for $\boldsymbol{W}$ and $\boldsymbol{Z}$ (where, for notational convenience, we have dropped the dependency of these on $\boldsymbol{b}_{r-1}$). In addition, we will also give expressions for U_j and the elements of $\mathbf{I}$, the information matrix.

The formula for U_j is

$$U_j = \frac{\partial l(\pi; \boldsymbol{y})}{\partial \beta_j} = \sum_{i=1}^{n} \left[\frac{\partial l_i}{\partial \mu_i} \frac{\partial \mu_i}{\partial \eta_i} \frac{\partial \eta_i}{\partial \beta_j} \right]$$

and we note that

$$\frac{\partial l_i}{\partial \mu_i} = \frac{n_i(y_i - \mu_i)}{\mu_i(n_i - \mu_i)} = \frac{(y_i - n_i \pi_i)}{n_i \pi_i (1 - \pi_i)}$$

$$\frac{\partial \eta_i}{\partial \mu_i} = \frac{n_i}{\mu_i(n_i - \mu_i)} = \frac{1}{n_i \pi_i (1 - \pi_i)}$$

and

$$\frac{\partial \eta_i}{\partial \beta_j} = x_{ij}.$$

Then

$$U_j = \sum_{i=1}^{n} \left[\frac{n_i(y_i - \mu_i)}{\mu_i(n_i - \mu_i)} \frac{\mu_i(n_i - \mu_i)}{n_i} x_{ij} \right]$$

$$= \sum_{i=1}^{n} (y_i - \mu_i)x_{ij}$$

$$= \sum_{i=1}^{n} (y_i - n_i\pi_i)x_{ij}.$$

The (j, k)th element of $\mathbf{I}$ is

$$\sum_{i=1}^{n} \frac{x_{ij}x_{ik}}{\text{Var}[Y_i]} \left(\frac{\partial\mu_i}{\partial\eta_i}\right)^2$$

$$= \sum_{i-1}^{n} \frac{x_{ij}x_{ik}}{n_i\pi_i(1 - \pi_i)} (n_i\pi_i(1 - \pi_i))^2$$

$$= \sum_{i=1}^{n} n_i\pi_i(1 - \pi_i)x_{ij}x_{ik}.$$

That is, the ith diagonal element of $\boldsymbol{W}$ is

$$W_{ii} = n_i\pi_i(1 - \pi_i).$$

The ith element of $\boldsymbol{Z}$ is

$$Z_i = \sum_{k=1}^{q} x_{ik}b_k + \frac{(y_i - \mu_i)}{n_i\pi_i(1 - \pi_i)}.$$

Restoring to our vectors and matrices their dependency on $\boldsymbol{b}_{r-1}$, we see that at the rth iteration $\boldsymbol{b}_r$ is obtained by a weighted regression that has response vector $\boldsymbol{Z}(\boldsymbol{b}_{r-1})$, model matrix $\boldsymbol{X}$ and weights $\boldsymbol{W}(\boldsymbol{b}_{r-1})$. The ith element of $\boldsymbol{Z}(\boldsymbol{b}_{r-1})$ can be written as

$$Z_{i,r-1} = \hat{\eta}_{i,r-1} + \frac{(y_i - n_i\hat{\pi}_{i,r-1})}{n_i\hat{\pi}_{i,r-1}(1 - \hat{\pi}_{i,r-1})}$$

where

$$\hat{\eta}_{i,r-1} = \sum_{k=1}^{q} x_{ik}b_{k,r-1}$$

is the estimated linear predictor obtained by using $\boldsymbol{b}_{r-1}$ and $\hat{\pi}_{i,r-1}$ is the corresponding estimate of π_i. The ith diagonal element of $\boldsymbol{W}(\boldsymbol{b}_{r-1})$ is $n_i\hat{\pi}_{i,r-1}(1 - \hat{\pi}_{i,r-1})$, the estimated variance of Y_i obtained using $\boldsymbol{b}_{r-1}$.

Example 10.3 (*continued*)

For the Poisson distribution the log-likelihood is

$$l(\boldsymbol{\mu}; \boldsymbol{y}) = -\sum_{i=1}^{n} \mu_i + \sum_{i=1}^{n} y_i \ln(\mu_i) + \ln\left(\prod_{i=1}^{n} y_i!\right)$$

and

$$\mathrm{E}[Y_i] = \mu_i = \mathrm{Var}[Y_i].$$

The canonical link is $g(\mu_i) = \ln(\mu_i)$ and let us assume that

$$\eta_i = g(\mu_i) = \beta_1 + \beta_2 x_i.$$

In other words the relationship between μ_i and x_i is multiplicative:

$$\mu_i = \mathrm{e}^{\beta_1 + \beta_2 x_i}.$$

The adjusted dependent variable and weights required for IWLS are, respectively

$$Z_i = \hat{\eta}_i + (y_i - \hat{\mu}_i)\frac{1}{\hat{\mu}_i}$$

and

$$W_{ii} = \hat{\mu}_i$$

where

$$\hat{\mu}_i = \mathrm{e}^{\hat{\beta}_1 + \hat{\beta}_2 x_i}.$$

10.4 Deciding if the model fits the data

One important aspect of fitting a model is to determine if it adequately describes the observed data. When using glms we determine the adequacy of a model by comparing the likelihood of the fitted model with the likelihood of the *saturated* model (assuming ϕ is known). The saturated model is of the same form as the fitted model except that it has as many parameters as there are observations, i.e. $q = n$. This model is a perfect fit to the data and is therefore generally of little use from a model fitting point of view. It is useful, however, as a means of obtaining a measure of how well the fitted model describes the data. To compare two glms, in this case the saturated and fitted models, we use the familiar likelihood ratio test statistic (Section 4.6).

We define

$$S = -2[l(\hat{\boldsymbol{\beta}}; \boldsymbol{y}, \phi) - l(\hat{\boldsymbol{\beta}}_{\mathrm{sat}}; \boldsymbol{y}, \phi)]$$

where $l(\hat{\boldsymbol{\beta}}_{\mathrm{sat}}; \boldsymbol{y}, \phi)$ is the log-likelihood evaluated using the saturated model and $l(\hat{\boldsymbol{\beta}}; \boldsymbol{y}, \phi)$ is the log-likelihood for the model under consideration evaluated using the maximum likelihood estimate $\hat{\boldsymbol{\beta}}$. The statistic S is the *scaled deviance*. (Confusingly, this is often simply called the deviance in the statistical literature.) If we express the model in canonical form and let $\hat{\theta}_i$ and $\tilde{\theta}_i$ denote the estimates of the ith canonical parameter under the fitted and saturated models, respectively, then S can be written as

$$S = 2\sum_{i=1}^{n} [y_i(\tilde{\theta}_i - \hat{\theta}_i) - b(\tilde{\theta}_i) + b(\hat{\theta}_i)]/a_i(\phi).$$

If $a_i(\phi) = \phi/w_i$, we can write

$$S = D(\mathbf{y}; \hat{\boldsymbol{\mu}})/\phi$$

where

$$D(\mathbf{y}; \hat{\boldsymbol{\mu}}) = 2\sum_{i=1}^{n} w_i[y_i(\tilde{\theta}_i - \hat{\theta}_i) - b(\tilde{\theta}_i) + b(\hat{\theta}_i)]$$

is known as the (unscaled) *deviance*.

When ϕ is known, the scaled deviance measures 'how far the fitted model deviates from the data'. If S is large the model is a poor fit. Conversely, a small value of S indicates a relatively good fit. In order to decide if S is large or not we use the fact that, as with likelihood ratio tests in general, S has an asymptotic χ^2 distribution if the model is a good fit. The degrees of freedom for this distribution equal the difference in the number of fitted parameters in the two models, i.e. $(n - q)$ in this case.

Example 10.2 (*continued*)

Let $\hat{\pi}_i$ be the maximum likelihood estimate of π_i under the fitted model. Then

$$l(\hat{\boldsymbol{\pi}}; \mathbf{y}) = \sum_{i=1}^{n} \left[\ln\binom{n_i}{y_i} + y_i \ln(\hat{\pi}_i) + (n_i - y_i)\ln(1 - \hat{\pi}_i)\right].$$

Under the saturated model π_i will be estimated by the observed proportion y_i/n_i, $i = 1, 2, \ldots, n$, giving

$$l(\hat{\boldsymbol{\pi}}_{\text{sat}}; \mathbf{y}) = \sum_{i=1}^{n} \left[\ln\binom{n_i}{y_i} + y_i \ln\left(\frac{y_i}{n_i}\right) + (n_i - y_i)\ln\left(\frac{n_i - y_i}{n_i}\right)\right].$$

The deviance (and also the scaled deviance, because $\phi = 1$) is then

$$\begin{aligned} D &= -2\sum_{i=1}^{n} \left[y_i \ln\left(\frac{n_i\hat{\pi}_i}{y_i}\right) + (n_i - y)\ln\left(\frac{n_i(1 - \hat{\pi}_i)}{n_i - y_i}\right)\right] \\ &= -2\sum_{i=1}^{n} \left[y_i \ln\left(\frac{\hat{y}_i}{y_i}\right) + (n_i - y_i)\ln\left(\frac{n_i - \hat{y}_i}{n_i - y_i}\right)\right] \end{aligned}$$

where $\hat{y}_i = n_i\hat{\pi}_i = \hat{\mu}_i$.

This latter expression shows that D is a measure of the difference between the observed and predicted values of y_i. It should be noted that for binary data, i.e. when $n_i = 1$, for all i, the deviance is uninformative, because it depends only on the fitted values. (See Collett, 1991, Section 3.8.1 and Exercise 10.4.)

Formulae to calculate the deviances of some well-known distributions are given in Table 10.2.

Table 10.2 Deviance formulae for some well-known distributions

Distribution	Deviance formula
Normal	$\sum_{i=1}^{n} (y_i - \hat{\mu}_i)^2$
Binomial	$2 \sum_{i=1}^{n} \left[y_i \ln\left(\frac{y_i}{\hat{\mu}_i}\right) + (n_i - y_i) \ln\left(\frac{n_i - y_i}{n_i - \hat{\mu}_i}\right)\right]$
Poisson	$2 \sum_{i=1}^{n} \left[y_i \ln\left(\frac{y_i}{\hat{\mu}_i}\right) - (y_i - \hat{\mu}_i)\right]$
Gamma	$2 \sum_{i=1}^{n} \left[-\ln\left(\frac{y_i}{\hat{\mu}_i}\right) + \frac{(y_i - \hat{\mu}_i)}{\hat{\mu}_i}\right]$

As noted above, the sampling distribution of S is only approximately χ^2. In fact, it is known (see Firth, 1991, for example) that the approximation may not be a good one. One special case where the result is exact is when $Y_1, Y_2, \ldots, Y_n$ are independent, normal random variables with common known variance $\sigma^2 = \phi$ and means $\mu_1, \mu_2, \ldots, \mu_n$, respectively. This result is given in Example 10.4 below.

Example 10.4

Suppose $Y_1, Y_2, \ldots, Y_n$ are independent, normally distributed random variables, each with variance σ^2 and means $\mu_1, \mu_2, \ldots, \mu_n$, respectively. Here $w_i = 1$, $i = 1, 2, \ldots, n$ and $\phi = \sigma^2$.

The log-likelihood function is

$$l(\boldsymbol{\mu}, \sigma^2; \boldsymbol{y}) = -\frac{1}{2\sigma^2} \sum_{i=1}^{n} (y_i - \mu_i)^2 - \frac{n}{2} \ln(2\pi\sigma^2)$$

and in the saturated model $\hat{\mu}_i = y_i$. In the simple model which sets $\mu_i = \mu$ for all i, $\hat{\mu}_i = \hat{\mu} = \bar{y}$.

Therefore we have

$$l(\hat{\boldsymbol{\beta}}, \sigma^2; \boldsymbol{y}) = -\frac{1}{2\sigma^2} \sum_{i=1}^{n} (y_i - \bar{y})^2 - \frac{n}{2} \ln(2\pi\sigma^2)$$

and

$$l(\hat{\boldsymbol{\beta}}_{\text{sat}}, \sigma^2; \boldsymbol{y}) = -\frac{n}{2} \ln(2\pi\sigma^2).$$

The scaled deviance S is then

$$S = -2\left[-\frac{1}{2\sigma^2} \sum_{i=1}^{n} (y_i - \bar{y})^2 \right]$$

$$= \sum_{i=1}^{n} (y_i - \bar{y})^2/\sigma^2,$$

which has a χ^2_{n-1} distribution.

For the exponential family we may write the deviance as (Firth, 1991, p. 67)

$$D(\boldsymbol{y}; \hat{\boldsymbol{\mu}}) = \sum_{i=1}^{n} d_i(y_i; \hat{\mu}_i)$$

where

$$d_i(y_i; \hat{\mu}_i) = -2 \int_{y_i}^{\hat{\mu}_i} \frac{w_i(y_i - t)}{V(t)} \, \mathrm{d}t.$$

Using this formula, the deviance for the Poisson distribution is derived in Example 10.5.

Example 10.5

Suppose $Y_1, Y_2, \ldots, Y_n$ are independent Poisson random variables with means $\mu_1, \mu_2, \ldots, \mu_n$, respectively. The deviance is

$$D(\boldsymbol{y}; \hat{\boldsymbol{\mu}}) = \sum_{i=1}^{n} d_i(y_i; \hat{\mu}_i)$$

where

$$\begin{aligned} d_i(y_i; \hat{\mu}_i) &= -2 \int_{y_i}^{\hat{\mu}_i} \frac{y_i - t}{t} \, \mathrm{d}t \\ &= -2[y_i \ln(t) - t]_{y_i}^{\hat{\mu}_i} \\ &= -2[y_i \ln(\hat{\mu}_i) - \hat{\mu}_i - y_i \ln(y_i) + y_i] \\ &= -2\left[y_i \ln\left(\frac{\hat{\mu}_i}{y_i}\right) - (\hat{\mu}_i - y_i)\right] \\ &= 2\left[y_i \ln\left(\frac{y_i}{\hat{\mu}_i}\right) - (y_i - \hat{\mu}_i)\right] \end{aligned}$$

as given in Table 10.2.

Although the scaled deviance is of limited value in assessing the goodness of fit of a model, it is useful for comparing the fits of two nested models. To explain the idea of nested models let us consider two models, M1 and M2, say. Model M2 has q_2 parameters and model M1 has q_1 parameters, where $q_2 > q_1$. Let $\boldsymbol{X}_1$ denote the model matrix for M1 and $\boldsymbol{X}_2$ the model matrix for M2. Model M1 is nested in model M2 if the columns of $\boldsymbol{X}_1$ are contained within the linear span of the columns of $\boldsymbol{X}_2$. Let $\hat{\boldsymbol{\beta}}_1$ and $\hat{\boldsymbol{\beta}}_2$ be the maximum likelihood estimates of the parameters in models M1 and M2, respectively. If $\hat{\boldsymbol{\beta}}_{\text{sat}}$ denotes the parameter estimates (i.e. the individual observations) of the saturated model, then

$$D_1 = -2\phi[l(\hat{\boldsymbol{\beta}}_1; \boldsymbol{y}) - l(\hat{\boldsymbol{\beta}}_{\text{sat}}; \boldsymbol{y})]$$

and

$$D_2 = -2\phi[l(\hat{\beta}_2; y) - l(\hat{\beta}_{sat}, y)].$$

As both D_1/ϕ and D_2/ϕ have asymptotic χ^2 distributions if each model is a good fit, and D_2 and $D_1 - D_2$ are independent, the difference $(D_1 - D_2)/\phi$ also has an asymptotic χ^2 distribution. The degrees of freedom for $(D_1 - D_2)/\phi$ equal $q_2 - q_1$.

For some distributions, like the binomial and Poisson, ϕ is known ($\phi = 1$) and so the scaled deviance is just the deviance. For some other distributions, however, ϕ has to be estimated. An estimator of ϕ was defined in the previous section in equation (10.13). When working with deviances it is convenient to use an alternative estimator of ϕ. Provided a reasonable model has been fitted to the data we may estimate ϕ by the mean deviance, i.e. by

$$\hat{\phi} = D/(n - q)$$

where D is the deviance and $(n - q)$ is the number of degrees of freedom for the deviance. The reasoning behind this is that under a reasonable model, $E(S) = (n - q)$, where $S = D/\phi$ is the scaled deviance. It will be recalled that the distribution of S is asymptotically (or exactly for the normal distribution) $\chi^2_{(n-q)}$. We should note, that $\hat{\phi}$ is not in general a consistent estimator. It may be used in practice, however, if it is known that $(1/\phi) \times \min(w_1, w_2, \ldots, w_n)$ is reasonably large, where $a_i(\phi) = \phi/w_i$ (Jorgensen, 1992, Section 3.5.4).

If ϕ has to be estimated, we can replace the χ^2 tests by F tests to determine if any parameters can be dropped from a model. If model M1 is nested in model M2, and if M1 is just as reasonable a fit as M2, then approximately

$$\frac{(D_1 - D_2)/(q_2 - q_1)}{D_2/(n - q_2)} \sim F_{(q_2 - q_1), (n - q_2)}$$

where $F_{(q_2 - q_1), (n - q_2)}$ is the central F-distribution on $[(q_2 - q_1), (n - q_2)]$ degrees of freedom. For normally distributed data this result is, of course, exact.

Although one model may be a better fit to the data than another we may still want to determine if all the parameters in the chosen model are needed. One way to do this is to drop terms from the model and, as described above, base the decision on whether the terms should be excluded by using tests based on the deviance. An alternative approach is to use the *Wald test* (Section 4.7.2). In our context, let $\boldsymbol{\beta}^T = (\boldsymbol{\beta}_1^T, \boldsymbol{\beta}_2^T)$ and suppose that we wish to test the null hypothesis H_0: $\boldsymbol{\beta}_2 = \boldsymbol{\beta}_{02}$, where the dimensions of $\boldsymbol{\beta}_1$ and $\boldsymbol{\beta}_2$ are $(r_1 \times 1)$ and $(r_2 \times 1)$, respectively. Further, let $\mathbf{I}_{22}$ denote the $(r_2 \times r_2)$ submatrix of the information matrix $\mathbf{I}$ corresponding to $\boldsymbol{\beta}_2$. On the assumption that ϕ is known, the *Wald statistic*, WD is

$$WD = (\hat{\boldsymbol{\beta}}_2 - \boldsymbol{\beta}_{02})^T \mathbf{I}_{22} (\hat{\boldsymbol{\beta}}_2 - \boldsymbol{\beta}_{02})$$

and has an asymptotic χ^2 distribution on r_2 d.f. if the null hypothesis is true. This result is equivalent to stating that $(\hat{\boldsymbol{\beta}}_2 - \boldsymbol{\beta}_{02})$ has an asymptotic multivariate normal distribution with mean vector $\mathbf{0}$ and variance–covariance matrix $\mathbf{I}_{22}^{-1}$. The Wald statistic can be used to determine if a set of parameters, held in the vector $\boldsymbol{\beta}_2$, can be dropped from the model by setting $\boldsymbol{\beta}_{02} = \mathbf{0}$. The Wald statistic can also be used to generate a multidimensional confidence interval for $\boldsymbol{\beta}_2$. For a single parameter, β_p, say, using the Wald statistic is equivalent to using the familiar standard normal test statistic

$$Z = \frac{\hat{\beta}_p}{\sqrt{\text{Var}[\hat{\beta}_p]}}.$$

On the null hypothesis that $\beta_p = 0$, Z has an asymptotic $N(0, 1)$ distribution. As when using the deviance we should be aware that the asymptotic normality is only approximate and the values of WD and Z should be used only as a guide. An approximate 95% confidence interval for β_p is $\hat{\beta}_p \pm 1.96\sqrt{\text{Var}[\hat{\beta}_p]}$. If the independent random variables $Y_1, Y_2, \ldots, Y_n$ have a normal distribution, the asymptotic results become exact, i.e. WD has a χ^2 distribution on r_2 d.f. and Z has a $N(0, 1)$ distribution. If ϕ is estimated we would assume Z has the t-distribution and WD has the F-distribution. The degrees of freedom (ν, say) for t would be the number corresponding to the deviance used to estimate ϕ. The degrees of freedom for F would be (r_2, ν).

The variances of the estimated parameters can be estimated using either the diagonal elements of $\mathbf{I}^{-1}(\hat{\boldsymbol{\beta}})$ or $-\boldsymbol{H}(\hat{\boldsymbol{\beta}})$. Although the sequence of estimates from both the Newton–Raphson and the method of scoring procedures converge to the maximum likelihood estimate of $\boldsymbol{\beta}$, the variances obtained from $\mathbf{I}^{-1}(\hat{\boldsymbol{\beta}})$ can, in some circumstances, differ from those obtained from $-\boldsymbol{H}(\hat{\boldsymbol{\beta}})$, although when n is large this difference is usually small. However, $-\boldsymbol{H}(\hat{\boldsymbol{\beta}})$ is generally preferable because it is not based on an average (i.e. an integrated) measure.

A disadvantage of using the Wald test is that it assumes the distribution of the test statistic is normal and hence symmetric. This may not be true, particularly in small samples, and then the test can be untrustworthy. An alternative approach is to base the tests and confidence intervals on the scaled deviance. (This is similar to the approach discussed in Section 5.2.3.)

Suppose we wish to test the hypothesis that $\beta_p = \beta_{0p}$. The likelihood ratio test statistic for this is

$$\lambda = S(\beta_{0p}) - S(\hat{\beta}_p)$$

where $S(\beta_{0p})$ is the scaled deviance calculated using $\beta_p = \beta_{0p}$ and $S(\hat{\beta}_p)$ is the scaled deviance calculated using the maximum likelihood estimate of β_p.

The hypothesis is rejected at significance level α if

$$\lambda > \chi^2_\alpha$$

where χ^2_α is the upper 100α% point of the χ^2 distribution on 1 d.f.

This is equivalent to the condition that

$$S(\beta_{0p}) > S(\hat{\beta}_p) + \chi^2_\alpha.$$

In other words, the hypothesis is *not* rejected if the scaled deviance evaluated at β_{0p} is not greater than χ^2_α units above the minimum scaled deviance at $\hat{\beta}_p$. The values of

β_{0p} that satisfy this requirement are a $100(1-\alpha)\%$ scaled deviance- or likelihood-based confidence interval for β_p.

Example 10.6

Gilligan (1983) described a test for randomness of infection of plant roots by soil-borne pathogens. In a series of experiments, fixed numbers of roots were planted in samples of soil infected with different densities of an inoculum. The higher the density the greater the chance of infection. The test required knowing only whether each root was infected or not, rather than having to count the number of infections per root, as had been necessary with previous tests for randomness.

If X is the number of observed infections per root, it is reasonable to assume that X has a Poisson distribution with some mean μ. In an experiment the observation for each soil sample is Y, the number of infected roots out of the total number, n say, of roots planted in that soil sample. The random variable Y has a binomial distribution $B(n, \pi)$, where π depends on the density of inoculum.

We might assume, for example, that for a particular density d the mean number of infections per root is proportional to d, i.e. $\mu = k \times d$, for some constant k. If π is the probability of a root becoming infected then, using the assumption that $X \sim P(\mu)$, we have that $\pi = 1 - e^{-kd}$ because e^{-kd} is the Poisson probability of no infections. That is,

$$-\ln(1-\pi) = kd$$

or

$$\ln[-\ln(1-\pi)] = \ln(k) + \ln(d).$$

Therefore in a glm that relates π to d, it is natural to consider using the complementary log-log (c-l-l) link function.

A more general form of the above equation is

$$\ln[-\ln(1-\pi)] = \ln(k) + \beta_1 \ln(d).$$

Here the link function is linear in the logarithm of the density. This implies that $\mu = kd^{\beta_1}$, i.e. the mean number of infections is proportional to a power (β_1) of the density.

We should note that we often have to exercise a choice when using the c-l-l link function because without our Poisson argument above we could either let π refer to the probability of a root being infected or the probability of a root being uninfected. The lack of symmetry in the c-l-l function means that the models for each choice are not equivalent. (See Exercise 10.5.)

In a particular experiment, not reported in Gilligan (1983), Gilligan obtained the results given in Table 10.3. In this experiment two different soils, conducive (labelled as C) and suppressive (labelled as S) were used in combination with five increasing densities of inoculum and each density was used in three independent samples of each soil type.

Table 10.3 Infected roots data

Observation number	Soil type	Inoculum density	Number of infected roots	Total number of roots in sample	Proportion infected
1	C	1	6	23	0.26
2	C	2	9	24	0.38
3	C	4	20	25	0.80
4	C	8	24	24	1.00
5	C	16	21	22	0.95
6	C	1	1	23	0.04
7	C	2	14	25	0.56
8	C	4	17	24	0.71
9	C	8	21	23	0.91
10	C	16	24	24	1.00
11	C	1	11	24	0.46
12	C	2	14	24	0.58
13	C	4	20	26	0.77
14	C	8	19	23	0.83
15	C	16	22	23	0.96
16	S	1	4	23	0.17
17	S	2	9	21	0.43
18	S	4	15	24	0.63
19	S	8	23	23	1.00
20	S	16	24	24	1.00
21	S	1	7	23	0.30
22	S	2	16	26	0.62
23	S	4	23	25	0.92
24	S	8	17	18	0.94
25	S	16	25	25	1.00
26	S	1	5	22	0.23
27	S	2	8	24	0.33
28	S	4	23	25	0.92
29	S	8	23	23	1.00
30	S	16	26	26	1.00

We begin the analysis of these data by first plotting the observed or empirical values of the c-l-l function against $\ln(d)$, the natural logarithm of the density. In a way similar to that used for the empirical logit transformation (Cox and Snell, 1989, p. 32), we define the *empirical c-l-l* to be

$$ec_i = \ln\left(-\ln\left[1 - \frac{(y_i + \frac{1}{2})}{(n_i + 1)}\right]\right)$$

where y_i is the observed number of roots infected out of a sample size of n_i, $i = 1, 2, \ldots, 30$. In Figure 10.1 we have plotted ec_i vs $\ln(d_i)$ for both soil types, where d_i is the

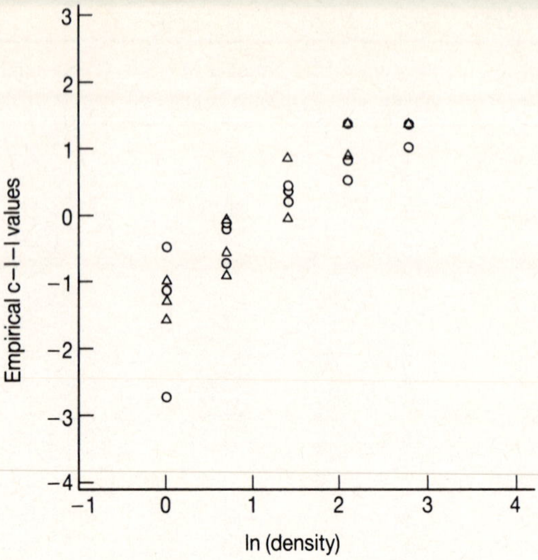

Fig. 10.1 Plot of empirical c-l-l values.

density for the *i*th observation. Except for one small value, it can be seen that there is a similarity in the shapes of the plots for each soil type and a linear relationship between ec_i and $\ln(d_i)$ is suggested.

If we fit a different fourth-order (i.e. quartic) equation to each soil type, then we will have fitted the 'maximal' model. That is, we will have accounted for the maximum number (10) of degrees of freedom for the five density levels for each soil type. The remaining 20 d.f. are for 'pure error, i.e. they correspond to the total number of d.f. for differences between the three replicate observations at each combination of soil type and density. If this maximal model is a good fit, then its mean deviance, i.e. its deviance/d.f., should be close to 1, the scale parameter for the binomial distribution. Values for the deviance, etc. for this model and some simpler models are given in Table 10.4. It can be seen that $\hat{\phi}$, the mean deviance for the maximal model, is 2.00, which suggests that the data are overdispersed (i.e. have a variance larger than that expected under the assumptions of the fitted model). As we have seen, a simple way to deal with this is to estimate ϕ and base any hypothesis testing on F-tests rather than χ^2-tests, and we will do this here. In the next section we will describe and use methods for checking the goodness of fit of a model. There, a possible explanation for the overdispersion in the data will be given and an alternative way of dealing with it will be described. In Section 10.6 we will return to the topic of overdispersion when we consider quasi-likelihood. The data will then be analysed again.

For more general discussions of overdispersion see Lindsey (1993, Chapter 5), Morgan (1992, Chapter 6), Collett (1991, Chapter 6) and Anderson (1988).

The aim of our analysis is to obtain the simplest model that reasonably explains the variation in the data. From Figure 10.1, if we ignore the one extreme value, it

Table 10.4 Deviances for some models for the roots data

Model	Deviance	Degrees of freedom
Quartic maximal model	40.057	20
Linear (L2) different line for each soil	42.958	26
Linear (L1) same line for each soil	53.432	28

seems sensible to try fitting (1) a different straight line to each soil type and (2) the same straight line to each soil type. Let us refer to these two models as L2 and L1, respectively. The deviances obtained when these models are fitted are also given in Table 10.4.

To test if model L2 is a reasonable fit we calculate

$$F = \frac{(42.958 - 40.057)/6}{40.057/20} = 0.24$$

on (6, 20) degrees of freedom. This is not significant at any standard significance level and so we can drop all quadratic and higher terms from our model. Next we need to determine if the simpler model L1 is reasonable. Comparing this to the maximal model gives a value of

$$F = \frac{(53.432 - 40.057)/8}{40.057/20} = 0.83$$

on (8, 20) degrees of freedom, which again is not significant at any standard significance level.

Our conclusion, therefore, is that the same straight-line adequately describes the relationship between the c-l-l transformation of π and ln(*density*) for each soil type. The equation of this line is

$$\ln[-\ln(1 - \pi)] = \beta_1 + \beta_2 \ln(d),$$

where π is the proportion of roots infected at dose d of inoculum. Taking $\phi = 2.003$ gives the estimates and standard errors of the parameters as in Table 10.5.

Table 10.5 Fitted parameters for model L1

Parameter	Estimate	Standard error
β_1	−1.096	0.163
β_2	1.021	0.109

A 95% confidence interval for β_2, based on the t-distribution is $1.021 \pm 2.086 \times 0.109$, i.e. (0.79, 1.25), where 2.086 is the upper 2.5% point of the t-distribution on 20 d.f. The scaled deviance-based confidence interval is the set of values of β_2 which give scaled deviances that are within 3.84 of the minimum value of 53.43/2.003, where 3.84 is the upper 5% point of the χ^2 distribution on 1 d.f. and 53.43 is the deviance for model L1. Calculating the scaled deviance over a range of values of β_2 leads to (0.83, 1.23) as the 95% scaled deviance-based confidence interval.

Arguably, we could have tested (with more power) the hypothesis that $\beta_2 = 0$ by comparing the fit of model L1 with that of L2. If we do this the hypothesis is still not rejected, though the result is less clear-cut.

10.5 Model checking for glms

10.5.1 *Normal data*

We first consider the special case of the linear model for independent, normally distributed data as defined in Section 10.1, where

$$\mathrm{E}[\boldsymbol{Y}] = \boldsymbol{X\beta} \quad \text{and} \quad \mathrm{Var}[\boldsymbol{Y}] = \sigma^2\mathbf{I}. \tag{10.14}$$

The maximum likelihood estimator of $\boldsymbol{\beta}$ is also the ordinary least squares estimator

$$\hat{\boldsymbol{\beta}} = (\boldsymbol{X}^{\mathrm{T}}\boldsymbol{X})^{-1}\boldsymbol{X}^{\mathrm{T}}\boldsymbol{Y}$$

and the fitted values are

$$\begin{aligned}\hat{\boldsymbol{Y}} &= \boldsymbol{X}\hat{\boldsymbol{\beta}} \\ &= \boldsymbol{X}(\boldsymbol{X}^{\mathrm{T}}\boldsymbol{X})^{-1}\boldsymbol{X}^{\mathrm{T}}\boldsymbol{Y} = \boldsymbol{HY}\end{aligned}$$

where

$$\boldsymbol{H} = \boldsymbol{X}(\boldsymbol{X}^{\mathrm{T}}\boldsymbol{X})^{-1}\boldsymbol{X}^{\mathrm{T}}.$$

The matrix $\boldsymbol{H}$ is known as the 'hat' matrix because it puts a hat on $\boldsymbol{Y}$, i.e. converts $\boldsymbol{Y}$ to $\hat{\boldsymbol{Y}}$.

We will use $\boldsymbol{y}$ to denote the vector of observed values of $\boldsymbol{Y}$. The ith elements of $\boldsymbol{y}$ and $\boldsymbol{Y}$, will be denoted by y_i and Y_i, respectively.

The difference between a response Y_i and its fitted value $\hat{Y}_i$ is known as the *residual* for that response. The vector of residuals is denoted by $\mathbf{R}$, where

$$\mathbf{R} = \boldsymbol{Y} - \hat{\boldsymbol{Y}} = (\mathbf{I} - \boldsymbol{H})\boldsymbol{Y}.$$

We note that

$$\mathrm{E}[\mathbf{R}] = 0 \quad \text{and} \quad \mathrm{Var}[\mathbf{R}] = (\mathbf{I} - \boldsymbol{H})\sigma^2$$

where $\mathbf{0}$ is an $n \times 1$ vector of zeros.

The observed value of R_i is denoted by r_i and referred to as the *raw residual*. We might think that we could obtain residuals with unit variance by dividing them by

s, where s^2 is the residual mean square obtained when fitting model (10.14). However, from the expression for Var[**R**] given above, it is clear that the variance of the residuals is not constant but changes with $\boldsymbol{x}_i$. Therefore, to obtain residuals with unit variance the residuals have to be standardized further to give what we will refer to here as the studentized residual r_i', i.e.

$$r_i' = \frac{y_i - \hat{y}_i}{s\sqrt{(1 - h_i)}}$$

is the *studentized residual*, where h_i is the ith diagonal element of $\boldsymbol{H}$.

It is usual to regard the r_i' as having an approximate $N(0, 1)$ distribution, at least for reasonably large n. Large values of r_i' (i.e. greater than 2) indicate that y_i is not predicted well by $\hat{y}_i$ and then y_i is termed an *outlier*. Normality of the r_i' can be checked by drawing a *normal probability plot*, i.e. by sorting the r_i' into ascending order and plotting them against their corresponding expected normal order statistics. These statistics can be approximated by, $\Phi^{-1}[(i - 3/8)/(n + 1/4)]$, $i = 1, 2, \ldots, n$. Non-linearity of the plot is evidence that the r_i' are not normally distributed. Alternatively, a *half normal plot* can be drawn by plotting the sorted absolute values of the r_i' against $\Phi^{-1}[(i + n - 1/8)/(2n + 1/2)]$. (See Collett, 1991, p. 129, for example.)

If a statistical test of normality is needed, the test described by Shapiro and Wilk (1965) can be used. (See Cook and Weisberg (1982), Section 2.3.4, or Wetherill (1986), Chapter 8, for further details.)

A simple way of checking the overall adequacy of the model is to plot the r_i' against their corresponding fitted values $\hat{y}_i$. Any pattern in this plot is suggestive of model

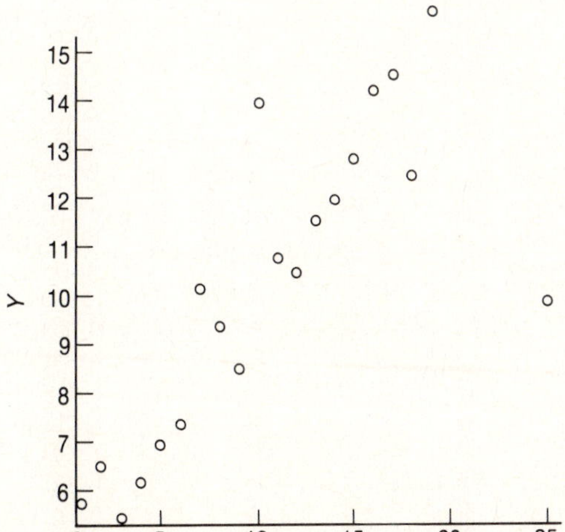

Fig. 10.2 Plot of *Y* vs x.

inadequacies. (See Draper and Smith (1981), Chapter 3, for further details and examples.)

Example 10.7

To illustrate the calculation and use of residuals, we will use the simulated data given in Table 10.6. These data represent observed values of a response variable Y at fixed values of an explanatory variable x. The data are plotted in Figure 10.2. It is clear that, apart from one or two unusual points, the data fall close to a straight line. If we assume that the relationship between Y and x is such that $E[Y_i] = \alpha + \beta x_i$ and $\text{Var}[Y_i] = \sigma^2$, the least squares estimates of the unknown parameters are $\hat{\alpha} = 6.078$, $\hat{\beta} = 0.384$ and $\hat{\sigma}^2 = s^2 = 4.264$.

The fitted values, residuals and some other useful quantities obtained from the data in Table 10.6 are given in Table 10.7.

A normal probability plot of the studentized residuals is given in Figure 10.3. On this plot we have also drawn the 45° degree line. If the data are normally distributed the points should lie close to this line. We can see that this is the case with the exception of observation 20, whose studentized residual has a value of -3.4, i.e. this observation is an outlier.

The *leverage* value h_i is a measure of the remoteness in the space of the x-variables of the ith observation from the remaining $n-1$ observations. Consequently, the leverage values are a useful diagnostic when looking for observations that might have an undue influence on the fit of the model. The values of the h_i are such that $0 \leqslant h_i \leqslant 1$

Table 10.6 Simulated data

Observation	x	Y
1	1	5.725
2	2	6.487
3	3	5.419
4	4	6.130
5	5	6.917
6	6	7.359
7	7	10.114
8	8	9.352
9	9	8.475
10	10	13.930
11	11	10.748
12	12	10.451
13	13	11.513
14	14	11.928
15	15	12.764
16	16	14.181
17	17	14.501
18	18	12.426
19	19	15.802
20	25	9.836

Table 10.7 Residuals and other diagnostics for the simulated data

Obs.	x	Y	Fitted value	Raw residual	Studentized residual	Deletion residual	Leverage	Modified Cook statistic
i	x_i	y_i	$\hat{y}_i$	r_i	r'_i	r^*_i	h_i	C_i
1	1	5.725	6.461	−0.736	−0.392	−0.382	0.171	0.521
2	2	6.487	6.845	−0.358	−0.188	−0.183	0.148	0.228
3	3	5.419	7.229	−1.810	−0.938	−0.935	0.127	1.068
4	4	6.130	7.613	−1.483	−0.760	−0.751	0.108	0.785
5	5	6.917	7.996	−1.079	−0.549	−0.538	0.092	0.514
6	6	7.359	8.380	−1.021	−0.515	−0.504	0.079	0.443
7	7	10.114	8.764	1.350	0.677	0.667	0.068	0.540
8	8	9.352	9.148	0.204	0.102	0.099	0.060	0.075
9	9	8.475	9.531	−1.056	−0.526	−0.515	0.054	0.369
10	10	13.930	9.915	4.015	1.996	2.198	0.051	1.524
11	11	10.748	10.299	0.449	0.223	0.217	0.050	0.150
12	12	10.451	10.683	−0.232	−0.115	−0.112	0.052	0.079
13	13	11.513	11.066	0.447	0.223	0.217	0.056	0.159
14	14	11.928	11.450	0.478	0.239	0.233	0.063	0.182
15	15	12.764	11.834	0.930	0.468	0.458	0.073	0.385
16	16	14.181	12.218	1.963	0.994	0.994	0.085	0.910
17	17	14.501	12.601	1.900	0.970	0.968	0.100	0.967
18	18	12.426	12.985	−0.559	−0.288	−0.281	0.117	0.307
19	19	15.802	13.369	2.433	1.268	1.292	0.137	1.543
20	25	9.836	15.671	−5.835	−3.400	−5.523	0.309	11.083

and $\Sigma_{i=1}^{n} h_i = q$. The maximum value of h_i is attained when $\hat{y}_i = y_i$. As the mean of the h_i values is q/n, a useful rule of thumb is to consider any observation that has $h_i > (2q)/n$ as potentially influential and worthy of closer scrutiny. Values of h_i are given in Table 10.7 and for our data $2q/n = 0.2$. We see that $h_{20} > 0.2$ and so observation 20 is potentially influential.

If $\boldsymbol{X}_{(i)}$ is the matrix $\boldsymbol{X}$ with its ith row deleted and $\boldsymbol{Y}_{(i)}$ is the corresponding vector of observations without Y_i, then $\hat{\boldsymbol{\beta}}_{(i)}$ is defined as

$$\hat{\boldsymbol{\beta}}_{(i)} = (\boldsymbol{X}_{(i)}^{\mathrm{T}}\boldsymbol{X}_{(i)})^{-1}\boldsymbol{X}_{(i)}^{\mathrm{T}}\boldsymbol{Y}_{(i)}.$$

If $\boldsymbol{x}_i^{\mathrm{T}}$ is the ith row of $\boldsymbol{X}$, the ith *deletion residual* r_i^* is defined as

$$r_i^* = \frac{y_i - \boldsymbol{x}_i^{\mathrm{T}}\hat{\boldsymbol{\beta}}_{(i)}}{s_{(i)}\sqrt{1 + \boldsymbol{x}_i^{\mathrm{T}}(\boldsymbol{X}_{(i)}^{\mathrm{T}}\boldsymbol{X}_{(i)})^{-1}\boldsymbol{x}_i}}$$

where $s_{(i)}^2$ is the residual mean square obtained when model (10.14) is fitted using all the data except those for the ith observation. Such residuals are useful for deciding if the deletion of the ith observation has a marked effect on prediction. Each r_i^* has a t-distribution on $(n - q - 1)$ degrees of freedom.

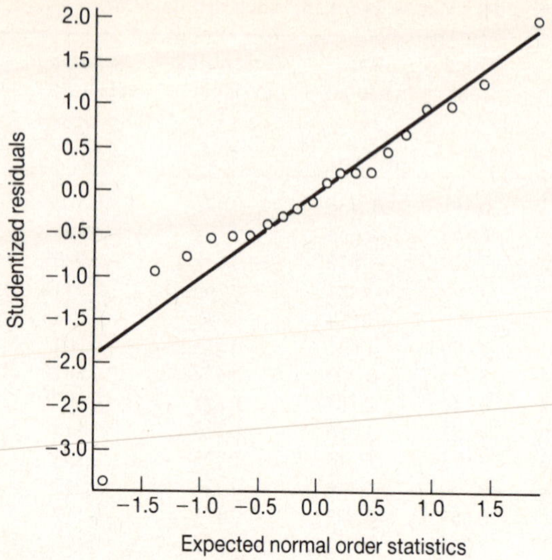

Fig. 10.3 Normal probability plot for simulated data.

The above formula can be simplified (see Exercises 10.7 and 10.8) to give

$$r_i^* = \frac{sr_i'}{s_{(i)}} = \frac{r_i'}{\sqrt{\left(\dfrac{n-q-r_i'^2}{n-q-1}\right)}}.$$

In order to determine in a more direct way which observations are influential on the values of the parameter estimates, we can use the statistic proposed by Cook (1977). For the i observation this statistic is

$$DC_i = \frac{(\hat{\boldsymbol{\beta}}_{(i)} - \hat{\boldsymbol{\beta}})^{\mathrm{T}}\mathbf{X}^{\mathrm{T}}\mathbf{X}(\hat{\boldsymbol{\beta}}_{(i)} - \hat{\boldsymbol{\beta}})}{qs^2}.$$

The statistic DC_i is a measure of the distance between $\hat{\boldsymbol{\beta}}_{(i)}$ and $\boldsymbol{\beta}$ and can be written as

$$DC_i = \frac{(r_i')^2 h_i}{q(1-h_i)}.$$

(See Atkinson (1985, p. 25) and Exercise 10.9.)

For graphical purposes Atkinson (1981) suggested using the modified *Cook statistic*

$$C_i = |r_i^*|\sqrt{\left(\frac{n-q}{q}\right)\left(\frac{h_i}{1-h_i}\right)}.$$

Atkinson (1985) describes and illustrates many useful plots of r_i^* and C_i. A particularly useful plot is the half-normal plot with simulated envelopes. We will illustrate this type of plot and others when we consider non-normal data in the next sub-section.

Example 10.7 (*continued*)

In Table 10.7 we have also given the values of r_i^* and C_i. The values for observation 20 are large, indicating that this observation is influential. If this observation is removed from the data set and the model is refitted, the estimated values of the parameters α, β and σ^2 change from 6.078, 0.384 and 4.264, respectively to 4.879, 0.537 and 1.616. Calculation of the studentized residuals will also reveal that observation 10 is an outlier, but is not influential. Its removal does, however, produce a large reduction in the residual sum of squares (see Exercise 10.10).

10.5.2 *Non-normal data*

As might be expected, the model-checking methods for non-normal data attempt to mirror those for normal data. However, because the variance of the response is a function of the mean, some modifications are necessary. To be consistent with the notation used in the literature, we will denote $\hat{y}_i$ by $\hat{\mu}_i$.

By analogy with the normal case the *studentized Pearson residual* for the ith observation is

$$r_i^P = \frac{y_i - \hat{\mu}_i}{\sqrt{\widehat{\mathrm{var}}[\hat{\mu}_i]}}$$

but here, for historical reasons, it is called the Pearson residual.

For the same reasons as in the normal case, we should use the *standardized Pearson residual*

$$r_i^{PS} = \frac{r_i^P}{\sqrt{(1 - h_i)}}$$

where h_i is the ith diagonal element of

$$\boldsymbol{H} = \boldsymbol{W}^{1/2}\boldsymbol{X}(\boldsymbol{X}^{\mathrm{T}}\boldsymbol{W}\boldsymbol{X})^{-1}\boldsymbol{X}^{\mathrm{T}}\boldsymbol{W}^{1/2}.$$

and $\boldsymbol{W}$ is the weight matrix defined in Section 10.3, below equation (10.11). We note that $\boldsymbol{W}$ depends not only on $\boldsymbol{X}$ but on $\hat{\boldsymbol{\beta}}$ as well.

For binomial data the (unstandardized) Pearson residual is

$$r_i^P = \frac{y_i - n_i\hat{\pi}_i}{\sqrt{n_i\hat{\pi}_i(1 - \hat{\pi}_i)}}$$

and for Poisson data is

$$r_i^P = \frac{y_i - \hat{\mu}_i}{\sqrt{\hat{\mu}_i}}.$$

Perhaps a more natural residual to consider for glms is the (*scaled*) *deviance residual*

$$r_i^D = \text{sgn}(y_i - \hat{\mu}_i)\sqrt{S_i}$$

where S_i is the contribution to the scaled deviance made by observation i with ϕ estimated, and $\text{sgn}(y_i - \hat{\mu}_i)$ is the sign of $(y_i - \hat{\mu}_i)$. Here we note that $\Sigma_{i=1}^n (r_i^D)^2 = S$, the scaled deviance for the fitted model under consideration.

The *standardized deviance residual* is

$$r_i^{DS} = \frac{r_i^D}{\sqrt{(1 - h_i)}}.$$

The so-called *likelihood residual* is defined as

$$r_i^L = \text{sgn}(y_i - \hat{\mu}_i)\sqrt{h_i(r_i^{PS})^2 + (1 - h_i)(r_i^{DS})^2}.$$

The usefulness of this residual lies in the fact that the change in the scaled deviance that results when the ith observation is deleted can be approximated by (Williams, 1987; Pregibon, 1981)

$$\Delta_i = h_i(r_i^{PS})^2 + (1 - h_i)(r_i^{DS})^2.$$

The sensitivity of the fit of a model to the removal of the ith observation can be assessed by looking at the size of r_i^L.

Using r_i^L we can define a modified Cook statistic as

$$C_i = |r_i^L| \sqrt{\frac{(n - q)h_i}{q(1 - h_i)}}.$$

Although outside the scope of this book, we should also note that many useful model checking procedures are based on score tests (see Chapter 4, Section 4.7.1). Further details are given by Pregibon (1982).

For a review of regression model diagnostics see Davison and Tsai (1992). For methods of testing whether a suitable link function has been chosen see Pregibon (1980). Collett (1991, Chapter 5) gives much useful advice and a number of examples of model checking for binomial and binary data.

Example 10.6 (*continued*)

To illustrate the calculation and use of the various sorts of residuals and diagnostics described in this subsection, we will return to the analysis of Gilligan's data which were presented in Table 10.3. We will ignore the fact that we suspected these data might be overdispersed and proceed to consider the fit of the quartic (maximal) model. The standardized residuals and other diagnostics for this model are given in Table 10.8.

Table 10.8 Residuals and other diagnostics for Gilligan's data

Obs. No. i	No. of infected roots y_i	Total no. of roots n_i	Fitted value $\hat{\mu}_i$	Leverage h_i	Standardized Pearson residual r_i^{PS}	Standardized deviance residual r_i^{DS}	Modified Cook statistic C_i
1	6	23	5.91	0.33	0.04	0.04	0.03
2	9	24	12.16	0.33	−1.11	−1.12	1.11
3	20	25	19.00	0.33	0.41	0.41	0.41
4	24	24	21.94	0.34	1.31	1.81	1.69
5	1	22	21.36	0.32	−0.39	−0.36	0.36
6	1	23	5.91	0.33	−2.02	−2.37	2.24
7	14	25	12.67	0.34	0.46	0.46	0.47
8	17	24	18.24	0.32	−0.51	−0.50	0.49
9	21	23	21.03	0.33	−0.02	−0.02	0.02
10	24	24	23.30	0.35	0.74	1.04	0.98
11	11	24	6.17	0.34	1.97	1.85	1.93
12	14	24	12.16	0.33	0.65	0.65	0.64
13	20	26	19.76	0.35	0.10	0.10	0.10
14	19	23	21.03	0.33	−1.30	−1.16	1.20
15	22	23	22.33	0.33	−0.36	−0.33	0.34
16	4	23	5.41	0.34	−0.60	−0.62	0.62
17	9	21	9.76	0.30	−0.28	−0.28	0.26
18	15	24	19.78	0.32	−2.21	−1.99	2.02
19	23	23	22.64	0.36	0.53	0.75	0.72
20	24	24	24.00	0.14	0.01	0.02	0.01
21	7	23	5.41	0.34	0.68	0.66	0.67
22	16	26	12.09	0.37	1.37	1.37	1.47
23	23	25	20.61	0.34	1.09	1.20	1.18
24	17	18	17.72	0.28	−1.13	−0.89	0.85
25	25	25	25.00	0.14	0.01	0.02	0.01
26	5	22	5.18	0.32	−0.08	−0.08	0.07
27	8	24	11.16	0.34	−1.12	−1.13	1.14
28	23	25	20.61	0.34	1.09	1.20	1.18
29	23	23	22.64	0.36	0.53	0.75	0.72
30	26	26	26.00	0.15	0.01	0.02	0.01

A normal probability plot of the standardized deviance residuals is given in Figure 10.4. On this plot we have again added the 45° line. The shape of the plot does not suggest that the residuals are anything other than normally distributed. A similar conclusion is obtained from a plot of the standardized Pearson residuals. If the Shapiro–Wilk test for normality is applied to the standardized Pearson residuals or the standardized deviance residuals, there is no evidence to reject the null hypothesis that they are normally distributed. Looking at the standardized residuals r_i^{PS} and r_i^{DS} in Table 10.8, we can see that the observations numbered as 6 and 18 are outliers. None of the observations has an unusually large leverage (h_i), though those for soil S and density 16 (observations 20, 25 and 30) have relatively small leverage. Turning to the values of the modified Cook statistic we see that, relatively speaking, observations 6, 11 and 18 are influential.

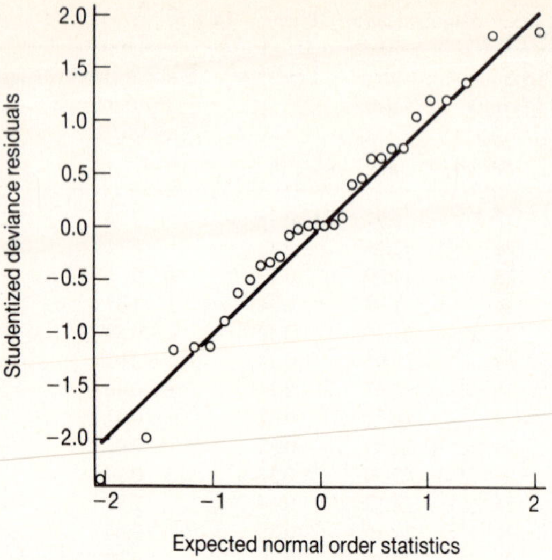

Fig. 10.4 Normal probability plot of standardized deviance residuals.

If we believe that observations 6 and 18 are in some way in error, we should attempt to discover their correct values by returning to the original records of the experiment. If they are indeed correct, then the sort of analysis we did in the previous section might be appropriate. In the next section we will consider these data from the point of view that they may be overdispersed. For now, however, we will take it that observations 6 and 18 are in error and we will reanalyze the data without them. The mean deviance for the maximal model now reduces to 1.16 and so the apparent overdispersion can be attributed to these aberrant observations. The residuals for the fit of the maximal model without observations 6 and 18 are given in the left-hand side of Table 10.9. From these it can be seen that observation 4 is now an outlier. When this observation is also removed the mean deviance reduces to 0.91 and the residuals and modified Cook statistic from the fit of the maximal model are as given in the right-hand side of Table 10.9. Now none of the residuals is exceptionally large. From the normal probability plots of the standardized Pearson and deviance residuals (not shown) there is no suggestion that the residuals are not normally distributed.

For the purposes of illustration we also give, in Figure 10.5, for the data set with observations 6, 18 and 4, removed, the half-normal plot of the standardized deviance residuals (assuming $\phi = 1$) with simulated envelopes. For this plot 19 sets of simulated data were generated to give a chance of 1 in 20 that the observed value of the largest residual lies above the envelope. (See Atkinson, 1985, pp. 35–36.) In this plot potential outliers are any points that fall above the upper edge of the envelope. (See Atkinson, 1985, Chapter 4 for examples where points fall below the lower envelope.) The lower and upper edges of the envelope are identified by the solid lines. The simulated mean

Table 10.9 Residuals and other diagnostics for Gilligan's data with some observations removed

Obs No.	Observations 6 and 18 removed			Observations 6, 18 and 4 removed		
i	Standardized Pearson residual r_i^{PS}	Standardized deviance residual r_i^{DS}	Modified Cook statistic C_i	Standardized Pearson residual r_i^{PS}	Standardized deviance residual r_i^{DS}	Modified Cook statistic C_i
1	−1.31	−1.34	1.74	−1.47	−1.51	1.91
2	−1.46	−1.47	1.38	−1.65	−1.66	1.51
3	0.53	0.54	0.51	0.60	0.61	0.56
4	1.72	2.38	2.11	–	–	–
5	−0.52	−0.48	0.45	−0.58	−0.54	0.49
6	–	–	–	–	–	–
7	0.61	0.61	0.59	0.69	0.69	0.65
8	−0.67	−0.65	0.61	−0.75	−0.74	0.66
9	−0.02	−0.02	0.02	0.92	0.97	1.23
10	0.97	1.37	1.22	1.10	1.54	1.34
11	1.31	1.29	1.78	1.48	1.45	1.95
12	0.85	0.85	0.80	0.96	0.96	0.88
13	0.13	0.13	0.12	0.14	0.14	0.14
14	−1.71	−1.52	1.49	−0.92	−0.88	1.17
15	−0.47	−0.44	0.43	−0.53	−0.50	0.47
16	−0.79	−0.82	0.78	−0.89	−0.92	0.85
17	−0.37	−0.37	0.32	−0.42	−0.42	0.35
18	–	–	–	–	–	–
19	0.70	0.99	0.90	0.79	1.11	0.99
20	0.02	0.03	0.02	0.03	0.04	0.02
21	0.89	0.87	0.84	1.00	0.98	0.92
22	1.80	1.80	1.83	2.02	2.03	2.00
23	0.00	0.00	0.00	0.00	0.00	0.00
24	−1.50	−1.16	1.06	−1.69	−1.31	1.16
25	0.02	0.03	0.02	0.03	0.04	0.02
26	−0.10	−0.10	0.09	−0.11	−0.11	0.10
27	−1.47	−1.49	1.43	−1.66	−1.68	1.56
28	0.00	0.00	0.00	0.00	0.00	0.00
29	0.70	0.99	0.90	0.79	1.11	0.99
30	0.02	0.03	0.02	0.03	0.04	0.02

values of the residuals are indicated by the dotted line. The absolute values of the standardized deviance residuals are marked by the circles. It is clear from this plot there are no unusual points.

The estimated values and standard errors of the parameters for model L1 (the same straight line for each soil type and assuming $\phi = 1$), obtained from the data set without observations 6, 18 and 4, are given in Table 10.10. Compared to the values in Table 10.5, we see that removing the outliers has changed the parameter estimates slightly and reduced their standard errors. An examination of the fit of this model will reveal some outliers. We will not pursue this further here, but leave it as an exercise (see Exercise 10.11). With such a small data set, we would not advise the

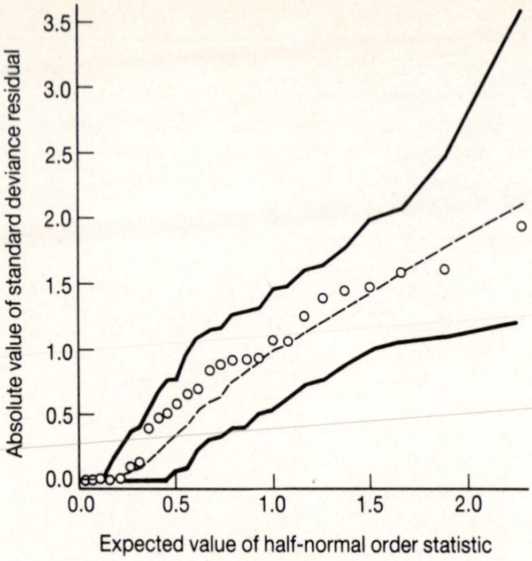

Fig. 10.5 Half-normal probability plot of standardized deviance residuals (observations 4, 6 and 18 removed).

removal of more than three observations in an attempt to make the model fit the data. In the next section we will attempt to improve the fit of the model by considering an alternative way of accounting for the unusually large variability in the data.

10.6 Quasi-likelihood

Sometimes we may not know the exact form of the p.d.f. or probability function of our response variable Y, but may be able to specify formulae for its mean and variance. This information may have come from theoretical considerations or from previous experiments or surveys. In such a situation we cannot use the complete-likelihood techniques described in Sections 10.2, 10.3 and 10.4 because the likelihood is not fully specified. One way of dealing with this situation is to use *quasi-likelihood estimation.* Here the only assumptions we shall make about the vector of responses $\boldsymbol{Y}$ are that

$$\mathrm{E}[\boldsymbol{Y}] = \boldsymbol{\mu}(\boldsymbol{\beta}) \quad \text{and} \quad \mathrm{Var}[\boldsymbol{Y}] = \phi V(\boldsymbol{\mu}) \tag{10.15}$$

Table 10.10 Fitted parameters for model L1 using depleted data set

Parameter	Estimate	Standard error
β_1	−0.982	0.116
β_2	0.955	0.078

where $\boldsymbol{\mu}(\boldsymbol{\beta})$ is the $n \times 1$ vector whose elements are $\mu_1, \mu_2, \ldots, \mu_n$, with $\mathrm{E}[Y_i] = \mu_i$, $i = 1, 2, \ldots, n$, ϕ is a known scale parameter and $V(\boldsymbol{\mu})$ is the matrix of functions that define the variances and covariances of the responses. We shall also assume, as usual, that $Y_1, Y_2, \ldots, Y_n$ are independent and so $V(\boldsymbol{\mu})$ is a diagonal matrix. The elements of $\boldsymbol{\mu}(\boldsymbol{\beta})$ depend on the unknown vector of regression parameters $\boldsymbol{\beta} = (\beta_1, \beta_2, \ldots, \beta_q)^{\mathrm{T}}$ in the usual way through the link function

$$g(\boldsymbol{\mu}) = \boldsymbol{X\beta}.$$

To appreciate the motivation for considering quasi-likelihood estimation let us consider the score U_j, defined in equation (10.6) in Section 10.3. This is

$$\begin{aligned} U_j &= \frac{\partial l(\boldsymbol{\theta}; \boldsymbol{y})}{\partial \beta_j} \\ &= \sum_{i=1}^{n} \frac{(y_i - \mu_i)}{a_i(\phi)V(\mu_i)} x_{ij}\left(\frac{\partial \mu_i}{\partial \eta_i}\right) \\ &= \sum_{i=1}^{n} \frac{(y_i - \mu_i)}{\phi V(\mu_i)} \frac{\partial \eta_i}{\partial \beta_j} \frac{\partial \mu_i}{\partial \eta_i} \qquad \text{(for the special case } a_i(\phi) = \phi \text{ considered here)} \\ &= \sum_{i=1}^{n} \frac{(y_i - \mu_i)}{\phi V(\mu_i)} \frac{\partial \mu_i}{\partial \beta_j}. \end{aligned}$$

Written in this way it can be seen that the only way the assumption that Y has a distribution that belongs to the exponential family enters into the definition of U_j is through μ_i and $V(\mu_i)$. (In fact, for ϕ known, these uniquely define the distribution.) If we did not wish to, or could not, state that the distribution of Y belonged to the exponential family, a score U_j could nevertheless be constructed. To emphasize that we are not working with a complete-likelihood specification we will denote the vector of such scores as $\boldsymbol{U}_q$, the *quasi-score* vector, where the jth element of $\boldsymbol{U}_q$ is U_{qj} and

$$U_{qj} = \sum_{i=1}^{n} \frac{1}{\phi} \frac{\partial \mu_i}{\partial \beta_j} \frac{1}{V(\mu_i)} (y_i - \mu_i).$$

In matrix notation the vector $\boldsymbol{U}_q$ can be written as

$$\boldsymbol{U}_q = \frac{1}{\phi} \boldsymbol{D}^{\mathrm{T}} \boldsymbol{W}(\boldsymbol{y} - \boldsymbol{\mu}) \qquad (10.16)$$

where $\boldsymbol{D}^{\mathrm{T}}$ is the matrix whose (i, j)th element is $\partial\mu_i/\partial\beta_j$ and $\boldsymbol{W} = \boldsymbol{V}^{-1}(\boldsymbol{\mu})$, where $\boldsymbol{V}(\boldsymbol{\mu})$ is the diagonal matrix with $V(\mu_i)$ as its ith diagonal element. We can see that $\boldsymbol{U}_q$ behaves just like the vector of ordinary scores, as defined in Section 10.3, in the sense that

$$\mathrm{E}[\boldsymbol{U}_q] = \boldsymbol{0}$$

$$\mathrm{E}[\boldsymbol{U}_q \boldsymbol{U}_q^{\mathrm{T}}] = \mathrm{Var}[\boldsymbol{U}_q] = \frac{1}{\phi} \boldsymbol{D}^{\mathrm{T}} \boldsymbol{W} \boldsymbol{D}$$

and

$$\mathrm{E}\left[\frac{\partial \boldsymbol{U}_q}{\partial \boldsymbol{\beta}}\right] = -\frac{1}{\phi}\boldsymbol{D}^{\mathrm{T}}\boldsymbol{W}\boldsymbol{D} = -\mathrm{E}[\boldsymbol{U}_q\boldsymbol{U}_q^{\mathrm{T}}].$$

The quasi-likelihood estimator $\hat{\boldsymbol{\beta}}$ of $\boldsymbol{\beta}$ is defined to be the solution of

$$\boldsymbol{D}^{\mathrm{T}}\boldsymbol{W}(\boldsymbol{Y} - \boldsymbol{\mu}(\boldsymbol{\beta})) = \boldsymbol{0}. \qquad (10.17)$$

McCullagh (1986, 1991) argues that equation (10.17) is the appropriate one to use when extending generalized least squares to the setting defined in equations (10.15). The term 'quasi-likelihood' dates back to Wedderburn (1974).

If $\hat{\boldsymbol{\beta}}$ is the solution to (10.17) then asymptotically $\hat{\boldsymbol{\beta}}$ has a multivariate normal distribution with mean vector $\boldsymbol{\beta}$ and variance–covariance matrix

$$\mathrm{Var}[\hat{\boldsymbol{\beta}}] = \phi(\boldsymbol{D}^{\mathrm{T}}\boldsymbol{W}\boldsymbol{D})^{-1}.$$

We also note that $\hat{\boldsymbol{\beta}}$ is a consistent estimator of $\boldsymbol{\beta}$.

If ϕ is unknown we replace it by a suitable estimate, e.g. the mean deviance or the degrees-of-freedom corrected estimate defined in equation (10.13) in Section 10.3.

Because $V(\boldsymbol{\mu})$ is a diagonal matrix with elements $V(\mu_i)$, $i = 1, 2, \ldots, n$, on the diagonal, we can construct the *log-quasi-likelihood* $Q(\boldsymbol{\beta}; \boldsymbol{y})$ which is defined as

$$Q(\boldsymbol{\beta}; \boldsymbol{y}) = \sum_{i=1}^{n} \int_{y_i}^{\mu_i} \frac{(y_i - t)}{\phi V_i(t)}\,\mathrm{d}t \qquad (10.18)$$

where $V_i(.)$ is the variance function for the ith observation. This behaves like the usual log-likelihood in the sense that its derivative with respect to β_j is the quasi-score, i.e.

$$\frac{\partial Q(\boldsymbol{\beta}; \boldsymbol{y})}{\partial \beta_j} = \sum_{i=1}^{n} \frac{(y_i - \mu_i)}{\phi V(\mu_i)}\frac{\partial \mu_i}{\partial \beta_j}, \qquad j = 1, 2, \ldots, q.$$

Using matrix notation, we see that this can be written as

$$\frac{\partial Q(\boldsymbol{\beta}; \boldsymbol{y})}{\partial \beta_j} = \frac{1}{\phi}\boldsymbol{D}^{\mathrm{T}}\boldsymbol{W}(\boldsymbol{y} - \boldsymbol{\mu}) = \boldsymbol{U}_q$$

the quasi-score vector.

By analogy with the complete-likelihood case, we can define the *quasi-deviance* as

$$D(\boldsymbol{y}; \boldsymbol{\mu}) = -2\phi Q(\boldsymbol{\beta}; \boldsymbol{y}).$$

Example 10.8

Suppose $Y_1, Y_2, \ldots, Y_n$ are independent with $\mathrm{E}[Y_i] = \mu_i$ and $\mathrm{Var}[Y_i] = \mu_i$, i.e. $\phi = 1$ and $V(\mu_i) = \mu_i$. The quasi-deviance is

$$Q(\boldsymbol{\beta}; \boldsymbol{y}) = -2\sum_{i=1}^{n} \int_{y_i}^{\mu_i} \frac{(y_i - t)}{t}\,\mathrm{d}t.$$

$$= -2 \sum_{i=1}^{n} [y_i \ln(t) - t]_{y_i}^{\mu_i}$$

$$= -2 \sum_{i=1}^{n} [y_i \ln(\mu_i) - \mu_i - y_i \ln(y_i) + y_i]$$

$$= -2 \sum_{i=1}^{n} \left[y_i \ln\left(\frac{\mu_i}{y_i}\right) - (\mu_i - y_i) \right]$$

$$= 2 \sum_{i=1}^{n} \left[y_i \ln\left(\frac{y_i}{\mu_i}\right) - (y_i - \mu_i) \right]$$

which we recognize from Table 10.2 in Section 10.4 as the deviance for a Poisson random variable. In fact, because $\phi = 1$, this had to be the case because the mean-variance relationship uniquely defined the distribution.

One reason we might want such a quasi-deviance is to calculate likelihood ratio statistics when comparing two nested models. Suppose model M1 is nested in model M2 and $\boldsymbol{\beta}_1$ of size $q_1 \times 1$ and $\boldsymbol{\beta}_2$ of size $q_2 \times 1$ are the parameter vectors, respectively, for M1 and M2. The quasi-likelihood ratio statistic, for ϕ known, is

$$\Lambda_q = -2[Q(\hat{\boldsymbol{\beta}}_1; \boldsymbol{y}) - Q(\hat{\boldsymbol{\beta}}_2; \boldsymbol{y})].$$

On the null hypothesis that the simpler model M1 is adequate, Λ_q has an approximate χ^2 distribution on $(q_2 - q_1)$ degrees of freedom.

Example 10.9 (McCullagh, 1986)

Let $Y_1, Y_2, \ldots, Y_n$ be uncorrelated random variables such that $\mathrm{E}[Y_i] = \mu_i < \infty$ and $\mathrm{Var}[Y_i] = \sigma^2\mu_i^2$. Suppose that $g(\mu_i) = \ln(\mu_i)$, where

$$\ln(\mu_i) = \beta_1 + \beta_2(x_i - \bar{x}), \qquad i = 1, 2, \ldots, n.$$

The estimating equations (10.17) may be written as

$$\sum_{i=1}^{n} \frac{(y_i - \hat{\mu}_i)}{\hat{\mu}_i} = 0 \tag{10.19}$$

and

$$\sum_{i=1}^{n} \frac{(x_i - \bar{x})(y_i - \hat{\mu}_i)}{\hat{\mu}_i} = 0. \tag{10.20}$$

The log-quasi-likelihood $Q(\boldsymbol{\beta}; \boldsymbol{y})$ is

$$Q(\boldsymbol{\beta}; \boldsymbol{y}) = -2 \sum_{i=1}^{n} \int_{y_i}^{\mu_i} \frac{(y_i - t)}{\sigma^2 t^2} \,\mathrm{d}t.$$

$$= \frac{-2}{\sigma^2} \sum_{i=1}^{n} \left[\frac{-y_i}{t} - \ln(t) \right]_{y_i}^{\mu_i}$$

$$= \frac{-2}{\sigma^2} \sum_{i=1}^{n} \left[\frac{-y_i}{\mu_i} - \ln(\mu_i) + 1 + \ln(y_i) \right]$$

$$= \frac{-2}{\sigma^2} \sum_{i=1}^{n} \left[\frac{-y_i}{\mu_i} + \ln\left(\frac{y_i}{\mu_i}\right) + 1 \right]$$

and the quasi-likelihood estimates for β_1 and β_2 are obtained by solving equations (10.19) and (10.20).

The solving of equation (10.17) to obtain the quasi-likelihood estimate of $\boldsymbol{\beta}$ can, of course, be done using the Method of Scoring via equations (10.8). The only change that will be required is that an appropriate formula is substituted for $\mathrm{Var}[Y_i]$. This leads to the important practical result that if the distribution of $\boldsymbol{Y}$ belongs to the (linear) exponential family with $\mathrm{E}[\boldsymbol{Y}] = \boldsymbol{\mu}$ and $\mathrm{Var}[\boldsymbol{Y}]$ proportional to $V(\boldsymbol{\mu})$, then the solution of (10.17) for $\boldsymbol{\beta}$ is the same as the one that would be obtained by solving the usual maximum likelihood equations for that exponential family. The standard errors of the estimates will need to be adjusted for the proportionality constant.

Apart from using the quasi-likelihood as a way of deciding if parameters can be dropped from a model, we can also use Wald or score tests. To calculate these tests we would use the information matrix $\mathbf{I}$, where

$$\mathbf{I} = \frac{1}{\phi} (\boldsymbol{D}^{\mathrm{T}} \boldsymbol{W} \boldsymbol{D}).$$

When computing standard errors using $\mathbf{I}$ we would substitute the quasi-likelihood estimate $\hat{\boldsymbol{\beta}}$ for $\boldsymbol{\beta}$. If ϕ is unknown we could estimate it using the mean deviance or the degrees of freedom corrected estimate defined in equation (10.13) in Section 10.3.

The description so far has been quite general and for that reason it is worth pointing out some simple cases. When $\mathrm{Var}[Y_i] = \phi$, i.e. $V(\mu_i) = 1$, $i = 1, 2, \ldots, n$, quasi-likelihood estimation reduces to ordinary least squares estimation. If $\mathrm{Var}[Y_i] = a_i\phi$ and the a_i are known constants, we get weighted least squares estimation. The practical result concerning exponential families, just mentioned, makes it easy to account for overdispersion in the binomial, Poisson and exponential distributions using quasi-likelihood if the following variance formulae are used: $\mathrm{Var}[Y_i] = \phi\mu_i$ for the Poisson, $\mathrm{Var}[Y_i] = \phi n_i \pi_i (1 - \pi_i)$ for the binomial and $\mathrm{Var}[Y_i] = \phi\mu_i^2$ for the exponential. The parameter estimates are obtained as in the complete-likelihood case but the standard errors of the parameter estimates are modified by multiplying them by $\phi^{1/2}$, if known, or its estimate $\hat{\phi}^{1/2}$ otherwise. This is what we did at the end of Section 10.4 when analyzing Gilligan's data.

In the presence of a modest amount of overdispersion, Cox (1983) showed that quasi-likelihood estimation was highly efficient for the estimation of a moment parameter of an exponential family distribution.

Example 10.10

The data given in Table 10.11 were first given by Bissell (1972) and have since been analyzed by Hinde (1982), Nelder (1985) and Lindsey (1993, Section 5.2.2). These data are the number of faults observed in rolls of fabric of various lengths. It seems reasonable that the distribution of faults might be approximated by a Poisson distribution. Hinde (1982) used ln(length) as an explanatory variable and we will do the same here. The link function we assume is

$$\ln(\mu_i) = \beta_1 + \beta_2 \ln(l_i)$$

where μ_i is the mean number of faults in a roll of length l_i metres, $i = 1, 2, \ldots, 32$.

If we fit the above model assuming a standard Poisson distribution with log-link, the estimate of the scale parameter ϕ, as given by the mean deviance, is $\hat{\phi} = 64.54/30 = 2.15$. As ϕ equals 1 for the Poisson distribution, there is evidence of

Table 10.11 Number of faults in rolls of fabric

Observation number	Number of faults	Length of roll
1	6	551
2	4	651
3	17	832
4	9	375
5	14	715
6	8	868
7	5	271
8	7	630
9	7	491
10	7	372
11	6	645
12	8	441
13	28	895
14	4	458
15	10	642
16	4	492
17	8	543
18	9	842
19	23	905
20	9	542
21	6	522
22	1	122
23	9	657
24	4	170
25	9	738
26	14	371
27	17	735
28	10	749
29	7	495
30	3	716
31	9	952
32	2	417

overdispersion. When we set $\beta_2 = 0$, the deviance increases to 103.71 on 31 d.f. As far as testing the hypothesis that $\beta_2 = 0$ is concerned, we can allow for the overdispersion by using an F-test as we did at the end of Section 10.4. That is, we assume that $\text{Var}[Y_i] = \phi\mu_i$ and use standard IWLS to get the quasi-likelihood estimates of β_1 and β_2. This gives

$$F = \frac{(39.18)/(1)}{2.15} = 18.2$$

on (1, 30) d.f., which is clearly highly significant.

The parameter estimates and their standard errors (multiplied by $\hat{\phi}^{1/2}$, where $\hat{\phi}$ is the mean deviance) are $\hat{\beta}_1 = -4.17$ (s.e. = 1.67) and $\hat{\beta}_2 = 1.00$ (s.e. = 0.26).

We will not proceed to examine in detail the various residuals and diagnostics obtained from the above fitted model, although this should be done. An examination of the standardized deviance residuals does, however, suggest that observations 13 and 26 are outliers. If they are removed, the mean deviance drops to 1.5. If observation 19 is also removed the mean deviance reduces even further to 1.2. It appears that much of the overdispersion could be due to these apparently aberrant observations. However, we will not pursue this line of argument any further but will reconsider the data from the point of view that they are overdispersed.

Hinde (1982) analyzed these data using a compound Poisson model that assumed the mean, μ, of the Poisson distribution was itself a log-normally distributed random variable with variance σ^2, i.e. he fitted the model

$$\ln(\mu_i) = \beta_1 + \beta_2 \ln(l_i) + \sigma Z$$

where $Z \sim N(0, 1)$.

Using a combination of the EM algorithm, numerical integration and IWLS, Hinde (1982) obtained maximum likelihood estimates of the parameters. For his model the deviance is 50.96 on 29 d.f., i.e. the mean deviance is 1.76, indicating an improved fit as compared to the standard model.

Nelder (1985) suggested that the data in Table 10.11 might be analyzed conveniently using quasi-likelihood estimation on the assumption that the variance function was $V(\mu_i) = \mu_i + \sigma^2\mu_i^2$. As σ^2 is unknown this approach requires, additionally, the estimation of σ^2. Breslow (1984), following suggestions that Williams (1982) gave for analyzing overdispersed binomial data, showed that quasi-likelihood estimates could be obtained using the standard IWLS procedure if the observations were suitably weighted. (See Example 10.6 (continued) below for further explanation concerning such weighting of the observations.) The weights were defined to be $w_i = (1 + \sigma^2\mu_i)^{-1}$, where, in the standard notation, given in Section 10.2, $a_i(\phi) = \phi/w_i$. Here, of course, we are assuming that $\phi = 1$. Breslow (1984) also described how the estimation of σ^2 could be included in the iterative procedure. Using the GLIM macros given by Breslow (1984), we obtained the results given in Table 10.12 which also displays the results obtained from fitting Hinde's (1982) Poisson-normal model. Obviously, there is a good deal of agreement between the two sets of results.

Table 10.12 Parameter estimates and standard errors for models fitted to the fabric data

Parameter	Poisson-normal		Quasi-likelihood	
	Estimate	Standard error	Estimate	Standard error
β_1	−3.31	1.53	−3.78	1.50
β_2	0.85	0.24	0.93	0.23
σ	0.36	—	0.37	—

For information on diagnostics for overdispersion see Ganio and Schafer (1992). Breslow (1989, 1990) describes score tests and other types of hypothesis test in overdispersed Poisson regression models.

We will end this chapter by illustrating the use of quasi-likelihood as a way of dealing with the suspected overdispersion in Gilligan's roots data.

Example 10.6 (*continued*)

Williams (1982) allowed for overdispersion in binomial data $Y_1, Y_2, \ldots, Y_n$, by introducing unobservable continuous random variables $P_1, P_2, \ldots, P_n$ which were independently distributed on the range (0, 1) with $\mathrm{E}_p[P_i] = \pi_i$ and $\mathrm{Var}_P[P_i] = \phi\pi_i(1-\pi_i)$. He then assumed that conditional on a particular value p_i of P_i, Y_i had a $B(n_i, p_i)$ distribution. To take his arguments further we need the following standard results concerning the expectations and variances of two random variables (called A and B, here):

$$\mathrm{E}[A] = E_B[E_A(A \mid B)]$$

and

$$\mathrm{Var}[A] = \mathrm{E}_B[\mathrm{Var}_A[A \mid B]] + \mathrm{Var}_B[\mathrm{E}_A[A \mid B]].$$

Using these results for our binomial random variables, the unconditional expectation of Y_i is

$$\begin{aligned}\mathrm{E}[Y_i] &= \mathrm{E}_P[\mathrm{E}_Y[Y_i \mid P_i = p_i]] \\ &= \mathrm{E}_P[n_i P_i] = n_i \mathrm{E}_P[P_i] = n_i \pi_i.\end{aligned}$$

The unconditional variance is

$$\mathrm{Var}[Y_i] = \mathrm{E}_P[\mathrm{Var}_Y[Y_i \mid P_i = p_i]] + \mathrm{Var}_P[\mathrm{E}_Y[Y_i \mid P_i = p_i]].$$

Looking at each part of the right-hand side separately we have that

$$\begin{aligned}\mathrm{E}_P[\mathrm{Var}_Y[Y_i \mid P_i = p_i]] &= \mathrm{E}_P[n_i P_i(1 - P_i)] \\ &= n_i[\mathrm{E}_P(P_i) - \mathrm{E}_P(P_i^2)] \\ &= n_i[\mathrm{E}_P(P_i) - \mathrm{Var}_P[P_i] - (\mathrm{E}_P[P_i])^2] \\ &= n_i[\pi_i - \phi\pi_i(1-\pi_i) - \pi_i^2] \\ &= n_i\pi_i(1-\pi_i)(1-\phi).\end{aligned}$$

$$\mathrm{Var}_P[E_Y(Y_i \mid P_i = p_i)] = \mathrm{Var}_P[n_i P_i]$$

$$= n_i^2 \, \mathrm{Var}_P[P_i] = n_i^2 \phi \pi_i(1 - \pi_i).$$

Putting these together gives

$$\mathrm{Var}[Y_i] = n_i \pi_i (1 - \pi_i)(1 + (n_i - 1)\phi).$$

Hence for $\phi > 0$, $\mathrm{Var}[Y_i]$ is greater than it would be in the standard binomial case, i.e. Y_i is overdispersed.

Here we are assuming that, for a given soil type and inoculum density, the probability of a root being infected in a sample of n_i roots is not constant, but is a realized value of a random variable. Our definition of this random variable ensures that, if $\phi > 0$, our data will be overdispersed. It should be noted, however, that we have not fully specified the distribution of each Y_i. All we have assumed is a particular form for the mean and variance of each Y_i. In other words we will need to use quasi-likelihood. A full-likelihood analysis could be done by fully specifying the distribution of each P_i. For example, we might assume that P_i has a beta distribution. Then each Y_i would have the beta-binomial distribution (see Crowder (1978), Brooks (1984) and Moore (1987), for example).

In the absence of overdispersion (or underdispersion) our model would be the standard binomial model used at the end of Section 10.4. Williams (1982) referred to this type of model as Model I. It is worth recalling now that we can fit Model I using the standard IWLS Method of Scoring algorithm.

Williams (1982) used Model II to denote the above quasi-likelihood model. He showed that Model II could be fitted (for ϕ known) by using a slightly modified version of IWLS. To appreciate this modification we note that in the standard IWLS algorithm we iteratively solve the following equation (equation (10.12) in Section 10.3)

$$(\boldsymbol{X}^{\mathrm{T}}\boldsymbol{W}(\boldsymbol{b}_{r-1})\boldsymbol{X})\boldsymbol{b}_r = \boldsymbol{X}^{\mathrm{T}}\boldsymbol{W}(\boldsymbol{b}_{r-1})\boldsymbol{Z}(\boldsymbol{b}_{r-1}).$$

With a suitable redefinition of $\boldsymbol{W}$ and $\boldsymbol{Z}$ to take account of the special form of $\mathrm{Var}[Y_i]$, this is the same iterative equation that is used to obtain the quasi-likelihood estimate of $\boldsymbol{\beta}$. We can see from the formula for $\mathrm{Var}[Y_i]$ that it is the standard binomial variance multiplied by a factor $[(1 + (n_i - 1)\phi]$. A standard implementation of IWLS, as in GLIM for example, allows each observation to be weighted by some constant, c_i. So if we choose the weights to be $c_i = [1 + (n_i - 1)\phi]^{-1}$ and use IWLS to fit the standard binomial model to our data with the c-l-l link function, we will have solved the estimating equations needed to get the quasi-likelihood estimate of $\boldsymbol{\beta}$. This, of course, assumes ϕ is known. Williams (1982) also showed how the estimation of ϕ could be incorporated into the iterative fitting procedure.

Fitting Williams' Model II to Gilligan's full data set gives 0.0368 as the estimated value of ϕ. Using this we get the (quasi-likelihood) results for model L1 that are given in Table 10.13.

Table 10.13. Fitted parameters for model L1 using quasi-likelihood

Parameter	Estimate	Standard error
β_1	−1.101	0.156
β_2	1.021	0.104

We see that these results are similar to those given in Table 10.5, which were obtained by using the standard model and scaling the standard errors by a factor $\hat{\phi}^{1/2}$. The parameter estimates are virtually identical and the standard errors are almost the same. This is not surprising, because for the special case of all n_i equal the two methods are equivalent. For Gilligan's data the n_i are very nearly all equal, hence the reason for the similarity in the two sets of results.

Exercises

10.1 Derive the expressions given in Table 10.1 for μ, $a(\phi)$, $b(\theta)$ and $c(y, \phi)$ for the normal, binomial and Poisson distributions.

10.2 For the Poisson distribution with log link show that the score U_j satisfies

$$U_j = \sum_{i}^{n} (y_i - \mu_i)x_{ij}$$

and that the ith diagonal element of $\mathbf{W}$, the weight matrix defined in Section 10.3, is

$$W_{ii} = \mu_i.$$

10.3 Derive the deviance formulae given in Table 10.2 for the Poisson and gamma distributions.

10.4 Consider the glm for binary data (i.e. $n_i = 1$, $i = 1, 2, \ldots, n$), with logit link. Show that

$$\sum_{j=1}^{q} \beta_j \frac{\partial l(\pi; \mathbf{y})}{\partial \beta_j} = \sum_{i=1}^{n} (y_i - \pi_i) \ln\left(\frac{\pi_i}{1 - \pi_i}\right)$$

and hence that

$$\sum_{i=1}^{n} y_i \ln\left(\frac{\hat{\pi}_i}{1 - \hat{\pi}_i}\right) = \sum_{i=1}^{n} \hat{\pi}_i \ln\left(\frac{\hat{\pi}_i}{1 - \hat{\pi}_i}\right).$$

Use this result to show that the deviance is

$$D = -2 \sum_{i=1}^{n} \left[\hat{\pi}_i \ln\left(\frac{\hat{\pi}_i}{1 - \hat{\pi}_i}\right) + \ln(1 - \hat{\pi}_i)\right].$$

10.5 Refit the model in Example 10.6 but now defining π as the probability of a root being uninfected.

10.6 Refit the model in Example 10.6 using *dose*, rather than ln(*dose*), as the explanatory variable. Determine if a quadratic in *dose* is a reasonable fit to the data.

10.7 For the model considered in Section 10.14 show that

$$\boldsymbol{x}_i^{\mathrm{T}}(\boldsymbol{X}_{(i)}^{\mathrm{T}}\boldsymbol{X}_{(i)})^{-1}\boldsymbol{x}_i = \frac{h_i}{1-h_i}$$

and

$$\boldsymbol{x}_i^{\mathrm{T}}\hat{\boldsymbol{\beta}}_{(i)} = \hat{y}_i - \frac{h_i r_i}{1-h_i}.$$

Hence show that

$$r_i^* = \frac{s r_i'}{s_{(i)}}.$$

10.8 For the model considered in Section 10.14 show that

$$(n-q-1)s_{(i)}^2 = (n-q)s^2 - \frac{r_i^2}{1-h_i}$$

and, using the result in Exercise 10.7, show that

$$r_i^* = \frac{r_i'}{\sqrt{\left(\dfrac{n-q-(r_i')^2}{n-q-1}\right)}}.$$

10.9 For the model considered in Section 10.14 show that

$$\hat{\boldsymbol{\beta}}_{(i)} - \hat{\boldsymbol{\beta}} = \frac{-(\boldsymbol{X}^{\mathrm{T}}\boldsymbol{X})^{-1}\boldsymbol{x}_i r_i}{(1-h_i)}$$

and hence that

$$DC_i = \frac{r_i^2 h_i}{qs^2(1-h_i)^2} = \frac{(r_i')^2 h_i}{q(1-h_i)}.$$

10.10 Analyze the data given in Example 10.7 but with both observations 10 and 20 removed.

10.11 Investigate the fit of model L1 to the data in Example 10.6 with observations 4, 6 and 18 removed.

References

Agresti, A. (1992) A survey of exact inference for contingency tables. *Statist. Sci.* **7**, 131–53.

Akaike, H. (1973) Information theory and an extension of the maximum likelihood principle. In: Petrov, B.N. & Csáki, F. (eds.) *2nd International Symposium on Information Theory.* Académiai Kiadó, Budapest, pp. 267–81.

Andersen, E.B. (1980) *Discrete Statistical Models with Social Science Applications.* North-Holland, Amsterdam.

Anderson, D.A. (1988) Some models for overdispersed binomial data. *Aust. J. of Statistics* **3**, 125–48.

Andrews, D.F., Bickel, P.J., Hampel, F.R., Huber, P.J., Rogers, W.H. & Tukey, J.W. (1972) *Robust Estimates of Location: Survey and Advances.* Princeton University Press, Princeton.

Angus, J.E. & Schafer, R.E. (1984) Improved confidence statements for the binomial parameter. *Amer. Statistician* **38**, 189–91.

Atkinson, A.C. (1981) Two graphical displays for outlying and influential observations in regression, *Biometrika* **68**, 13–20.

Atkinson, A.C. (1985) *Plots, Transformations and Regression.* Clarendon Press, Oxford.

Barnett, V. (1982) *Comparative Statistical Inference* (2nd edn). Wiley, London.

Berger, J.O. (1985) *Statistical Decision Theory and Bayesian Analysis* (2nd edn). Springer-Verlag, New York.

Berger, J.O. & Berry, D.A. (1988) The relevance of stopping rules in statistical inference (with discussion). In: Gupta, S.S. & Berger, J.O. (eds.) *Statistical decision theory and related topics IV*, vol. 1. Springer-Verlag, New York, pp. 29–72.

Berger, J.O. & Sellke, T. (1987) Testing a point hypothesis: the irreconcilability of p values and evidence. *J. Amer. Statist. Assoc.* **82**, 112–22 (rejoinder 135–9).

Besag, J. & Diggle, P.J. (1977) Simple Monte Carlo tests for spatial pattern. *Appl. Statist.* **26**, 327–333.

Besag, J. & Green, P.J. (1993) Spatial statistics and Bayesian computation. *J.R. Statist. Soc.* **B55**, 25–37.

Bickel, P.J. & Doksum, K.A. (1977) *Mathematical Statistics; Basic Ideas and Selected Topics.* Holden-Day, San Francisco.

Bissell, A.F. (1972) A negative binomial model with varying element sizes. *Biometrika* **59**, 435–441.

Breslow, N.E. (1984) Extra-Poisson variation in log-linear models. *Appl. Statist.* **33**, 38–44.

Breslow, N.E. (1989) Score tests in overdispersed GLMs. In: Decarli, A., Francis, B., Gilchrist, R. & Seber, G.U.H. (eds.) *Statistical Modelling.* Springer-Verlag, Berlin, pp. 64–74.

Breslow, N.E. (1990) Tests of hypotheses in overdispersed Poisson regression and other quasi-likelihood models. *J. Amer. Statist. Assoc.* **85**, 565–71.

Brooks, R.J. (1984) Approximate likelihood ratio tests in the analysis of beta-binomial data. *Appl. Statist.* **33**, 285–9.

Buse, A. (1982) The likelihood ratio, Wald and Lagrange multiplier tests: an expository note. *Amer. Statistician* **36**, 153–157.

Casella, G. & Berger, R.L. (1987) Reconciling Bayesian and frequentist evidence in the one-sided testing problem. *J. Amer. Statist. Assoc.* **82**, 106–11 (rejoinder 133–5).

Catchpole, E.A. & Morgan, B.J.T. (1994) Boundary estimation in ring recovery models. *J. Roy. Statist. Soc.* **B56**, 385–91.

Cheng, R.C.H. & Iles, T.C. (1987) Corrected maximum likelihood in non-regular problems. *J.R. Statist. Soc.* **B49**, 95–101.

Cheng, R.C.H. & Traylor, L. (1995) Non-regular maximum likelihood problems (with discussion). *J.R. Statist. Soc.* **B57**, 3–44.

Christensen, R. & Hoffman, M.D. (1985) Bayesian point estimation using the predictive distribution. *Amer. Statistician* **39**, 319–21.

Cochran, W.G. (1963) *Sampling Techniques* (2nd edn). Wiley, London.

Collett, D. (1991) *Modelling Binary Data.* Chapman & Hall, London.

Cook, R.D. (1977) Detection of influential observations in linear regression. *Technometrics* **19**, 15–18.

Cook, R.D. & Weisberg, S. (1982) *Residuals and Influence in Regression.* Chapman & Hall, London.

Copas, J.B. (1983) Regression, prediction and shrinkage (with discussion). *J. Roy. Statist. Soc.* **B45**, 311–54.

Cordeiro, G.M., De Paula Ferrari, S.L. & Paula, G.A. (1993) Improved score tests for generalized linear models. *J.R. Statist. Soc.* **B55**, 661–74.

Cordeiro, G.M., Paula, A.P. & Botter, D.A. (1994) Improved likelihood ratio tests for dispersion models. *Int. Statist. Rev.* **62**, 257–74.

Cox, D.R. (1961) Tests of separate families of hypotheses. *Proc. Fourth Berkeley Symposium on Mathematical Statistics and Probability* **1**, 105–23.

Cox, D.R. (1975) Partial likelihood. *Biometrika* **62**, 269–76.

Cox, D.R. (1983) Some remarks on overdispersion. *Biometrika* **70**, 269–74.

Cox, D.R. & Hinkley, D.V. (1974) *Theoretical Statistics.* Chapman & Hall, London.

Cox, D.R. & Reid, N. (1987) Parameter orthogonality and approximate conditional inference. *J.R. Statist. Soc.* **B49**, 1–39.

Cox, D.R. & Reid, N. (1993) A note on the calculation of adjusted profile likelihood. *J.R. Statist. Soc.* **B55**, 467–71.

Cox, D.R. & Snell, E.J. (1989) *Analysis of Binary Data* (2nd edn). Chapman & Hall, London.

Cressie, N.A.C. (1993) *Statistics for Spatial Data.* Wiley, New York.

Crowder, M.J. (1978) Beta-binomial Anova for proportions. *Appl. Statist.* **27**, 34–7.

Cryer, J.D. (1986) *Time Series Analysis.* Duxbury Press, Boston.

Davis, L.J. (1986) Exact tests for 2 × 2 contingency tables. *Amer. Statistician* **40**, 139–41.

Davison, A.C. & Tsai, C.-L. (1992) Regression model diagnostics. *Int. Statist. Rev.* **60**, 337–53.

De Groot, M.H. (1970) *Optimal Statistical Decisions.* McGraw-Hill, New York.

De Groot, M.H. (1986) *Probability and Statistics* (2nd edn). Addison-Wesley, Reading, Massachusetts.

Dempster, A.P., Laird, N.M. & Rubin, D.B. (1977) Maximum likelihood estimation from incomplete data via the EM algorithm (with discussion). *J.R. Statist. Soc.* **B39**, 1–38.

Devroye, L. (1986) *Non-uniform Random Variate Generation.* Springer-Verlag, New York.

DiCiccio, T.J. & Martin, M.A. (1993) Simple modifications for signed roots of likelihood ratio statistics. *J. Roy. Statist. Soc.* **B55**, 305–16.

DiCiccio , T.J. & Stern, S.E. (1994) Frequentist and Bayesian Bartlett correction of test statistics based on adjusted profile likelihoods. *J. Roy. Statist. Soc.* **B56**, 397–408.

Doganaksoy, N. & Schmee, J. (1993) Comparisons of approximate confidence intervals for distributions used in life-data analysis. *Technometrics* **35**, 175–84.

Draper, N.R. & Smith, H. (1981) *Applied Regression Analysis* (2nd edn). Wiley, New York.

Easton, G.S. (1991) Compromise maximum likelihood estimators for location. *J. Amer. Statist. Assoc.* **86**, 1051–64.

Edgington, E.S. (1980) *Randomization Tests.* Marcel Dekker, New York.

Efron, B. (1979) Bootstrap methods: another look at the jackknife. *Annals Statist.* **7**, 1–26.

Efron, B. (1982) *The Jackknife, the Bootstrap and Other Resampling Plans.* Society for Industrial and Applied Mathematics, Philadelphia.

Efron, B. (1987) Better bootstrap confidence intervals (with discussion). *J. Amer. Statist. Assoc.* **82**, 171–200.

Efron, B. & Tibshirani, R.J. (1993) *An Introduction to the Bootstrap.* Chapman & Hall, London.

Engle, R.F. (1984) Wald, likelihood ratio, and Lagrange multiplier tests in econometrics. In: Griliches, Z. & Intriligator, M.D. (eds.) *Handbook of Econometrics*, Vol 2. North-Holland, Amsterdam, pp. 776–826.

Ericson, W.A. (1969) A note on the posterior mean of the population mean. *J.R. Statist. Soc.* **B31**, 332–4.

Everitt, B.S. (1987) *Introduction to Optimization Methods and their Application in Statistics.* Chapman & Hall, London.

Ferguson, T.S. (1967) *Mathematical Statistics; a Decision Theoretic Approach.* Academic Press, New York.

Firth, D. (1991) Generalized linear models. In: Hinkley, D.V., Reid, N. & Snell, E.J. (eds.) *Statistical Theory and Modelling.* Chapman & Hall, London, pp. 55–82.

Firth, D. (1993) Recent developments in quasi-likelihood methods. In: *Proceedings of the 49th Session of the ISI, Florence*, Book 2, pp. 341–58.

Francis, B., Green, M. & Payne, C. (1993) *The GLIM System, Release* 4 *Manual.* Clarendon Press, Oxford.

Ganio, L.M. & Schafer, D.W. (1992) Diagnostics for overdispersion. *J. Amer. Statist. Assoc.* **87**, 795–804.

Garthwaite, P.H. & Buckland, S.T. (1992) Generating Monte Carlo confidence intervals by the Robbins-Monro process. *Appl. Statist.* **41**, 159–71.

Gates, J. (1993) Testing for circularity of spatially located objects. *J. Appl. Statistics* **20**, 95–103.

Geisser, S. (1993) *Predictive Inference: An Introduction.* Chapman & Hall, New York.

Gelfand, A.E., Hills, S.E., Racine-Poon, A. & Smith, A.F.M. (1990) Illustration of Bayesian inference in normal data models using Gibbs sampling. *J. Amer. Statist. Assoc.* **85**, 972–85.

Ghosh, J.K. & Mukerjee, R. (1994) Adjusted versus conditional likelihood: power properties and Bartlett-type adjustment. *J. Roy. Statist. Soc.* **B56**, 185–8.

Gibbons, J.D. (1985) *Nonparametric Statistical Inference* (2nd edn). Marcel Dekker, New York.

Gilks, W.R. & Wild, P. (1992) Adaptive rejection sampling for Gibbs sampling. *Appl. Statist.* **41**, 337–48.

Gilks, W.R., Clayton, D.G., Spiegelhalter, D.J., Best, N.G., McNeil, A.J., Sharples, L.D. & Kirby, A.J. (1993) Modelling complexity: applications of Gibbs sampling in medicine. *J.R. Statist. Soc.* **B55**, 39–52.

Gilligan, C.A. (1983) A test for randomness of infection by soilborne pathogens. *Phytopathology* **73**, 300–3.

Good, I.J. & Gaskins, R.A. (1980) Density estimation and bump-hunting by the penalized likelihood method exemplified by scattering and meteorite data. *J. Amer. Statist. Assoc.* **75**, 42–73.

Green, P.J. (1990) On use of the EM algorithm for penalized likelihood estimation. *J.R. Statist. Soc.* **B52**, 443–52.

Green, P.J. & Silverman, B.W. (1994) *Nonparametric Regression and Generalized Linear Models.* Chapman & Hall, London.

Gunst, R.F. & Mason, R.L. (1980) *Regression Analysis and its Applications.* Marcel Dekker, New York.

Hall, P. (1992) *The Bootstrap and Edgeworth Expansion.* Springer-Verlag, New York.

Hampel, F.R. (1974) The influence curve and its role in robust estimation. *J. Amer. Statist. Assoc.* **69**, 383–93.

Hampel, F.R., Ronchetti, E.M., Rousseeuw, P.J. & Stahel, W.A. (1986) *Robust Statistics: The Approach Based on Influence Functions.* Wiley, New York.

Harville, D.A. (1977) Maximum likelihood approaches to variance component estimation and to related problems. *J. Amer. Statist. Assoc.* **72**, 320–40.

Hinde, J.P. (1982) Compound Poisson regression models. In: Gilchrist, R. (ed.), *GLIM82, Proceedings of the International Conference on Generalized Linear Models.* Springer-Verlag, Berlin, pp. 109–21.

Hjorth, J.S.U. (1993) *Computer Intensive Statistical Methods.* Chapman & Hall, London.

Hoaglin, D.C., Mosteller, F. & Tukey, J.W. (1983) *Understanding Robust and Exploratory Data Analysis.* Wiley, New York.

Hogg, R.V. & Craig, A.T. (1970) *Introduction to Mathematical Statistics* (3rd edn). Macmillan, New York.

Hogg, R.V. & Tanis, E.A. (1993) *Probability and Statistical Inference* (4th edn). Macmillan, New York.

Hora, S.C. & Iman, R.L. (1988) Asymptotic relative efficiencies of the rank-transformation procedure in randomized complete block designs. *J. Amer. Statist. Assoc.* **83**, 462–70.

Hosking, J.R.M. & Wallis, J.R. (1987) Parameter and quantile estimation for the generalized Pareto distribution. *Technometrics* **29**, 339–49.

Huber, P. (1964) Robust estimation of a location parameter. *Ann. Math. Statist.* **35**, 73–101.

Huber, P.J. (1981) *Robust Statistics.* Wiley, New York.

James, W. & Stein, C. (1961) Estimation with quadratic loss. *Proc. Fourth Berkeley Symposium on Mathematical Statistics and Probability* **1**, 361–79.

Jeffreys, H. (1961) *Theory of Probability* (3rd edn). Oxford University Press, London.

Jennison, C. & Turnbull, B.W. (1989) Interim analysis: the repeated confidence interval approach (with discussion). *J.R. Statist. Soc.* **B51**, 305–61.

Jorgensen, B. (1987) Exponential dispersion models (with discussion). *J. Roy. Statist. Soc.* **B49**, 127–62.

Jorgensen, B. (1992) *The Theory of Exponential Dispersion Models and Analysis of Deviance* (2nd edn). Instituto de Matematica Pura e Aplicada, Rio de Jeneiro.

Juola, R.C. (1993) More on shortest confidence intervals. *Amer. Statistician* **47**, 117–19, letter **48**, 176–7.

Kendall, M.G., Stuart, A. & Ord, J.K. (1987) *Kendall's Advanced Theory of Statistics* (5th edn), Vol. 1. Griffin, London.

Korn, E.L. (1990) Projecting power from a previous study: maximum likelihood estimation.

Amer. Statistician **44**, 290–2.

Lehmann, E.L. (1983) Estimation with inadequate information. *J. Amer. Statist. Assoc.* **78**, 624–7.

Lehmann, E.L. (1986) *Testing Statistical Hypotheses* (2nd edn). Wiley, New York.

Lehmann, E.L. (1993) The Fisher, Neyman–Pearson theories of testing hypotheses: one theory or two? *J. Amer. Statist. Assoc.* **88**, 1242–9.

Lehmann, E.L. & Scheffé, H. (1950) Completeness, similar regions and unbiased estimation. *Sankhya* **10**, 305–40.

Levy,M.S.(1985) A note on nonunique MLEs and sufficient statistics. *Amer. Statistician* **39**, 66.

Lindley, D.V. & Smith, A.F.M. (1972) Bayes estimates for the linear model (with discussion). *J.R. Statist. Soc.* **B34**, 1–41.

Lindsey, J.K. (1993) *Models for Repeated Measurements.* Clarendon Press, Oxford.

Lindsey, J.K. (1995) *Parametric Statistical Inference.* Clarendon Press, Oxford.

Lütkepohl, H. (1991) *Introduction to Multiple Time Series Analysis.* Springer-Verlag, Berlin.

McCullagh, P. (1986) Quasi-likelihood. In: Kotz, S. & Johnson, N.L. (eds.) *Encyclopaedia of Statistical Sciences*, Vol. 7. Wiley, New York, pp. 464–7.

McCullagh, P. (1991) Quasi-likelihood and estimating functions. In: Hinkley, D.V., Reid, N. & Snell, E.J. (eds.) *Statistical Theory and Modelling.* Chapman & Hall, London, pp. 265–86.

McCullagh, P. & Nelder, J.A. (1989) *Generalized Linear Models* (2nd edn). Chapman & Hall, London.

Mak, T.K. (1993) Solving non-linear estimation equations. *J. Roy. Statist. Soc.* **B55**, 945–55.

Manly, B.F.J. (1991) *Randomization and Monte Carlo Methods in Biology.* Chapman & Hall, London.

Maritz, J.S. & Lwin, T. (1989) *Empirical Bayes Methods* (2nd edn). Chapman & Hall, London.

Martz, H.F. & Zimmer, W.J. (1992) The risk of catastrophic failure of the solid rocket boosters on the space shuttle. *Amer. Statistician* **46**, 42–7.

Meeden, G. (1987) Estimation when using a statistic that is not sufficient. *Amer. Statistician* **41**, 135–6.

Meeden, G. & Vardeman, S. (1985) Bayes and admissible set estimation. *J. Amer. Statist. Assoc.* **80**, 465–71.

Mikulski, P.W. (1982) Efficiency. In: Kotz, S. & Johnson, N.L. (eds.) *Encyclopaedia of Statistical Sciences*, Vol. 2. Wiley, New York, pp. 469–72.

Miller, R.G. (1966) *Simultaneous Statistical Inference.* McGraw-Hill, New York.

Moore, D.F. (1987) Modelling the extraneous variance in the presence of extra-binomial variation. *Appl. Statist.* **36**, 8–14.

Morgan, B.J.T. (1978) A simple comparison of Newton–Raphson and the method of scoring. *Int. J. Math. Educ. Sci. Technol.* **9**, 343–8.

Morgan, B.J.T. (1992) *Analysis of Quantal Response Data.* Chapman & Hall, London.

Mosteller, F. & Tukey, J.W. (1977) *Data Analysis and Regression.* Addison Wesley, Reading, Massachusetts.

Muirhead, R.J. (1985) Estimating a particular function of the multiple correlation coefficient. *J. Amer. Statist. Assoc.* **80**, 923–5.

Nelder, J.A. (1985) Quasi-likelihood and GLIM. In: Gilchrist, R., Francis, B. & Whittaker, J. (eds.) *Generalised linear models.* Lecture notes in statistics, Vol. 32. Springer-Verlag, Berlin, pp. 120–7.

Osborne, M.R. (1992) Fisher's method of scoring. *Int. Statist. Rev.* **60**(1), 99–117.

Pal, N. & Berry, J.C. (1992) On invariance and maximum likelihood estimation. *Amer. Statistician* **46**, 209–12.

Patterson, H.D. & Thompson, R. (1971) Recovery of the inter-block information when block sizes are unequal. *Biometrika* **58**, 545–54.

Payne, R.W., Lane, P.W., Digby, P.G.N., Harding, S.A., Leech, P.K., Morgan, G.W., Thompson, R., Todd, A.D., Tunnicliffe Wilson, G., Welham, S.J. & White, R.P. (1993) *GENSTAT 5 Release 3 Reference Manual.* Clarendon Press, Oxford.

Pierce, D.A. & Peters, D. (1992) Practical use of higher order asymptotics for multiparameter exponential families. *J.R. Statist. Soc.* **B54**, 701–37.

Pratt, J.W. & Gibbons, J.D. (1981) *Concepts of Nonparametric Theory.* Springer Verlag, New York.

Pregibon, D. (1980) Goodness of link tests for generalized linear models. *Appl. Statist.* **29**, 15–24.

Pregibon, D. (1981) Logistic regression diagnostics. *Ann. Statist.* **9**, 705–24.

Pregibon, D. (1982) Score tests in GLIM with applications. In: Gilchrist, R. (ed.) *GLIM82. Proceedings of the International Conference on Generalized Linear Models.* Springer-Verlag, Berlin, pp. 87–97.

Press, W.H. (1992) *Numerical Recipes in Fortran: the Art of Statistical Computing* (2nd edn). Cambridge University Press, Cambridge.

Quenouille, M.H. (1949) Approximate tests of correlation in time series. *J.R. Statist. Soc.* **B11**, 68–84.

Quenouille, M.H. (1956) Notes on bias in estimation. *Biometrika* **43**, 353–60.

Randles, R.H. & Wolfe, D.A. (1979) *Introduction to the Theory of Nonparametric Statistics.* Wiley, New York.

Rao, C.R. (1973) *Linear Statistical Inference and its Applications* (2nd edn). Wiley, New York.

Ravishanker, N., Melnick, E.L. & Tsai, C.-L. (1990) Differential geometry of ARMA models. *J. Time Series Analysis* **11**, 259–74.

Ridout, M.S. (1994) A comparison of confidence interval methods for dilution series experiments. *Biometrics* **50**, 289–96.

Robbins, H. & Monro, S. (1951) A stochastic approximation method. *Ann. Math. Statist.* **22**, 400–7.

Shapiro, S.S. & Wilk, M.B. (1965) An analysis of variance test for normality (complete samples). *Biometrika* **52**, 591–611.

Silverman, B.W. (1986) *Density Estimation for Statistics and Data Analysis.* Chapman & Hall, London.

Silvey, S.D. (1970) *Statistical Inference.* Penguin, Harmondsworth, England.

Smith, A.F.M. & Roberts, G.O. (1993) Bayesian computation via the Gibbs sampler and related Markov chain Monte Carlo methods. *J.R. Statist. Soc.* **B55**, 3–23.

Sprent, P. (1989) *Applied Nonparametric Statistical Methods.* Chapman & Hall, London.

Stein, C. (1960) Multiple regression. In: Olkin, I. (ed.) *Contributions to Probability and Statistics: Essays in Honour of Harold Hotelling.* Stanford University Press, Stanford, pp. 424–43.

Still, A.W. & White, A.P. (1981) The approximate randomization test as an alternative to the F test in analysis of variance. *Brit. J. Math. Statist. Psychol.* **34**, 243–52.

Stone, M. (1974) Cross-validatory choice and assessment of statistical predictions. *J.R. Statist. Soc.* **B36**, 111–47.

Stone, M. (1977) An asymptotic equivalence of choice of model by cross-validation and Akaike's criterion. *J.R. Statist. Soc.* **B39**, 44–7.

Stuart, A. & Ord. J.K. (1991) *Kendall's Advanced Theory of Statistics* (5th edn), Vol. 2. Edward Arnold, London.

Thompson, G.L. & Ammann, L.P. (1989) Efficacies of rank-transform statistics in two-way models with no interaction. *J. Amer. Statist. Assoc.* **84**, 325–30.

Tierney, L. (1991) Markov chains for exploring posterior distributions. *Technical Report 560*. School of Statistics, University of Minnesota, Minneapolis.

Tiku, M.L., Tan, W.Y. & Balakrishnan, N. (1986) *Robust Inference*. Marcel Dekker, New York.

Tingley, M. & Li, C. (1993) A note on obtaining confidence intervals for discrete parameters. *Amer. Statistician* **47**, 20–3.

Vardeman, S.B. (1992) What about the other intervals? *Amer. Statistician* **46**, 193–7 (correction **47**, 238).

Vinod, H.D. & Ullah, A. (1981) *Recent Advances in Regression Methods*. Marcel Dekker, New York.

Wakefield, J.C. Gelfand, A.E. & Smith, A.F.M. (1991) Efficient generation of random variates via the ratio-of-uniforms method. *Statist. Comput.* **1**, 129–33.

Wald, A. (1947) *Sequential Analysis*. Wiley, New York.

Wedderburn, R.W.M. (1974) Quasi-likelihood functions, generalized linear models, and the Gauss–Newton method. *Biometrika* **61**, 439–47.

Weerahandi, S. (1993) Generalized confidence intervals. *J. Amer. Statist. Assoc.* **88**, 899–905 (correction **89**, 726).

Wetherill, G.B. (1963) Sequential estimation of quantal response curves (with discussion). *J.R. Statist. Soc.* **B25**, 1–48.

Wetherill, G.B. (1975) *Sequential Methods in Statistics* (2nd edn). Chapman & Hall, London.

Wetherill, G.B. (1986) *Regression Analysis with Applications*. Chapman & Hall, London.

Williams, D.A. (1982) Extra-binomial variation in logistic linear models. *Appl. Statist.* **31**, 144–8.

Williams, D.A. (1987) Generalized linear model diagnostics using the deviance and single case deletions. *Appl. Statist.* **36**, 181–91.

Winkler, R.L. (1972) *Introduction to Bayesian Inference and Decision*. Holt, Rinehart & Winston, New York.

Wu, C.F. (1986) Jackknife, bootstrap and other resampling plans in regression analysis (with discussion). *Annals Statist.* **14**, 1261–1350.

Zacks, S. (1971) *The Theory of Statistical Inference*. Wiley, New York.

Index